# Women in Science

# Women in Science

*With An Introductory Chapter on Woman's Long Struggle For Things of The Mind*

H J MOZANS

Nine BOOKS

वाणी प्रकाशन
*Presents*

Nine
BOOKS

*4695, 21-A, Daryaganj, New Delhi 110 002*

—— Branch ——

*Ashok Rajpath, Patna 800 004*

Phone : +91 11 23273167 Fax : +91 11 23275710

www.vaniprakashan.in
sales@vaniprakashan.in
vaniprakashan@gmail.com

WOMEN IN SCIENCE
*by* H J MOZANS

ISBN: 978-93-5299-021-5
Criticism/Feminism


First Edition

Price : ₹ 595



*Printed at Printland Digital, New Delhi-110 020*

*Vani Prakashan's signature motif is created by Artist Maqbool Fida Husain*

To

Mrs. Charles M Schwab

As A Slight Tribute

To Her Charming Personality

Goodness Of Heart And Nobility Of Soul.

This Volume

Is Respectfully Dedicated

With The Best Wishes Of

The Author.

# Preface

THE FOLLOWING PAGES are the outcome of studies begun many years ago in Greece and Italy. While wandering through the famed and picturesque land of the Hellenes, rejoicing in the countless beauties of the islands of the Ionian and Aegean seas or scaling the heights of Helicon and Parnassus, all so redolent of the storied past, I saw on every side tangible evidence of that marvellous race of men and women whose matchless achievements have been the delight and inspiration of the world for nearly three thousand years. But it was especially while contemplating, from the portico of the Parthenon, the magnificent vista which there meets the charmed vision, that I first fully experienced the spell of the favoured land of Hellas, so long the home of beauty and of intellect. The scene before me was indeed enchanting beyond expression; for, every ruin, every marble column, every rock had its history, and evoked the most precious memories of men of godlike thoughts and of

> A thousand glorious actions that may claim
> Triumphal laurels and immortal fame.

It was a tranquil and balmy night in midsummer. The sun, leaving a gorgeous afterglow, had about an hour before disappeared behind the azure-veiled mountains of Ithaca, where, in the long ago, lived and loved the hero and the heroine of the incomparable Odyssey. The full moon, just rising above the plain of Marathon, intensified the witchery of that memorable spot consecrated by the valour of patriots battling victoriously against the invading hordes of Asia. Hard by was the Areopagus, where St Paul preached to the "superstitious" Athenians on "The Unknown God." Almost adjoining it was the Agora, where Socrates was wont to hold converse with noble and simple on the sublimest questions which can engage the human mind. Not distant was the site of the celebrated "Painted Porch," where Zeno developed his famous system of ethics. In another quarter were the shady walks of the Lyceum, where Aristotle, "the master of those who know," lectured before an admiring concourse of students from all parts of Hellas. Farther afield, on the banks of the Cephissus, was the grove of Academus, where the divine Plato expounded that admirable idealism which, with Aristotelianism, has controlled the progress of speculative thought for more than twenty centuries, and enunciated those admirable doctrines which have become the common heritage of humanity.

But where, in this venerable city — "the eye of Greece, mother of arts and eloquence" — was the abode of Aspasia, the wife of Pericles and the inspirer of the noblest minds of the Golden Age of Grecian civilisation? Where was that salon, renowned these four and twenty centuries as the most brilliant court of culture the world has ever known, wherein this gifted and accomplished daughter of Miletus gathered about her the most learned men and women of her time? Whatever the location, there it was that the wit and talent of Attica found a congenial trysting-place, and human genius burst into fairest blossom. There it was that poets, sculptors, painters, orators, philosophers, statesmen were all equally at home. There Socrates discoursed on philosophy; there Euripides and Sophocles read their plays; there Anaxagoras dilated upon the nature and constitution of the universe; there Phidias, the greatest sculptor of all time, and Ictinus and Callicrates unfolded their plans for that supreme creation of architecture, the temple of Athena Parthenos on the Acropolis. Like Michaelangelo, long centuries afterwards, who "saw with the eyes and acted by the inspiration" of Vittoria Colonna, these masters of Greek architecture and sculpture saw with the eyes and acted by the sublime promptings of Aspasia, who was the greatest patron and inspirer of men of genius the world has ever known.

I felt then, as I feel now, that this superb monument to the virgin goddess of wisdom and art and science was in great measure a monument to the one who by her quick intelligence, her profound knowledge, her inspiration, her patronage, her influence, had so much to do with its erection — the wise, the cultured, the richly dowered Aspasia.

This thought it was that started the train of reflections on the intellectual achievements of women which eventually gave rise to the idea of writing a book on woman's work in things of the mind.

The following day, as I was entering the University of Athens, I noticed above the stately portal a large and beautiful painting which, on inspection, proved, to my great delight, to be nothing less than a pictorial representation of my musings the night before on the portico of the Parthenon. For there was Aspasia, just as I had fancied her in her salon, seated beside Pericles, and surrounded by the greatest and the wisest men of Greece. "This," I exclaimed, "shall be the frontispiece of my book; it will tell more than many pages of text." Nor did I rest till I had procured a copy of this excellent work of art.

Shortly after my journey through Greece I visited the chief cities and towns of Italy. I traversed the whole of Magna Graecia and, to enjoy the local colour of things Grecian and breathe, as far as might be, the atmosphere which once enveloped the world's greatest thinkers, I stood on the spot in Syracuse where Plato discoursed on the true, the beautiful and the good, before enthusiastic audiences of men and women, and wandered through the land inhabited by the ancient Bruttii, where Pythagoras has his famous school of science and philosophy — a school which was continued after the founder's death by his celebrated wife, Theano. For in Crotona, as well as in Athens, and in Alexandria in the time of Hypatia, women were teachers as well as scholars, and attained to marked distinction in every branch of intellectual activity.

As I visited, one after the other, what were once the great centres of learning and culture in Magna Graecia, the idea of writing the book aforementioned appealed to me more strongly from day to day, but it did not assume definite form until after I had tarried for some weeks or months in each of the great university towns of Italy. And as I wended my way through the almost deserted streets of Salerno, which was for centuries one of the noblest seats of learning in Christendom, and recalled the achievements of its gifted daughters — those wonderful *mulieres Salernitanae*, whose praises were once sounded throughout Europe, but whose names have been almost forgotten — I began to realise, as never before, that women of intellectual eminence have received too little credit for their contributions to the progress of knowledge, and should have a sympathetic historian of what they have achieved in the domain of learning.

But it was not until after I had visited the great university towns of Bologna, Padua and Pavia, had become more familiar with their fascinating histories and traditions, and surveyed there the scenes of the great scholastic triumphs of women as students and professors, that I fully realised the importance, if not the necessity, of such a work as I had in contemplation. For then, as when standing in silent meditation on the pronaos of the Parthenon, the past seemed to become present, and the graceful figures of those illustrious daughters of *Italia la Bella*, who have conferred such honour on both their country and on womankind throughout the world, seemed to flit before me as they returned to and from their lecture halls and laboratories, where their discourses, in flowing Latin periods, had commanded the admiration and the applause of students from every European country, from the Rock of Cashel to the Athenian Acropolis.

Only then did the magnitude and the difficulty of my self-imposed task begin to dawn upon me. I saw that it would be impossible, if I were to do justice to the subject, to compass in a single volume anything like an adequate account of the contributions of women to the advancement of general knowledge. I accordingly resolved to restrict my theme and confine myself to an attempt to show what an important role women have played in the development of those branches of knowledge in which they are usually thought to have had but little part.

The subject of my book thus, by a process of elimination, narrowed its scope to woman's achievements in science. Many works in various languages had been written on what women had accomplished in art, literature, and statecraft, and there was, therefore, no special call for a new volume on any of these topics. But, with the exception of a few brief monographs in German, French and Italian, and an occasional magazine article here and there, practically nothing had been written about woman in science. The time, then, seemed opportune for entering upon a field that had thus far been almost completely neglected; and, although I soon discovered that the labour involved would be far greater than I had anticipated, I never lost sight of the work which had its virtual inception in the peerless sanctuary of Pallas Athena in the "City of the Violet Crown."

Duties and occupations innumerable have retarded the progress of the work. But not the least cause of delay has been the difficulty of locating the material essential to the production of a volume that would do even partial justice to the numerous topics requiring treatment. My experience, *parva componere magnis,* was not unlike that of Dr Johnson, who tells us in the preface to his *Dictionary of the English Language,* "I saw that one inquiry only gave occasion to another, that book referred to book, that to search was not always to find, and that thus to pursue perfection was, like the first inhabitants of Arcadia, to chase the sun, which, when they reached the hill where he seemed to rest, was still beheld at the same distance from them."

Although I have endeavoured to give a place in this work to all women who have achieved special distinction in science, it is not unlikely that I may have inadvertently overlooked some, particularly among those of recent years, who were deserving of mention. Should this be the case, I shall be grateful for information which will enable me to correct such oversights and render the volume, should there be a demand for more than one edition, more complete and serviceable. And, although I have striven to be as accurate as possible in all my statements, I can scarcely hope, in traversing so broad a field, to have been wholly successful. For all shortcomings, whether through omission or commission,

Quas aut incuria fudit,
Aut humana parum cavit natura,

I crave the reader's indulgence, and trust that the present volume will have at least the merit of stimulating some ambitious young Whewell to explore more thoroughly the interesting field that I have but partially reconnoitred, and give us ere long an adequate and comprehensive history of the achievements of woman, not only in the inductive but in all the sciences.

# Contents

Le donne son venute in excellenza
Di ciascun'arte, ove hanno posta cura;
E qualunque all'istorie abbia avvertenza,
Ne sente ancor la fama non oscura.

What art so deep, what science so high,
But worthy women have thereto attained?
Who list in stories old to look may try,
And find my speech herein not false nor
fain'd.

*Ariosto, Orlando*
*Furioso, Canto XX,*
*Strophe 2.*

Ad omnem igitur doctrinam..... muliebres
animos natura comparavit.

*Maria Gaetana Agnesi.*

# Woman's Long Struggle for Things of the Mind

*WOMAN AND EDUCATION IN ANCIENT GREECE*

I PROPOSE TO review the progress and achievements of woman in science from her earliest efforts in ancient Greece down to the present time. I shall relate how, in every department of natural knowledge, when not inhibited by her environment, she has been the colleague and the emulatress, if not the peer, of the most illustrious men who have contributed to the increase and diffusion of human learning. But a proper understanding of this subject seems to require some preliminary survey of the many and diverse obstacles which, in every age of the world's history, have opposed woman's advancement in general knowledge. Without such preliminary survey it is impossible to realise the intensity of her age-long struggle for freedom and justice in things of the mind or fully to appreciate the comparative liberty and advantages she now enjoys in almost every department of intellectual activity. Neither could one understand why woman's achievements in science, compared with those of men, have been so few and of so small import, especially in times past, or why it is that, as a student of nature or as an investigator in the various realms of pure and applied science, we hear so little of her before the second half of the nineteenth century.

To exhibit the nature of the difficulties woman has had to contend with in every age and in every land, in order to secure what we now consider her inalienable rights to things of the mind, it is not necessary to review the history of female education, or to enter into the details of her gradual progress forward and upward in the

New and Old Worlds. But it is necessary that we should know what was the attitude of mankind toward woman's education during the leading epochs of the world's history and what were, until almost our own day, the opinions of men — scholars and rulers included — respecting the nature and the duties of woman and what was considered, almost by all, her proper sphere of action. Understanding the numerous and cruel handicaps which she had so long to endure, the opposition to her aspirations which she had to encounter, even during the most enlightened periods of the world's history, and that, too, from those who should have been the first to extend to her a helping hand, we can the better appreciate the extent of her recent intellectual enfranchisement and of the value of the work she has accomplished since she has been free to exercise those God-given faculties which were so long held in restraint.

The first great bar to the mental development of woman was the assumed superiority of the male sex, the opinion, so generally accepted, that, in the scheme of creation, woman was but "an accident, an imperfection, an error of nature;" that she was either a slave conducing to man's comfort, or, at best, a companion ministering to his amusement and pleasure.

From the earliest times she was regarded as man's inferior and relegated to a subordinate position in society. She was, so it was averred, but a diminutive man — a kind of mean between the lord of creation and the rest of the animal kingdom. By some she was considered a kind of half man; by others, as was cynically asserted, she was looked upon as a *mas occasionatus* — a man marred in the making. She was, both mentally and physically, what Spencer would call a man whose evolution had been arrested, while man, as in the modern language of Darwin, was a woman, whose evolution had been completed.

When such views prevailed, it was inevitable that, so long as physical force was the *force majeure*, a woman should be relegated to the position of a slave or to that of "a mere glorified toy." Every man then said, in effect, if not in words, of the woman who happened to be in his power what Petruchio said of Katherine:

I will be master of what is mine own,
She is my goods, my chattels; she is my house,
My household stuff, my field, my barn,
My horse, my ox, my ass, my everything.

Even after civilisation had superseded savagery and barbarism, it was still inevitable, so long as such views found acceptance, that woman should continue to be held in vassalage and ignorance and to suffer all the disabilities and privations of "the lesser man." She was studiously excluded from civic and social functions and compelled to pass her life in the restricted quarters of the harem or gyneceum. This was the case among the Athenians, as well as among other peoples; for, during the most brilliant period of their history, women, when not slaves or hetaerae, were considered simply child-bearers or housekeepers.[1] A girl's education, when she received any at all, was limited to reading, writing and music, and for a knowledge of these subjects she was dependent on her mother. From her earliest years the Athenian maiden was made to realise that the great fountains of knowledge, which were always available for her brothers, were closed to her. Her duty was to become proficient in the use of the needle and the distaff, and, later on, to learn how to embroider, to ply the loom and make garments for herself and for the other members of her family.

Until she was seven years old, she was brought up with her brothers under the eye of her mother. During this period of childhood she had a certain amount of freedom, but, after her seventh year, she was kept in the gyneconitis — women's quarters — "under the strictest restraint, in order," as Xenophon informs us in his *Oeconomicus,* "that she might see as little, hear as little and ask as few questions as possible." On rare occasions she was permitted to be a spectator at a religious procession, or to take part in certain of the choral dances that constituted so important a part in the religious ceremonies of ancient Greece. Whether in public or in private, silence was always considered an imperative duty for a woman.

But more than this. Not only was she expected to observe silence herself, but she was also expected so to conduct herself that no one would have occasion to speak about her. Pericles, in a celebrated discourse, gave expression to the prevailing opinion regarding this phase of female excellence when, on a notable occasion, he addressed to a certain number of women the following words: "Great will be your glory in not falling short of your natural character; and greatest will be hers who is least talked of among men whether for good or for evil."[2]

From the foregoing observations it will be seen that the general attitude of the Athenians toward woman was anything but favourable to her intellectual development, or to her exerting any influence beyond the limits of her own household. And what is said of the Greeks can be affirmed, with still greater emphasis, of the other nations of antiquity. Indeed, it can be safely asserted that, had they all entered into a solemn compact systematically to discredit woman's mental capacity and to repress all her noblest aspirations, they could not have succeeded more effectually than by the methods they severally adopted. In ancient Greece the condition of woman was little better than it is in India today under the law of Manu, where the husband, no matter how unworthy he may be, must be regarded by the wife as a god.

And yet, notwithstanding the dominant force of public opinion and the strange traditional prejudices that possessed for the majority of people all the semblance and commanding power of truth, woman was here and there able to break through the barriers that impeded her progress in her quest of knowledge and to defy the social conventions that precluded her from being seen or heard in the intellectual arena.

One of the first and most notable of Greek women to assert her independence and to emerge from the intellectual eclipse which had so long kept her sex in obscurity, was the Lesbian Sappho, who, as a lyric poet, stands, even today, without a superior. So great was her renown among the ancients that she was called "The Poetess," as Homer was called "The Poet." Solon, on hearing one of her songs sung at a banquet, begged the singer to teach it to him at once that he might learn it and die. Aristotle did not hesitate to endorse a judgement that ranked her with Homer and Archilochus, while Plato, in his Phaedrus, exalts her still higher by proclaiming her "the tenth Muse." Horace and Ovid and Catullus strove to reproduce her passionate strains and rhythmic beauty; but their efforts were little better than paraphrase and feeble imitation. Her features were stamped on coins, "though she was but a woman," and, after her death, altars were raised and temples erected in honour of this "flower of the Graces," of

> That mighty songstress, whose unrivalled powers
> Weave for the Muse a crown of deathless flowers.

Second only to the "violet-crowned, pure, sweetly-smiling Sappho," as her rival, Alcaeus, calls her, were Gorgo, Andromeda and Corinna. The last of these was the teacher of Pindar, the celebrated lyric poet, whom she defeated five times in poetic contests in Thebes.[3] She was one of the nine lyrical muses, corresponding to "the celestial nine," who dwelt on the sacred slopes of Helicon.[4] Telesilla and Praxilla were two others. The last named was by her countrymen ranked with Anacreon.

Scarcely inferior to Corinna were those ardent pupils of Sappho, who had flocked from the sunny isles of the Aegean and the laurel-crowned hills of Greece around "the fair-haired Lesbian" in her island home, which was, at the same time, a school of poetry and music. The most gifted of these were Danophila, the Pamphylian, and Erinna, whose hexameters were said by the ancients to reveal a genius equal to that of Homer. She died at the early age of nineteen and has always excited a pathetic interest because, like so many others of her sex since her time — women and maidens of the loftiest spiritual aspirations — she was condemned to the spindle and the distaff when she wished to devote her life to the service of the Muses. The following is her own epitaph:

> These are Erinna's songs, how sweet, though slight!
> For she was but a girl of nineteen years.

Yet stronger far than what most men can write;
Had death delayed, whose fame had equalled hers?

Never before nor since did such a wave of feminine genius pass over the fragrant valleys and vine-clad plains of Greece. Never in any other place or time shone so brilliant a galaxy of women of talent and imagination; never was there a more perfect flowering of female intelligence of the highest order. According to tradition, there appeared in the favoured land of Hellas, when the entire population of the country was not equal to that of a fair-sized modern city, within the brief space of a century, no fewer than seventy-six women poets. When we remember that the Renaissance produced only about sixty female poets, though in a more extended territory and with a much larger population, and that none of them could approach the incomparable Sappho, or even many of her pupils, in the perfection of their work, we can realise the splendour of the achievements of the female intellect in the Hellenic world during the golden age of feminine poetic art.[5]

One would think that this phenomenal outburst of mental vigour, and especially the marvellous achievements of Sappho, Corinna and those of their pupils and followers, would have compelled the world for all subsequent time to recognise the innate power of the female mind, and perceive the wisdom — not to say justice — of according to women the same advantages for the development of their inborn gifts as were afforded to men. They had proved that, under favourable conditions, there was essentially no difference between the male and the female intellect, and that genius knows no sex. And this they demonstrated not only in poetry, but also in philosophy and in other branches of human knowledge as well.

Among those who had especially distinguished themselves were Hipparchia, the wife of the philosopher Crates; Themista, the wife of Leon and a correspondent of Epicurus, who was pronounced "a sort of female Solon;" Perictione, a disciple of Pythagoras, who distinguished herself by her writings on *Wisdom* and *The Harmony of Woman,* and Leontium, a disciple and companion of Epicurus, who wrote a work against Theophrastus, which was pronounced by Cicero a model of style.

And was not the school of Pythagoras at Crotona continued after his death by his daughter and his wife, Theano? And did not this fact alone manifest woman's capacity for abstract thought, as effectively as the Lesbian school had demonstrated her talent for consummate verse?[6]

But it was all to no purpose. The comparative freedom and advantages which Sappho, Corinna and their friends had enjoyed was soon — for some reason scarcely comprehensible by us — taken from all the women of Greece except the peculiar class known in history as *hetaerae* — companions. These we should now rank among the *demimonde,* but the Greek point of view was different from ours. The hetaerae were the friends and companions of the men who spent most of their time in public resorts, and they accompanied them to the gymnasium, to banquets, the games, to the theatre and other similar assemblies from which the wives and daughters of the Athenians, during the golden age of Greece, were rigorously excluded. For so great was the seclusion in which the wives of the Greeks then lived that they never attended public spectacles and never left the house, unless accompanied by a female slave. They were not permitted to see men except in the presence of their husbands, nor could they have a seat even at their own tables, if their husbands happened to have male guests.

It was by reason of this strict seclusion and the enforced ignorance to which they were subjected that we hear very little of the virtuous women of this period of Greek history. We have records of a few instances of filial and conjugal affection, but, outside of this, the names of the wives and daughters of even the most distinguished citizens have long since passed into oblivion. Only the hetaerae attracted public notice, and only among them, during the period to which reference is now made, do we find any women who achieved distinction by their intellectual attainments, or by the influence which they exerted over those with whom they were associated.

But strange as it may appear, these extra-matrimonial connections, far from incurring the censure which they would now provoke, received the cordial recognition of both legislators and moralists, and even those who were considered the most virtuous among men openly entered into these relations without exposing themselves to the slightest stigma or reproach. Many of the hetaerae, contrary to what is sometimes thought, were "of highly moral character, temperate, thoughtful and earnest, and were either unattached or attached to one man, and to all intents and purposes married. Even if they had two or three attachments but behaved in other respects with temperance and sobriety, such was the Greek feeling in regard to their peculiar position that they did not bring down upon themselves any censure from even the sternest of the Greek moralists."[7]

The most famous men of Greece, married as well as unmarried, had their "companions," many of whom were as distinguished for their accomplishments as for their wit and beauty. Thus Epicurus had Leontium, Menander Glycera, Isocrates Metaneira, Aristotle Herpyllis, and Plato Archlanassa, while Aristippus, the philosopher, Diogenes, the cynic, and Demosthenes, the great orator, each had a companion bearing the name of Lais.[8] More than this. So strongly had many of the hetaerae impressed themselves on the aesthetic sense of the beauty-loving Greeks that not a few of them had statues erected in their honour, especially in Athens and Corinth, and thus shared in the honour that hitherto had been reserved exclusively for the goddess of beauty and love, fair Aphrodite.

The hetaerae from Ionia and Aetolia were particularly conspicuous for their intelligence and culture. And all of them, whencesoever they came, enjoyed unrestricted liberty and, unlike the wives of the citizens of Athens, had free access to the Portico and the Academy and the Lyceum, and were permitted to attend the lectures of the philosophers on the same footing as the men. Thus, to mention only a few, Thais was a pupil of Alciphron, Nicarete of Stilpo, and Lasthenia of Plato.

And so keen were their intellects and so marked was their

progress in the most abstract studies, that many of them were recognised as the most distinguished pupils of their masters. This accounts, in part, for the popularity of their salons, at which were gathered the most eminent statesmen, poets, artists, philosophers and orators of the day. The nearest approach in modern times to such trysting-places, where beauty, wit and talent found a congenial atmosphere, were the celebrated salons of Ninon de Lenclos, Mlle de l'Espinasse and Mme du Deffand. At these reunions were discussed, not only the news of the day, but also, and especially, art, science, literature and politics, and always to the advantage of both guests and hostesses.

Possessing such freedom and enjoying such splendid opportunities for culture and intellectual advancement, it is not surprising that the hetaerae played so remarkable a role in the social and civic life of Greece, and that they were able to wield such influence over their associates, and that they often attained even the highest royal honours. Nor is it surprising to read in Plato's *Symposium* the splendid tribute which Socrates renders to Diotima of Mantinea, when, in discussing the true nature of divine and eternal beauty, he speaks of her as his teacher.

Many of the hetaerae were not only the models but also the inspirers of the most famous painters and sculptors of antiquity. Thus, Lais was the companion and inspirer of Apelles, the most noted painter of Greece, while Phryne, said to have been the most beautiful woman who ever lived, was the inspirer of the peerless Praxitiles, who, in reproducing her form, succeeded in bequeathing to the world what was undoubtedly the most lovely representation of "the human form divine" that ever came from a sculptor's chisel.[9]

On account of the relations of the hetaerae, especially those of the fourth and fifth centuries BC, with the greatest men of their time, the writers of antiquity thought them of sufficient importance to preserve their history. One author has left us an account of no fewer than one hundred and thirty-five of them. But, of all those whose names have come down to us, by far the most noted, accomplished and influential was the famous Aspasia of Miletus. In many respects she was the most remarkable woman Greece ever produced. Of rare talent and culture, of extraordinary tact and finesse, of a fascinating personality combined with the grace and sensibility of her sex, together with a masculine power of intellect, "this gracious Ionian," as has well been said, "stands with Sappho on the pinnacle of Hellenic culture, each in her own field the highest feminine representative of an aesthetic race."

At an early age she won the passionate love of the great statesman Pericles, after which she entered upon that marvellous career which secured for her a place in the front rank of the most eminent women of all time. "Her house became the resort of all the great men of Athens. Socrates was often there. Phidias and Anaxagoras were intimate acquaintances, and probably Sophocles and Euripides were in constant attendance. Indeed, never had any woman such a salon in the whole history of man. The greatest sculptor that ever lived, the grandest man of all antiquity, philosophers and poets, sculptors and painters, statesmen and historians, met each other and discussed congenial subjects in her rooms. And probably hence has arisen the tradition that she was the teacher of Socrates in philosophy and politics, and Pericles in rhetoric. Her influence was such as to stimulate men to their best, and they attributed to her all that was best in themselves. Aspasia seems especially to have thought earnestly on the duties and destiny of women. The cultivated men who thronged her assemblies had no hesitation in breaking through the conventionalities of Athenian society, and brought their wives to the parties of Aspasia; and she discussed with them the duties of wives. She thought they should be something more than mere mothers and housewives. She urged them to cultivate their minds, and be in all respects fit companions for their husbands."[10]

She is said to have written some of the best speeches of Pericles —among them his noted funeral oration over those who had died in battle before the walls of Potidaea. As to Socrates, he himself explicitly refers to her, in the *Memorabilia*, as his teacher. She is a notable character in the Socratic dialogues and appears several times in those of Aeschines, while there is every reason to believe that she strongly influenced the views of Plato, as expressed by him in the *Republic* respecting the equality of woman with man.

She was continually consulted regarding affairs of state, and her influence in social and political matters was profound and far-reaching. This is evidenced by the abuse heaped upon her by the comic dramatists of the time. Referring to the ascendancy which she had over Pericles, she was called Dejanira, the wife of Hercules; Hera, the queen of the gods and wife of the Olympian Jove. It was asserted by her enemies that the Samian war had been brought about at her instigation and that the Peloponnesian war had been undertaken to avenge an insult which had been offered her. These and similar statements which, when not absurd, were greatly exaggerated, show the boundless influence she wielded over Pericles, and what an important part she took in the government of Greece in the zenith of its glory.

But, however great her influence, we are warranted in asserting that it was never exercised in an illegitimate manner. She was ever, as history informs us, the good, the wise, the learned, the eloquent Aspasia. It was her goodness, her wisdom, her rare and varied accomplishments, her clear insight and noble purposes that gave her the wonderful power she possessed and which enabled her, probably more than any one person, to make the age of Pericles not only the most brilliant age of Greek history, but also the most brilliant age of all time.[11]

But, notwithstanding the beneficent influence which Aspasia ever exerted on those about her, notwithstanding the heroic efforts she had made to liberate her own sex from the restrictions that had so long harassed and degraded it, the wives and daughters of the citizens of Athens were still kept in almost absolute seclusion and denied the opportunities of mental culture which were so generously accorded the free-born hetaerae from Asia Minor and the islands of the Aegean. Socrates, as we learn from Xenophon, asserted woman's equality with man, while Plato taught that mentally there was no essential difference between man and woman. He concluded, accordingly, that women of talent should have the same educational advantages as men. In *The Republic* as well as in the *Laws*, when he refers to education — which he would make compulsory for "all and sundry, as far as possible" — his views are far in advance of those which have been entertained until the last half century. He would have girls as well as boys thoroughly instructed in music and gymnastic — "music for the mind and gymnastic for the body."[12]

In the *Laws* he contends that "women ought to share, as far as possible, in education and in other ways with men. For consider — if women do not share in their whole life with men, then they must have some other order of life."

Again he asserts "Nothing can be more absurd than the practice which prevails in our own country of men and women not following the same pursuits with all their strength and with one mind, for thus the state, instead of being a whole, is reduced to a half."[13]

In *The Republic* he expresses the same idea when he affirms that "the gifts of nature are alike diffused in both" — men and women — "all the pursuits of men are the pursuits of women."[14]

These opinions of Socrates and Plato are so at variance with those of their contemporaries, and so contrary to the custom that then obtained of excluding all but free-born hetaerae from the advantages of education and culture, that we cannot but think that they were due to the profound influence which had been exercised directly or indirectly by Aspasia on both of these great philosophers. Be this as it may, neither the efforts of Aspasia nor the teachings of Socrates and Plato were able to remove the bars to intellectual development from which the women of Greece had so long suffered. A change in customs and laws concerning the rigid, oriental seclusion of women did not come until much later, and then it was under a new regime — that of the Caesars — while complete equality of men and women in school and college was not recognised until long centuries afterward.

It is interesting to speculate regarding what Greece would have become had she developed her women as she developed her men. Never in the history of the world were there in any one city so many eminent men — poets, orators, statesmen, painters, scu[illegible] architects, philosophers — as in Athens, and yet not a single native born Athenian woman ever attained the least distinction in any department of art or science or literature. We cannot conceive for a moment that Greece's fertility in great men and barrenness in great women was due to the fact that the mothers of such illustrious men were ordinary housewives and entirely devoid of the talent and genius which gave immortality to their distinguished sons. The careers of Aspasia and the achievements of Sappho, Corinna, Myrtides, Erinna, Praxilla, Telesilla, Myrus, Anytae and Nossidis, Theano and her daughter, to mention no others, absolutely preclude such an assumption.

The women in Greece, there can be no doubt about it, were as richly endowed by nature as were the men, and only lacked the opportunities that men enjoyed to achieve, in every sphere of intellectual activity, a corresponding measure of success. They were extraordinary types, these women of ancient Greece; for among them we find the dignified Roman matron, the chatelaine of the Middle Ages, the brilliant woman of the Renaissance and the cultured mistress of the French *salon*. But all their talent, power and genius counted for naught.

Had the civilisation of Greece been a woman's civilisation, as well as a man's civilisation, had there been a federation of all the Greek states, as Aspasia seems to have striven for, instead of a number of small and independent city-states; had the women of Hellas been allowed the same liberty of action in intellectual work as was granted to the Italian women during and after the revival of letters, and had they been encouraged to develop all their latent powers that were so systematically suppressed, and to work in unison with the men for the welfare and advancement of a united nation, it is difficult to imagine what a dazzling intellectual zenith a supremely gifted people, "full summ'd in all their powers," would have attained. Their capacity for work and for achieving great things would have been doubled and their power as a political organisation would have been practically irresistible.

"We are the only women that bring forth men," said Gorgo, the wife of Leonidas. The Spartan mothers, who had more of liberty than their Athenian sisters, did, indeed, bring forth warriors of undying renown; but it was the mothers of Athens who, notwithstanding all their grievous disabilities, gave to the world all the greatest masters in art, literature, and philosophy — the men who through the ages have been the leaders and the teachers of humanity, and who seem destined to hold their exalted position until the end of time.

The failure of the men of Greece to avail themselves of the immense potential power, which they always kept latent in their women, was the occasion of a terrible nemesis in the end. For this failure, coupled with the frightful license introduced by a class of educated women, like the hetaerae, without legal status or domestic ties, and the wave of corruption that subsequently followed the advent of the countless dissolute women who flocked to the Hellenic cities from every part of the East, paved the way for the nation's downfall and for its ultimate conquest by the resistless Roman legions that swept the once glorious but ill-fated country of Pericles and Aspasia.

## Woman And Education In Ancient Rome

The condition of women in Rome, especially from 150 BC to 150 AD, was quite different from what it was in Athens, even during her palmiest days. Owing to the lack of authentic documents we know but little of the history of the Roman people during the first five hundred years of their existence, but we do know that during this period many and important changes were effected regarding the social and civil status of women.

In the first place the Roman matron had much more freedom than was accorded the Greek wife during the age of Pericles. Far from being kept in oriental seclusion, like her Athenian sister, she was at liberty to receive and dine with the friends of her husband, and to appear in public whenever she desired. She went to the theatre and the Forum; she took part in all reputable entertainment, whether public or private. Besides this, she had more and greater legal rights than Greek women had ever known, and was treated rather as the peer and companion of man than as his toy or his slave.

Besides this, foreign women were never so conspicuous in Rome as in Athens. Even after Greece had become a Roman province, and after *Graecia capta Romam cepit* — when Greek ideas and Greek customs were introduced into the capital of the Roman world — it was still the Roman matron that was supreme. And, although many Greek women, some of them of rare beauty and culture, found their way to Rome, especially under the empire, they were always kept in the background and never succeeded in achieving anything approaching the ascendancy which distinguished them during the time of Aspasia. Their influence in literature and politics was almost *nil*.

In the case of the women of Rome, on the contrary, it may well be questioned whether woman has ever wielded a greater influence than she did during the three centuries that followed the reign of Augustus. But she did not attain to this position of pre-eminence without a long and bitter struggle. Every advance toward the goal of social and intellectual equality was strenuously contested by the men, who wished to limit the activities of their wives to the spindle, the distaff and the loom and the other occupations of the household. For, as in Greece, the generally accepted view was that woman, in the language of Gibbon, "was created to please and obey. She was never supposed to have reached the age of reason or experience." And her noblest epitaph, it was averred, was couched in the following words:

She was gentle, pious, loved her husband, was skilful at the loom and a good housekeeper.[15]

As to her mental work, far from being considered on its own merits or as a factor in the world's growth, it was flouted as

> Mere woman's work
> Expressing the comparative respect

Which means the absolute scorn.

As early as 450 BC, when the laws of the Twelve Tables were promulgated, the girls of Rome received instruction in reading, writing and arithmetic. "Up before dawn, with a lamp to light the way, and an attendant to carry her satchel, the little Roman maiden of seven years, or over, would trudge off to the portico where the schoolmaster wielded his rod.[16] For some years this life continued, with but few holidays, and those far between, until she attained some proficiency in the rudiments. Then, most probably, her education in the scholastic sense came to an end. Her brothers and boy schoolmates, if their parents wished it, could proceed from the primary school to the secondary, where geography, history and ethics were taught; where the art of elocution was assiduously practiced and the works of the great Greek and Roman poets were carefully read and expounded; but it was enough for the girl to have learned how to read, write and cipher; she had then to learn her domestic duties."[17]

With the extension of the empire and the consequent enormous increase in wealth and the rapid progress in social and intellectual freedom, there was a notable change in the character of the education given to women, at least to those of the wealthier and patrician families. This was, in great measure, due to the wave of Hellenism which, shortly after the conquest of Greece, broke upon the Roman capital with such irresistible force. To the large and rapidly increasing number of women of keen intellect and lofty aspirations, whose minds had hitherto been confined to the comparatively barren field of Roman letters, the splendid creations of Greek genius came as a revelation. To become thoroughly versed in Greek poetry and proficient in the teachings of Greek philosophy was the ambition of scores of Roman women, who soon became noted for the extent and variety of their attainments, as well as for their rare culture and charming personality.

Among the pioneers of the intellectual movement in Rome, and one of the most beautiful types of the learned women of her time, was the celebrated daughter of the elder Scipio Africanus — Cornelia, mother of the Gracchi. She is famous on account of her devotion to her two sons, Tiberius and Caius. She was their teacher; and it was her educated and refined mind that, more than anything else, contributed to the formation of those splendid characters for which they were so highly esteemed by their countrymen. Plutarch informs us that these noble sons of a noble mother "were brought up by her so carefully that they became beyond dispute the most accomplished of Roman youth; and, thus, they owed perhaps more to their excellent upbringing than to their natural parts."[18] One is not surprised to learn that this noble lady was almost idolised by the Romans, and that they erected a statue to her with the inscription, "Cornelia, Mother of the Gracchi."

Scarcely less distinguished and accomplished was another Cornelia, the wife of Pompey, the Great. "Besides her youthful beauty," writes Plutarch, in his *Life of Pompey*, "she possessed other charms, for she was well versed in literature, in playing on the lyre, and in geometry, and she had been used to listen to philosophical discourses with profit. Besides this, she had a disposition free from all affectation and display of pedantry — blemishes which such acquirements usually breed in women."[19]

Then there was the cultured and devoted Aurelia, the mother of Julius Caesar. It is safe to say that this eminent man was as much indebted to his mother for his success and greatness as were Tiberius and Caius Gracchus to the benign influence and careful teachings of the gentle and virtuous Cornelia. Highly educated and of commanding personalities, both these women, like many others of their time, contributed much to the making of Roman history by the success they achieved in moulding the characters of some of the greatest men of their own or of any age.

It is a splendid tribute that Cicero, in his *Orator*, pays to Laelia when he tells of the purity of her language and the charm of her conversation. "When I listen," he declares, "to my mother-in-law, Laelia — for women preserve the traditional purity of accent the best because, being limited in their intercourse with the multitude, they retain their early impressions — I could imagine that I hear Plautus or Naevius speaking, the pronunciation is so plain and simple, so perfectly free from all affectation and display; from which I infer that such was the accent of her father and his ancestors — not harsh like the pronunciation to which I have just referred, not broad nor rustic nor rugged, but terse, smooth and flowing."[20]

These are a few of the cultured and learned women who shed glory on their country by the refining influence which they exerted in the quiet and unostentatious precincts of the family circle. But there were others who chose a wider field for their activities, and who, by reason of their unerring judgement, well-poised and highly cultivated minds, had so won the confidence of the nation's greatest leaders that they were frequently consulted on important affairs of state. Thus, Cicero tells us of an interview which he had at Antium with Brutus and Cassius. Besides the men, there were present on this occasion three women, who took an active part in the discussion. These were Servilia, the mother of Brutus, Porcia, the wife of Brutus and the daughter of Cato, and Tertulla, the wife of Cassius and sister of Brutus. The views of the women were not without effect, and so confident was Servilia of her power that she engaged to have a certain clause in one of the decrees of the Senate expunged. This is but one of many similar instances which might be adduced from the lives of the women of Rome who took an active part in politics. As we learn from Tacitus, their counsels and assistance were considered of peculiar value by the Commonwealth. For, when some of the sterner old moralists wished to exclude women from all participation in public affairs, the Senate, after a heated debate, decided by a large majority that the cooperation of women in questions of administration, far from being a menace, as some contended, was so beneficial to the state that it should be continued.

Among other noteworthy makers of Roman history, besides those just mentioned, is Livia, the wife of Augustus and the mother of Tiberius. So great was her influence and so persistent was her activity in government affairs, that it is sometimes asserted that she was the prime mover of most of the public acts of both these rulers.

This woman, whom Ovid describes as having the features of Venus and the manner of Juno, and who, he declares, "held her head above all vices," was credited with having the benevolence of Ceres, the purity of Diana and the wisdom and craft of Minerva — "a woman," as was said by one of her contemporaries, "in all things more comparable to the gods than to men, who knew how to use her power so as to turn away peril and advance the most deserving."

Then there was the gracious, the virtuous, the self-sacrificing Octavia, sister of the Emperor Augustus, who was so successful in composing grave differences between her brother and her husband, and who so exerted her influence for peace during the troublous times in which she lived that she lives in history as a peacemaker. In marked contrast to this gentle and sympathetic woman was the energetic and heroic Agrippina, the wife of Germanicus. In many respects she was the most commanding personality of her age, and exhibited in an eminent degree those sterling qualities which we are wont to associate with the strong, dignified, courageous women of ancient Rome, who gave to the world so many and so great men in every sphere of human endeavour. "She was," as Tacitus informs us, "a greater power in the army than legates and commanders, and she, a woman, had quelled a mutiny which the emperor's authority could not check."[21] She was, indeed, as has well been said, "a woman to whom one might address an epic but never a sonnet."

I have referred to these distinguished women because they are embodiments of the best types of the noble, patrician families who made the great Roman empire the admiration of all time, and because they exhibit the wonderful advance that had been made in the general status of women since the days of Pericles and Aspasia. I have referred to them, also, to show what women are capable of achieving in the difficult and complicated affairs of public life, when they are accorded the necessary freedom of action and when they are properly equipped for work by education and by association with men of learning and experience. Comparing the secluded and illiterate Greek wife with the free and highly accomplished Roman matron, we find almost as much difference between the two as there is between a child and a fully developed woman — all the difference there was between the unsophisticated young wife, not quite fifteen, of whom Xenophon gives us such a charming picture,[22] and the highly educated and competent mother of the Gracchi.

Of the Greek maiden we are told that, before her marriage she "had been most carefully brought up to see and hear as little as possible and to ask the fewest questions;" that her whole experience before her marriage "consisted in knowing how to take the wool and make a dress, and in seeing how her mother's handmaidens had their daily spinning tasks assigned to them." Cornelia, on the contrary, was not only, as we have seen, highly accomplished, but also one who, after her husband's death, was quite prepared, as Plutarch assures us, to undertake the management of the extensive property which he left his family, and who, we may well believe, would also have been qualified, had the occasion demanded it, to perform with distinction the same duties that fell to the lot of the gifted wives of Germanicus and Augustus.

Nothing in the history of Greek and Roman womanhood more strikingly illustrates than the two instances given the vast difference in the status of the wives of Greece and Rome, or exhibits more clearly the advantages accruing to early training and thorough mental development. If there was any difference in talent or intellect between the Greek and the Roman woman it was, so far as we can determine, in favour of the Greek. The sole reason, then, for such a marked difference in their capacity for work and for achieving distinction in intellectual and administrative fields of action arose from the lack of education in the Athenian wife and the fullest measure of educational freedom enjoyed by the Roman. That Aspasia, in spite of all the odds against her, was able to rise to such a pinnacle of glory does not prove that she was the superior of her countrywomen — the mothers of the greatest poets, artists and philosophers of all time — but it exhibits rather her good fortune in being able to effect a partnership with the greatest statesman of Greece, and one who was at the same time fully able to appreciate all her rare mental attainments and give her marvellous genius free scope for development by cooperating with him in making the period during which he held the reign of power the most brilliant one in the annals of human progress.

Plato, referring to the oriental seclusion to which Athenian wives were condemned, speaks of them as "a race used to living out of the sunshine," and that, too, among a people that habitually lived out of doors. We have already seen how much greater freedom Roman women enjoyed and how much more important was the role they played in public as well as private life; but we have not told all. They not only went to, but presided over, public games and religious ceremonies. They were admitted to aristocratic clubs and had, under the empire, a regular assembly or senate of their own, known as the *Conventus Matronarum*. Hortensia, the daughter of the great orator Hortensius, pleaded the cause of her sex before the tribunal of the triumvirs, and so eloquent and effective was her speech that she not only won her case, but also won the praise of the critic, Quintilian, for her splendid oratorical effort.

Yet more. A certain woman in the Roman possessions in Africa had so impressed her fellow citizens by her intellectual capacity and administrative ability that she was chosen as one of the two chief magistrates of the place. She is known in history as Messia Castula, *duumvira*. It is true that the men of the older school, who would limit woman's activities to the distaff and the loom, strongly objected to the increasing freedom and power of women, and endeavoured to counteract their influence; but all to no purpose. And it was the crabbed old Cato, the Censor, who growled in undisguised disgust: "We Romans rule over all men and our wives rule over us."

But great as were the freedom and educational advantages of the Roman women, the startling fact remains that, with the exception of a few fragmentary verses of slight merit and of questionable authenticity, we have absolutely no tangible evidence of the Roman woman's literary ability while under pagan influence. We have seen, in considering her intellectual attainments — especially after the introduction of Greek art and letters into the City of the Seven Hills

— that every woman who pretended to culture was obliged to be familiar with the Greek as well as with the Latin authors, that her education was deemed incomplete without a knowledge of Greek poetry, oratory, history and philosophy, but the fact is indisputable that Roman women were not producers like their Greek sisters, and that in no instance did their productions reach anything like the supreme excellence of the creations of a Corinna or a Sappho. There was, it is true, Sulpicia, of whom Martial writes: "Let every girl, whose wish it is to please a single man, read Sulpicia; let every man, whose wish it is to please a single maid, read Sulpicia;" but, if the few amatory verses that are credited to her represent the highest flights of the Roman women in the domain of poetry, then, indeed, were they far behind not only Sappho and Corinna, but also far behind scores of their pupils. Martial does indeed speak of a young maiden in whom were combined the eloquence of Plato with the austere philosophy of the Porch, and who wrote verses worthy of a chaste Sappho; but this was evidently a great exaggeration, for we have no other evidence of her existence.

The creative work of Roman women was, so far as we are able to judge, quite as limited in prose as it was in poetry. Agrippina, the mother of Nero, was one of the few prose writers whose name has come down to us. From her memoirs it was that Tacitus received much of the material incorporated in his *Annals*.

That some of the women had literary ability of a high order is indicated by a letter of Pliny to one of his correspondents, in which occurs the following passage:

> Pomponius Saturninus recently read me some letters which he averred had been written by his wife. I believed that Plautus or Terence was being read in prose. Whether they were really his wife's, as he maintains, or his own, which he denies, he deserves equal honour, either because he composes them or because he has made his wife, whom he married when a mere girl, so learned and so polished.[23]

Scarcely less distinguished for her taste in literature, and for her talent as a letter writer, was Pliny's wife, Calphurnia, who, at his request, wrote to him in his absence every day and sometimes even twice a day. According to Cicero, his daughter Tulia was "the best and most learned of women;" but her literary work, it is probable, did not extend much beyond her letters to her illustrious father. Nevertheless, what would we not give to possess these letters — to have as complete a collection of them as we have of those of the great orator and philosopher. They would be of inestimable value and would be absolutely beyond compare, except, possibly, with the letters of Mme du Deffand or of Elizabeth Barrett Browning of a much later age.

Considering the number of educated women that lived in the latter days of the Republic and during the earlier part of the Empire, and their well known culture and love of letters, it is reasonable to suppose that they may have written much in both prose and verse of which we have no record. Literary productions must have more than ordinary value to survive two thousand years, and especially two thousand years of such revolutions and upheavals as have convulsed the world since the time of the *Pax Romana*, when all the world was at peace under Augustus.

How much of the literary work of the women of today will receive recognition twenty centuries hence? Some of it may, it is true, find a place in the fireproof libraries of the time; but who, outside of a few antiquarians, will take the trouble to read it or estimate its value? A few anthologies containing our gems of prose and poetry will probably be all that our fortieth century readers will deem worthy of notice. In view of the chaotic condition of Europe for so many centuries, the wonder is not that we have so little of the literary remains of Greece and Rome, but rather that we have anything at all.

As one might expect, the literary women of Rome, as well as those who ventured to take part in public affairs, had their critics. The satirists of the time were as unsparing of their ridicule as they were long afterward when Moliere wrote his *Femmes Savantes* and his *Precieuses Ridicules*. And as for men of the old conservative type, a learned woman was as much an object of horror as is a militant suffragette in conservative England today. "No learned wife for me," exclaims Martial, "but rather a well-fed slave."[24]

And Juvenal had no more love for educated women than have some of our contemporaries for a blue-stocking housekeeper. He gives his opinion of them in the following characteristic fashion:

> That woman is a worse nuisance than usual who, as soon as she reclines on her couch, praises Virgil; makes excuses for doomed Dido; pits bards against one another and compares them, and weighs Homer and Mars in the balance. Teachers of literature give way, professors are vanquished, the whole mob is hushed, and so great is the torrent of words that no lawyer or auctioneer may speak, nor any other woman.[25]

But if learned women had their enemies and detractors they also had friends and defenders. Among these was the Stoic philosopher, C Musonius Rufus, who lived in the time of Nero. Like Plato, he contended that women should have the same training as men and that the faculties of both should be equally developed. The gist of his teaching is contained in the statement that:

> If the same virtues must pertain to men and women, it follows, necessarily, that the same training and education must be suitable for both.[26]

Our brief sketch of women's work in ancient Rome would be incomplete without some reference to the famous *Ecclesia Domestica* —Church of the Household—on the Aventine, and the distinguished women who were its chief ornaments. During the time of Pope Damasus, and not long before the sacking of Rome by Alaric, the *Ecclesia Domestica* was a kind of conventual home to which had retired, or in which were frequently gathered, some of the most noble and learned women of the city. Among the most notable of these were Marcella and her friends, Paula and Eustochium.

For beauty of character and nobility of purpose and rare mental endowments they recall the best traditions of a Cornelia or a

Calphurnia, while so great was their purity of life and so unbounded was their charity to the poor and suffering that they were honoured by being numbered among the saints of the early church. But what specially distinguished them among all the great women of the Roman world was their great and varied learning. In this respect they probably were far in advance of all their predecessors. For, in addition to a thorough knowledge of Latin and Greek literature, history and philosophy, they had, under the great theologian and orientalist, St Jerome, become proficient in Hebrew and deeply versed in Scripture.

Special mention should be made of Paula and her daughter Eustochium; for it is probable that, had it not been for their influence on Jerome, and their active cooperation in his great life work, we should not have the Latin version of the Scriptures that is today known as the Vulgate. This is evinced from the letters of the saint himself and from what we know of the lives of these two remarkable women, who, as St Jerome informs us in the epitaph which he had engraved on Paula's tomb in the Church of the Nativity in Bethlehem, were descended from the Scipios, the Gracchi and the Pauli on the mother's side, and on the father's side from the half-mythical kings of Sparta and Mycenae.[27]

They aided him not only by their sympathy and by purchasing for him, often at a great price, the manuscripts he needed for his colossal undertaking, but also assisted him by their thorough knowledge of Latin, Greek and Hebrew in translating the Sacred Books from the original Hebrew into Latin. So great was Jerome's confidence in their scholarship and so high was his appreciation of their ability and judgement that he did not hesitate to submit his translations to them for their criticism and approval. After he had completed his version of the first Book of Kings, he turned it over to them, saying: "Read my Book of Kings — read also the Latin and Greek translations and compare them with my version." And they did read and compare and criticise. And more than this, they frequently suggested modifications and corrections which the great man accepted with touching humility and incorporated in a revised copy.

More wonderful still, the Latin Psalter, as it has come down to us, is not, as is generally supposed, the translation from the Hebrew of Jerome, but rather a corrected version made from the Septuagint by his illustrious collaborators — Paula and Eustochium.

It is safe to say that no two women were ever engaged in a more important or more difficult literary undertaking — one requiring keener critical sense or more profound learning — than were Paula and Eustochium, or one in which their efforts were crowned with more brilliant success than were those of these two supreme exemplars of the grace, the knowledge, the culture, the refinement of Roman womanhood — the crowning glories of womanhood throughout the ages.

St Jerome showed his grateful recognition of the invaluable assistance received from his devoted and talented co-workers by dedicating to them a great number of his most important books. This scandalised the pharisaical men of the time, who looked askance at all learned women and resented particularly the pre-eminence given to Paula and her accomplished daughter. But their reproaches provoked a reply from the saint that was worthy of the most chivalrous champion of woman, and revealed, at the same time, all the nobility of soul of the roused "Lion of Bethlehem." It is not only a defence of his course, but also a splendid tribute to his two illustrious friends, and a tribute also to the great and good women of all time.

"There are people, O Paula and Eustochium," exclaims the Christian Cicero, vibrant with emotion and in a burst of eloquence that recalls one of the burning philippics of Marcus Tullius, "who take offence at seeing your names at the beginning of my works. These people do not know that Olda prophesied when the men were mute; that while Barach was atremble, Deborah saved Israel; that Judith and Esther delivered from supreme peril the children of God. I pass over in silence Anna and Elizabeth and the other holy women of the Gospel, but humble stars when compared with the great luminary, Mary. Shall I speak now of the illustrious women among the heathen? Does not Plato have Aspasia speak in his dialogues?

Does not Sappho hold the lyre at the same time as Alcaeus and Pindar? Did not Themista philosophise with the sages of Greece? And the mother of the Gracchi, your Cornelia, and the daughter of Cato, wife of Brutus, before whom pale the austere virtue of the father and the courage of the husband — are they not the pride of the whole of Rome? I shall add but one word more. Was not it women to whom our Lord first appeared after His resurrection? Yes, men could then blush for not having sought what the women had found."[28]

Time has spared a joint letter of Paula and Eustochium to their friend Marcella — a letter which exhibits so well the rare culture and literary ability of the writers that we cannot but lament that we have not more of the correspondence which was carried on between the learned inmates of the Church of the Household on the Aventine and Paula's convent home near the Church of the Nativity in Bethlehem. Such a collection would be beyond price, as it would complete the picture of the age so well sketched by St Jerome; and, as a contribution to the literary world, it would have a value not inferior to that of those exquisite classics of a later age — the letters of Madame Sevigne to her daughter.[29]

## Woman And Education During The Middle Ages

The period of nearly a thousand years intervening between the downfall of Rome in AD 476 and the taking of Constantinople by the Turks in 1453 is usually known in history as the Middle Ages. By some it is considered as synonymous with the Dark Ages, because of the decline of learning and civilisation during this long interval of time. The former designation seems preferable, for, as we shall see, the latter is more or less misleading. During the "wandering of the nations" in the fourth and fifth centuries, and the long and fierce struggles between the barbarian hordes from the north with the decadent peoples of the once great Roman empire, there was, no doubt, a partial eclipse of the sun of civilisation; but the consequent darkness was not so dense nor so general and long-continued as is sometimes imagined. The progress of intellectual culture was, indeed, greatly retarded, but there was no time when the light of learning was entirely extinguished. For even during the most troublous times there were centres of culture in one part of Europe or another. At one time the centre was in Italy, at another in Gaul, and, at still another, it was in Britain or Ireland or Germany.

But whether it was in the south, or the west or the north of Europe that letters flourished, it was always the convent or the monastery that was the home of learning and culture. Within these holy precincts the literary treasures of antiquity were preserved and multiplied. Here monks and nuns laboured and studied, always keeping lighted the sacred torch of knowledge — *Et quasi cursores vitaï lampada tradunt* — and passing it on to the generations that succeeded them. That any of the great literary masterpieces of Greece and Rome have come to us, in spite of the destructive agencies of time and the wreck of empires, is due wholly to the unremitting toil through long ages of the zealous and intelligent inmates of the cloister.

Of the monastic institutions for men there is no occasion to speak, except in so far as they contributed to the intellectual advancement of woman. In some cases the women of the cloister owed much to ecclesiastics for their literary training; but there are not wanting instances in which the nuns took the lead in education and had the direction of schools which gave to the church priests and bishops of recognised scholarship.

Practically the only schools for girls during the Middle Ages were the convents. Here were educated rich and poor, gentle and simple. And in these homes of piety and learning the inmates enjoyed a peace and a security that it was impossible to find elsewhere. They were free from the dangers and annoyances that so often menaced them in their own homes and were able to pursue their studies under the most favourable auspices.

Among the first convent schools to achieve distinction were those of Arles and Poitiers in Gaul, in the latter part of the sixth century. The Abbess of Poitiers is known to us as St Radegund. She not only had a knowledge of letters rare for her age, but wrote poems of such merit that they were until recently accepted as the productions of her master, the poet Fortunatus,[30] who subsequently became bishop of Poitiers.

Far more notable, however, than the convents of Arles and Poitiers was the celebrated convent of St Hilda at Whitby. Hilda, the foundress and first abbess of Whitby, was a princess of the blood-royal and a grand-niece of Edwin, the first Christian king of Northumbria. Her convent and adjoining monastery for monks soon became the most noted centre of learning and culture in Britain. And so great was her reputation for knowledge and wisdom that not only priests and bishops, but also princes and kings sought her counsel in important matters of church and state.

As to the monks subject to her authority, she inspired them with so great a love of knowledge, and urged them to so thorough a study of the Scriptures, that her monastery became, as Venerable Bede informs us, a school not only for missionaries but for bishops as well. He speaks in particular of six ecclesiastical dignitaries who were sent forth from this noble institution — all of whom were bishops. Five of them he describes as men of singular merit and sanctity — "*singularis meriti et sanctitatis viros*," while the sixth, he declared, was a man of rare ability and learning — "*doctissimus et excellentis ingenii.*" Of this number was St John of Beverly, who, we are told, "attained a degree of popularity rare even in England, where the saints of old were so universally and so readily popular."[31] Hilda governed her double monastery with singular wisdom and success; and, so great was the love and veneration she inspired among all classes that she merited the epithet of "Mother of her Country."

Celebrated, however, as Hilda was for her great educational work at Whitby, she is probably better known to the world as the one who first recognised and fostered the rare gifts of the poet Caedmon. "It is on the lips of this cowherd," as Montalembert beautifully expresses it, "that the Anglo-Saxon speech first bursts into poetry. Indeed, nothing in the whole history of European literature is more original or more religious than this first utterance of the English muse."[32]

As soon as Hilda discovered the extraordinary poetic faculty of Caedmon, she did not hesitate to regard it "as a special gift of God, worthy of all respect and of the most tender care." And, in order that she might the more readily develop the splendid talents of this literary prodigy, the keen discerning abbess received Caedmon into the monastery of monks, and had him translate the entire Bible into Anglo-Saxon. "As soon as the Sacred Text was read for him he forthwith," as Bede declares, "ruminated it as a clean animal ruminates its food, and transformed it into songs so beautiful that all who heard were delighted."

As his poetical faculty became more developed, his profoundly original genius became more marked, and his inspiration more earnest and impassioned. It was this Northumbrian cowherd, transformed into a monk of Whitby, who sang before the abbess Hilda the revolt of Satan and Paradise Lost, a thousand years earlier than Milton, in verses which may still be admired even beside the immortal poem of the British Homer. So remarkable, indeed, in some instances is the similarity in the productions of the two poets that F Palgrave, one of the most competent of English critics, does not hesitate to declare that certain of Caedmon's verses resembled so closely certain passages of the *Paradise Lost* that some of Milton's lines seem almost like a translation from the work of his distinguished predecessor. And M Taine, in his *History of English Literature*, referring to the "string of short, accumulated, passionate images, like a succession of lightning flashes," of the old Anglo-Saxon poet, asserts that "Milton's Satan exists in Caedmon's as the picture exists in the sketch."[33]

Well could Caedmon's first biographer, the Venerable Bede, say of him, "Many Englishmen after him have tried to compose religious poems, but no one has ever equalled the man who had only God for a master." And not without warrant does the eloquent Montalembert, in the masterly work just quoted, pen the following statement: "Apart from the interest which attaches to Caedmon from a historical and literary point of view, his life discloses to us essential peculiarities in the outward organisation and intellectual life of those great communities which in the seventh century studded the coast of Northumbria, and which, with all their numerous dependents, found often a more complete development under the crozier of such a woman as Hilda than under the superiors of the other sex."[34]

Space precludes my telling of other convents which were centres of literary activity, and of nuns who distinguished themselves by their learning and by the benign influence which they exerted far beyond the walls of the cloister. I cannot, however, refrain from referring to that group of learned English nuns who are chiefly known by their Latin correspondence with St Boniface, the Apostle of Germany, and by the assistance which they gave him in his arduous labours. Conspicuous among these was St Lioba, who, at the request of Boniface, left her home in England to found a convent at Bischopsheim in Germany, which, under the direction of its learned and zealous abbess, soon became the most important educational centre in that part of Europe. Teachers were formed here for other schools in Germany and Lioba's biographer tells us that there were few *monasteria feminarum* — monasteries of women — within the sphere of Boniface's missionary activities for which Lioba's pupils were not sought as instructresses.

Like her illustrious countrywoman, St Hilda, the abbess of Bischopsheim was the friend and counsellor of spiritual and temporal rulers. Charlemagne, that eminent patron of scholars, had a great admiration for her and gave her many substantial proofs of his esteem and veneration. "Princes," writes her biographer, "loved her, noblemen received her, and bishops gladly entertained her and conversed with her on the Scriptures and on the institutions of religion, for she was familiar with many writings and careful in giving advice. She was so bent on reading that she never laid aside her book except to pray or to strengthen her slight frame with food or sleep."[35] She was thoroughly conversant with the books of the Old and the New Testaments and was, at the same time, familiar with the writings of the Fathers. It is not surprising, then, that she was regarded as an oracle, and that all classes flocked to her as they did to the abbess of Whitby for guidance and assistance.

From what has been said of the accomplishments and achievements of the Anglo-Saxon nuns just mentioned, it is evident that they were, of a truth, women of exceptional worth and of sterling character. And it is equally clear that their pupils must have shared in the education and culture of their distinguished teachers.[36] Many of them, in addition to having a wide acquaintance with literature, sacred and profane, were also mistresses of several languages. A woman's education, at this time, was not complete unless she could write Latin and speak it fluently. The author of that most interesting early English work, *Ancren Riwle* — Rule of Anchoresses — presupposes in his auditors, for whose benefit his instructions were given, a knowledge of Latin and French, as well as of English. In certain convents Latin was almost the sole medium of communication — to such an extent, indeed, that a special rule was made prohibiting "the use of the Latin tongue except under special circumstances."

"As long as the conventual system lasted the only schools for girls in England were the convent schools where, says Robert Aske, 'the daughters of gentlemen were brought up in virtue.' From an educational point of view, the suppression of the convents was decidedly a blunder." Thus writes Georgiana Hill in her instructive work on *Women in English Life*, and there are, we fancy, but few readers of her instructive pages who will not be inclined to agree with her conclusions.[37] Lecky speaks of the dissolution of convents at the time of the Reformation as "far from a benefit to women or the world."[38] And Dom Gasquet declares "that destruction by Henry VIII of the conventual schools where the female population, the rich as well as the poor, found their only teachers, was the absolute extinction of any systematic education of women for a long period."[39]

But this is not all. The strangest and saddest result, consequent on the suppression of the convents, was that men were made to profit by the loss which women had sustained. The revenues of the houses that were suppressed had been intended for the sole use and behoof of women, and had been administered by them in this sense for centuries. When they were appropriated by Henry VIII, it never occurred to him or his ministers to make any provision for the education of women in lieu of that which had so ruthlessly been wrested from them. Thus the nunnery of St Radegund, together with its revenues and possessions, was transformed into Jesus College, Cambridge, while from the suppressed convents of Bromhall in Berkshire and Lillechurch in Kent funds were secured for the foundation and endowment of St John's College, also at Cambridge. Similarly, the properties of other nunneries, large and small, were appropriated for the foundation of collegiate institutions at Oxford, all of which were for the benefit of men.

And so it was that, in a few short years, the great work of centuries was undone and women were left little better educational facilities than when the Anglo-Saxon nuns began their noble work in a land that was enveloped in "one dark night of unillumined barbarism."

One would have thought that Elizabeth, who was so highly educated, and who did so much for the supremacy of her country on land and sea, would have bethought herself of the necessity of doing something for the education of her female subjects. But no. She did nothing for them, and the founders of the endowed grammar schools, during her reign, gave never a thought to the educational necessities of the girls. They made provision only for the boys. In this respect, however, the "Virgin Queen" was but following in the footsteps of the male sovereigns and legislators who had preceded her, and who, although affecting an interest in having women "sensible and virtuous, seem by their conduct toward the sex to have entered into a general conspiracy to order it otherwise."

The truth is, when anything was achieved for the intellectual advancement of women it was due either to private instruction or to the result of a protracted struggle on the part of women themselves for what they deemed their indefeasible rights. Had they relied on the spontaneous action of men and on legislation in favour of female education to which men had given the initiative, they would today be in the same condition of ignorance and seclusion and servitude as was the Athenian woman twenty-five centuries ago, and would occupy a status but little above that of the inmates of oriental harems and zenanas.

The Anglo-Saxon nuns were, as we have seen, specially distinguished for their learning and for the splendid work they performed for the education of their sex during the long period of the Middle Ages. But however great their pre-eminence in these respects, they were not without rivals. There were, besides the schools, already named, conducted by St Lioba and her companions, also flourishing schools in Germany under the direction of native nuns, whose success as educators was as marked as that of Lioba or Hilda, and who, in addition to their labours in the class-room, achieved distinction by their productive work. The Anglo-Saxon convents developed few writers, whereas those of Germany produced several who not only shed lustre on their sex but who also showed what woman is capable of accomplishing when accorded some measure of encouragement and full liberty of action.

One of the most noted writers of her age was the famous nun of Gandersheim, Hroswitha, who was born in the early part of the tenth century. She was the pupil of the abbess Gerberg, who was of royal lineage, and one of the most zealous promoters of learning and culture in Saxony during the forty-two years of her rule in the convent to which she and her favourite pupil gave undying renown.

Hroswitha's literary work consists of legends and contemporary history in metrical form and of her dramas written in the style of Terence. As a writer of history and legends she ranks with the best authors of her time, while as a writer of dramas she stands absolutely alone. Hers, indeed, were the first dramatic compositions given to the world during the long interval that elapsed between the last comedies of classic antiquity and the first of the miracle plays which had such a vogue between the twelfth and the sixteenth century.

Her dramas, which, of all her works, have attracted the most attention, are seven in number. They deal with the moral and mental conflicts which characterised the period of transition from heathendom to Christianity. Some of them exhibit poetic talent of a high order as well as the inspiration and courage of genius. They reveal also a wide acquaintance with the classic authors of Rome and Greece, besides a knowledge of many of the Christian writers. They are, likewise, distinguished by originality of treatment, complete mastery of the material used, as well as by genuine beauty of rhyme and rhythm. In form, all the plays preserve the simple directness of their model, Terence, while, in conception, they embody the noblest ideals of Christian teaching. In marked contrast to her model, who invariably exhibits the frailties and lapses of woman, Hroswitha's plays turn on the resistance of her sex to temptation, and on their steadfast adherence to duty and to vows voluntarily assumed. A recent English writer, W H Hudson, in an appreciative estimate of the work of this learned Benedictine nun expresses himself as follows:

> It is on the literary side alone that Hroswitha belongs to the classic school. The spirit and essence of her work belong entirely to the Middle Ages; for beneath the rigid garb of a dead language (— she wrote in Latin —) "beats the warm heart of a new era. Everything in her plays that is not formal but essential, everything that is original and individual, belongs wholly to the Christianised Germany of the tenth century. Everywhere we can trace the influence of the atmosphere in which she lived; every thought and every motive is coloured by the spiritual conditions of her time. The keynote of all her works is the conflict of Christianity with paganism; and it is worthy of remark that in Hroswitha's hands Christianity is throughout represented by the purity and gentleness of woman, while paganism is embodied in what she describes as the vigour of men — *virile robur.*[40]

Among her legends the one entitled *The Lapse and Conversion of Theophilus* has a special interest as being the precursor of the well-known legend of Faust.

In Hroswitha's time, as in our own, there were people who were strongly opposed to the higher education of women. There were others who would deny them even the elements of an education — who declared that they should be taught anything rather than reading and writing, which were a cause of temptation and sin — that their knowledge should be confined solely to the duties of an ordinary housewife, that their books should consist solely of thimble, thread and needles — "*Et leurs livres, un de, du fil et des aguilles.*" Some, it is true, were willing to make an exception in favour of nuns; but, as to all others, the less they knew the better it was for their spiritual, if not for their temporal, welfare also.[41] To those who were thus minded, Hroswitha pithily replied that it was not knowledge itself but the bad use of it that was dangerous — "*Nec scientia scibilis Deum offendit, sed injustitia scientis.*"

Among other women who were Hroswitha's equals in knowledge, if not in literary attainments, were several other nuns who illumined the closing centuries of the Middle Ages. Chief among these were St Hildegard, "the sybil of the Rhine;" Herrad, the noted author of the *Hortus Deliciarum — Garden of Delights* — and Matilda and Gertrude, those remarkable mystical writers, whose descriptions of heaven and hell so closely resemble those in the *Divina Commedia* that many writers are of the opinion that the great Florentine poet must have been familiar with the accounts which they gave of their visions.

St Hildegard was for a third of a century the abbess of the convent of St Rupert at Bingen. So great was her reputation for sanctity and for the extent and variety of her attainments that she was called "the marvel of Germany." She is without doubt one of the most beautiful and imposing as well as one of the greatest figures of the Middle Ages — great beside such eminent contemporaries as Abelard, Martin of Tours and Bernard of Clairvaux. People from all parts of the Christian world sought her counsel; and her convent at Bingen became a Mecca for all classes and conditions of men and women. But nothing shows better the immense influence which she wielded than her letters of which nearly three hundred have been preserved.

Among her correspondents were people of the humble walks

of life as well as the highest representatives of Church and State. There were simple monks and noble abbots; dukes, kings and queens; archbishops and cardinals and no fewer than four Popes. Letters came to her from the orient and the occident, from the patriarch of Jerusalem, from Queen Bertha of Greece, from Frederick Barbarossa, Philip the Count of Flanders, St Bernard, the professors of the University of Paris; from Henry II of England, and from his granddaughter Eleonora, "The Damsel of Brittany." It is safe to say that no woman during the Middle Ages exercised a wider or more beneficent influence than did this humble Benedictine abbess of Bingen on the Rhine and had unsought so large a number of distinguished correspondents. And, if we accept the criterion that influence is measured by the number and nature of one's relations, it would be difficult to find in any age relations that were more select or more cosmopolitan.

But her astonishing collection of letters is the slightest product of her intellectual activity. She is without doubt the most voluminous woman writer of the Middle Ages. Her works on theology, Scripture and science make no less than six or eight large octavo volumes. The Bollandists, than whom there is no more competent authority, express their amazement at the amount and quality of Hildegard's work. Witness the following language of one of their number: "Although we may not be surprised that our saint was interrogated regarding secret things by so many men eminent both by reason of their dignity and their learning, I am nevertheless forced to recognise with stupefaction that a woman without instruction, and who had not acquired knowledge by study, was consulted concerning the most difficult questions of theology and the most subtle of Holy Scriptures, and that she gave, without hesitation, the answers that were demanded by theology and Scripture."[42]

Is it, then, surprising that the famous William of Auxerre, after a critical examination of her works, should compare her with Peter Lombard, the celebrated "Master of the Sentences,"[43] and one of the most learned of the Schoolmen, and write that Hildegard is *Sententiarum Magistra* — Mistress of the Sentences — and that "in her works the words are not human but divine"? Has any woman writer ever received higher praise, and from one so competent to express an opinion as the scholarly divine of Auxerre?

Herrad, the gifted abbess of Hohenburg in Alsace, was a contemporary of Hildegard, and, like her, was noted for her culture and wide range of knowledge. She is chiefly known for her *Hortus Deliciarum*, a remarkable work, encyclopaedic in character, which she wrote for the nuns of her convent and which was designed to embody in words and in pictures the knowledge of her age.

Nothing that time has bequeathed to us gives us a clearer conception of the manifold activities of a medieval nunnery, of the industry, talents and enthusiastic love of learning of its inmates, than Herrad's wonderful *Garden of Delights*. Nor is there any other work that gives us a better knowledge of the manners, customs and ideals of the twelfth century, or one that, in its particular sphere, is of more value to the student of art, philology and archaeology. It exhibits Herrad's intense interest in the intellectual advancement of her nuns and pupils as well as her superior talent and acquirements. Unfortunately the manuscript copy of this work was destroyed at the time of the bombardment of Strasburg by the Germans in 1870, and our knowledge of it is limited to portions of it which had previously been transcribed or to accounts left of it by those who had examined it before its destruction. Of such exceptional value was this unique work that the editor of the great collection of pictures, which illustrates this remarkable book, does not hesitate to declare that "Few illuminated manuscripts had acquired a fame so well deserved as the *Hortus Deliciarum* of Herrad."[44]

No sketch, however brief, of the literary nuns of medieval Germany would be complete without some reference to the learned religious of the convent of Helfta, near Eisleben in Saxony. Of the abbess Gertrude we read that her enthusiasm for knowledge was so great that she not only inspired others with the same enthusiasm, but that she was an incessant collector of books, which she had her nuns transcribe. Among her most distinguished subjects were two religious by the name of Matilda, one of whom was her sister, and a third, who, to distinguish her from the abbess, is known as "Gertrude the Great."

The writings of these nuns were inspired by that great mystic movement which then prevailed in various parts of Europe and are among the most impassioned productions of the age. For this reason they still have a special claim on the attention of students of art and literature, as well as those of theology and mysticism. Impressed by the similarity of their ideas and descriptions as compared with those found in Dante's great masterpiece, there are not wanting scholars who contend that the prototype of the Matelda in the earthly paradise of the *Purgatorio* was none other than one of the Matildas of the famous convent of Helfta.[45]

The writings of Hroswitha, Hildegard, Herrad, Gertrude and the Matildas, to speak of no others, are the best evidence of the studious character of the nuns of medieval times, and of their devotion to the cause of education. They command, likewise, our admiration for the system of training which made such development possible, and show that, in certain departments, the schools as then conducted were on as high a plane as any we have today.[46] They show us, too, that nuns and convent-bred women of the age in question were of quite different mental calibre from that of the "gentle lady of chivalry living in her bower, playing upon her lute and waiting patiently for the return of her triumphant knight," and quite different, too, from that of the castle lady-loves — whose sole attractions were often no more than youth and beauty — who inspired the impassioned lyrics of troubadour and minnesinger.

A recent writer sums up in a few words the status and the accomplishments of the lady of the abbey in the following paragraph:

> No institution of Europe has ever won for the lady the freedom and development that she enjoyed in the convent in early days. The modern college for women only feebly reproduces it, since the college for women has arisen at a time when colleges in general are under a cloud. The lady-abbess, on the other hand, was part of the two great social

> forces of her time, feudalism and the Church. Great spiritual rewards and great worldly prizes were alike within her grasp. She was treated as an equal by the men of her class, as is witnessed by letters we still have from popes and emperors to abbesses. She had the stimulus of competition with men in executive capacity, in scholarship, and in artistic production, since her work was freely set before the general public; but she was relieved by the circumstances of her environment from the ceaseless competition in common life of woman with woman for the favour of the individual man. In the cloister of the great days, as on a small scale in the college for women today, women were judged by each other as men are everywhere judged by each other, for sterling qualities of head and heart and character.[47]

Nor is this all. Never was woman more highly honoured, never was her power and influence greater than during the period of conventual life extending from Hilda of Whitby to Gertrude and the Matildas of Helfta, and especially during that golden period of monasticism and chivalry when cloister and court were the radiant centres of learning and culture. Abbesses took part in ecclesiastical synods and councils and assisted in the deliberations of national assemblies. In England, they ranked with lords temporal and spiritual, and had the right to attend the king's council or to send proxies to represent them, while in Germany, where they held property directly from the king or emperor, they enjoyed the rights and privileges of barons and, as such, took part in the proceedings of the imperial diet either in person or through their accredited representatives. In Saxony, the abbesses had the right to strike coins bearing their own portraits, notably the abbesses of Gandersheim and Quedlinburg. In England they were invested with extraordinary powers, and in certain cases owed obedience to none save the Pope. In Kent abbesses, as representatives of religion, came immediately after bishops.

Possessing such power and prestige, it is not surprising to learn that abbesses wielded great influence in temporal as well as spiritual matters; that it pervaded politics and extended to the courts of kings and emperors. Thus, Matilda, the abbess of Quedlinburg, together with Adelheid, the mother of Otto III who was but three years old at the time of his father's death, practically ruled the empire. At a later period during the prolonged absence in Italy of Otto III, the control of affairs was entrusted to the abbess alone; and so successful was her administration, and so vigorous were the measures which she adopted against the invading Wends, that she commanded the admiration of all. In view of these facts, the learned authoress of *Woman under Monasticism* is fully warranted in declaring as she does "The career open to the inmates of convents in England and on the Continent was greater than any other ever thrown open to women in the course of modern European history."[48]

"The educational influence of convents during centuries," continues the same writer, "cannot be rated too highly. Not only did their inmates attain considerable knowledge but education in a nunnery, as we see from Chaucer and others, secured an improved standing for those who were not professed."[49] It prepared the way for, if it did not train, those highly educated women who appeared during the time of the transition between the Middle Ages and what is now designated as the Modern Period.

Among these were Christine de Pisan, who was a prolific writer on many subjects in both prose and verse, and who, it is said, was the first woman to earn a livelihood by her pen.[50] There were also some of those remarkable women who lectured on law in the University of Bologna, among whom were Bettina Gozzadini,[51] who, some writers will have it, occupied the chairs of law in her *alma mater* as early as 1236, and the celebrated Novella d'Andrea, of the following century, who frequently acted as a substitute for her father, a professor of canon law in the university, and who, by reason of her varied and profound knowledge, held a prominent place among the most learned men of her time. Both of these noted women were worthy prototypes of that long list of learned Italian women who, during the Renaissance, won such honour for themselves and such undying glory for their country. Not less remarkable were several women of the school of Salerno, who, during its palmiest days, distinguished themselves as teachers, writers and medical practitioners,[52] and the still more remarkable daughters of one Mangord, a professor of Paris, whose daughters taught Sacred Scripture.[53] There were few in number, it is true, but they were the worthy prototypes of those learned and brilliant women who achieved such distinction and glory for their sex during that most interesting period of history known as the Renaissance.

## Woman And Education During The Renaissance

By the Renaissance we understand not only a phase in the development of the nations of Europe but also that period of transition between the medieval and the modern world during which the latent spiritual energies of the Middle Ages developed into the intellectual forces and moral habits of thought which now pervade the civilised world. Various dates are assigned for its starting point. Among them is the fall of Constantinople in 1453, when there was a great influx of scholars from the famed metropolis on the Bosphorus to the Italian peninsula, who brought with them those forgotten treasures of science and literature which were so instrumental in producing that interesting phenomenon known in history as the Revival of Learning. But whatever date be assigned for the beginning of the Renaissance, whether it be the year when Constantinople fell into the hands of the Turk or the fateful millennial year which was to witness the termination of all things, there certainly was never at any period a distinct breach of historical continuity between the old order and the new.

This is particularly true of Italy where the Renaissance had its origin. For here, during the entire medieval period, there never was a time when the study of antiquity was completely neglected; when the traditions of the old Roman culture had died out, or when the art and the literature of the classical ages of the past had ceased to exert an influence on artists and scholars. Osanam was, then, right when he declared that the night of the Dark Age, which in Italy intervened between "the intellectual daylight of antiquity and the dawn of the Renaissance," was, in reality, like "one of those luminous nights in which the fading brightness of evening is prolonged into the first beaming of the morning."[54]

So much, indeed, was this the case that those who have made the most profound study of the Middle Ages recognise a first Renaissance in the twelfth century, which was not less real than the Renaissance *par excellence* of the fifteenth century, a renaissance which counts such masters of Latinity as Abelard, John of Salisbury and Hildebert of Tours, and such schools as that of Chartres, where classical Latin was taught with as much thoroughness as in the great universities of Europe during the brilliant age of the humanists. It was then, as Rashdall truly observes, that "a revival of architecture heralded, as it usually does, a wider revival of Art. The schools of Christendom became thronged as they were never thronged before. A passion for enquiry took the place of the old routine. The Crusades brought different parts of Europe into contact with one another and into contact with the new world of the East — with a new religion and a new philosophy, with the Arabic Aristotle, with the Arabic commentators on Aristotle, and eventually even with Aristotle in the original Greek."[55]

Roughly speaking, the Renaissance attained its culmination during the second half of the fifteenth century. It was during this period that gunpowder and printing with movable types were invented — the first completely revolutionising the methods of warfare and the second marvellously facilitating the diffusion of knowledge. And it was during the same period also that Vasco da Gama doubled the Cape of Good Hope, that Columbus crossed the Sea of Darkness and that Copernicus laid the foundation of modern astronomy.

But this wonderful half-century constituted only a small portion of the period embraced by the Renaissance. From the fall of Constantinople until it attained the highest phase of development in England, the Renaissance covers a period of nearly two centuries.

The progress of the intellectual and moral movement which it represented, from the land of its birth, to the northern and western parts of Europe, was comparatively slow. Thus, while Italy was exhibiting the full effulgence of the re-birth, England was still in the feudal condition of the Middle Ages. A striking illustration of this truth is seen in the fact that "a brother of the Black Prince banqueted with Petrarch in the palace of Galeazzo Visconti — that is to say, the founder of Italian humanism, the representative of Italian despotic statecraft, and the companion of Froissart's heroes met together at a marriage feast." "In Italy," as Symonds has shown, "the keynote was struck by the *Novella*, as in England by the drama."[56] The supreme exponents of the Renaissance as manifested in literature were, without doubt, Ariosto in Italy, Rabelais in France, Cervantes in Spain, Camoens in Portugal, Erasmus in the Netherlands and Shakespeare in England.

Considering the splendid achievements of men during the Renaissance in every department of intellectual activity, one would imagine that women also would have attained to a somewhat proportionate distinction, at least in literature and the arts. But, outside of Italy, this was far from being the case. In France, Spain, Portugal and England there were, it is true, a certain number of women who won distinction by their talents and learning, but these were the exceptions which but served to throw into greater relief the prevailing ignorance of the great mass of their sex, which had few, if any, of the advantages of instruction, even in the most elementary branches of knowledge.

The Italian women, as we have already seen, had commanded marked recognition for their talents and learning even before the close of the Middle Ages. The most famous of these were among those who, having obtained the doctorate, became lecturers and professors in the great university of Bologna. The existence and accomplishments of some of these may, perhaps, be more or less legendary, but there can be no doubt that many of them, some before the time of the Renaissance, had gained a European reputation for the breadth and variety of their attainments. But it was during the

Renaissance that the remarkable flowering of the intellect of the Italian woman was seen at its best. While the women in the other parts of Europe, especially in England and Germany, were suffering the ill effects consequent on the suppression of the convents, which, for centuries, had been almost the only schools available for girls, the women of Italy were taking an active part in the great educational movement inaugurated by the revival of learning, and winning the highest honours for their sex in every department of science, art and literature. Not since the days of Sappho and Aspasia had woman attained such prominence, and never were they, irrespective of class-condition, accorded greater liberty, privileges or honour. The universities, which had been opened to them at the close of the Middle Ages, gladly conferred upon them the doctorate, and eagerly welcomed them to the chairs of some of their most important faculties. The Renaissance was, indeed, the heyday of the intellectual woman throughout the whole of the Italian peninsula — a time when woman enjoyed the same scholastic freedom as men, and when Mme de Staël's dictum, *Le genie n'a pas de sexe*, expressed a doctrine admitted in practice and not an academic theory.

It would require a large volume, or rather many volumes, to do justice to the learned women of Italy who conferred such honour upon their sex during the period we are considering. Suffice it to mention a few of those who achieved special distinction and whose memories are still green in the land which had been made so illustrious by their talent and genius.

That which the modern reader finds the most surprising in the Italian women of the Renaissance is their enthusiasm for the *literse humaniores* — the Latin and Greek classics — and the proficiency which so many of them, even at an early age, attained in the literature and philosophy of antiquity. It was no uncommon thing for a girl in her teens to write and speak Latin, while many of them were almost equally familiar with Greek.[57] Thus Laura Brenzoni, of Verona, had such a mastery of these two languages that she wrote and spoke them with ease, while Alessandra Scala was so familiar with them that she employed them in writing poetry. Lorenza Strozzi, who was educated in a convent and eventually became a nun, was distinguished for her great versatility, for her profound knowledge of science and art, as well as for her proficiency in Latin and Greek. Her Latin poems were so highly valued that they were translated into foreign languages. Livia Chiavello, of Fabriano, was celebrated as one of the most brilliant representatives of the Petrarchan school. Her style was so pure and noble that, had Petrarch not lived, she alone would have upheld the honour of the vulgar tongue. So successful was Isotta of Rimini in the cultivation of the Muses that she was hailed as another Sappho. Cassandra Fedele, of Venice, deserved, according to Polizian, the noted Florentine humanist, to be ranked with that famous universal genius, Pico de la Mirandola. So extensive were her attainments that in addition to being a thorough mistress of Latin and Greek, she was likewise distinguished in music, eloquence, philosophy and even theology. Leo X, Louis XII of France, and Isabella of Spain were eager to have her as an ornament for their courts, but the Venetian senate was so proud of its treasure that it was unwilling to have her depart. Catarina Cibo, of Genoa, was another prodigy of learning; for, besides a knowledge of Latin and Greek, philosophy and theology, she was well acquainted with Hebrew. Donna Felice Rasponi, of Ravenna, devoted herself to the study of Plato and Aristotle, of Scripture and the Fathers. But, for the extent and variety of her attainments, Tarquinia Molza seems to have eclipsed all her contemporaries. She had as teachers the ablest scholars of an age of distinguished scholars. Not only did she excel in poetry and the fine arts, but she also had a rare knowledge of astronomy and mathematics, Latin, Greek and Hebrew. And so great was the esteem in which she was held that the senate of Rome conferred on her the singular honour of Roman citizenship, transmissible in perpetuity to her descendants. The Sovereign Pontiff and the flower of the Roman prelacy begged her to take up her residence in the Eternal City, but she could not be prevailed upon to leave the land of her birth.

In the arts of sculpture and painting the women of Italy, during the Renaissance, were no less illustrious than they were in science, literature and philosophy. Indeed, many of the treasures in the Italian churches and art galleries that still delight all lovers of the beautiful are from the chisel and the brush of women who achieved distinction between three and four centuries ago.[58]

Probably the most famous sculptress was Properzia de Rossi, whose ability was so remarkable that she excited the envy of the men who were her competitors.[59] Among painters there was Suor Plantilla Nelli, who was a nun and prioress in the convent of Santa Catarina in Florence. Both Lanzi and Vasari bestow high praise on her work and declare some of her productions to be of rare excellence. There were also Maria Angela Crisculo, of whose splendid work many examples are still preserved in the churches of Naples, and Lavinia Fontana of Bologna, who exhibited such extraordinary ability as an artist that some of her pictures passed for the work of her great contemporary, Guido Reni.[60] Still more remarkable were the achievements of four sisters of the noted family Anguisciola of Cremona. So admirable was the work of the eldest sister, Sofonisba, that Philip II invited her to his court in Spain, where she excited the amazement of every one by the splendid canvases which she executed for her illustrious patron and for the members of the royal family.

Of the fifty female poets who flourished in Italy during the Renaissance the most eminent were Gaspara Stampa, Veronica Gambara, and Vittoria Colonna. Of such merit and exquisite finish were the productions of their Muse that they are still read with never failing pleasure. So highly did Cardinal Bembo — the famous "dictator of letters" — value the scholarship and critical acumen of Veronica Gambara that he never published anything without previously submitting it to her judgement. But far more eminent as a poet was the noble and accomplished Marchesa of Pescara, Vittoria Colonna, who, on account of her talents and virtues, was named *La Divina*. The friend and adviser of scholars and the confidante of princes, she represented, as has truly been said, "the best phases of the Renaissance, its learning, its intelligence, its enthusiasm, its subtle Platonism, combined with a profound religious faith and the trace of the mysticism of a simpler age." The chorus of universal praise which was sung by her contemporaries is well echoed by Ariosto when he writes of her: "She has not only made herself immortal by her beautiful style, of which I have heard not better, but she can raise from the tomb those of whom she speaks or writes and make them live forever." But it was as the friend and inspirer of Michaelangelo that she is best known to us today. "Without wings," he writes to her, "I fly with your wings; by your genius I am raised to the skies; in your soul my thought is born."

Among those who specially distinguished themselves for their profound scholarship, as exhibited in the halls of universities, were Dorotea Bucca, who occupied a chair of medicine in the University of Bologna, where, by reason of her rare eloquence and learning, she had students from all parts of Europe; Laura Ceretta, of Brescia, who, during seven years, gave public lectures on philosophy; Battista Malatesta, of Urbino, who taught philosophy with such marked success that the most distinguished professors of the day were forced to recognise themselves as her inferiors; and Fulvia Olympia Morati, who "at the age of fourteen wrote Latin letters and dialogues in Greek and Latin in the style of Plato and Cicero," and who, when she was scarcely sixteen, "was invited to give lectures in the University of Ferrara on the philosophical problems of the *Paradoxes of Cicero*." So great, indeed, was her knowledge of the ancient languages that she was offered the professorship of Greek in the University of Heidelberg; but death cut short her brilliant career before she could enter upon her duties in this famed institution of learning. It was female professors of this type — masters of Greek and Latin letters, who in the words of a recent writer, "sent forth from Italy such students as Moritz von Spiegelberg and Rudolph Agricola, to reform the instruction of Deventer and Zwoll and prepare the way for Erasmus and Reuchlin."

In the preceding list of learned women — and but a few only have been named of the many who in every city of importance conferred undying glory on their sex — it is clear that the Renaissance in Italy was, indeed, the golden age of women. Never in history had they greater freedom of action in things of the mind; never were they, except probably in the case of the English and German abbesses of the Middle Ages, treated with more marked deference and consideration or fairness; never were their efforts more highly appreciated or more generously rewarded, and never was their success more highly and enthusiastically applauded. Temporal and spiritual rulers, princes and cardinals, Popes and emperors vied with one another in paying just tribute to woman's genius as well as to woman's virtue. The nun in the cloister as well as the lady in the palace shared in the general enthusiasm for learning, and they enjoyed throughout the peninsula the same opportunities as men and received the same recognition for their work. Everywhere the intellectual arena was open to them on the same terms as to men. Incapacity and not sex was the only bar to entrance.

But the men of those days, especially scholars of the type of Bembo, Politian and Ariosto, were liberal and broad-minded men, who never for a moment imagined that a woman was out of her sphere or unsexed because she wore a doctor's cap or occupied a university chair. And far from stigmatising her as a singular or strong-minded woman, they recognised her as one who had but enhanced the graces and virtues of her sex by the added attractions of a cultivated mind and a developed intellect. Not only did she escape the shafts of satire and ridicule, which are so frequently aimed at the educated woman of today, but she was called into the councils of temporal and spiritual rulers as well.

Woe betide the ill-advised misogynist who should venture to declaim against the inferiority of the female sex, or to protest against the honours which an appreciative and a chivalrous age bestowed upon it with so lavish a hand. The women of Italy, unlike those of other nations, knew how to defend themselves, and were not afraid to take, when occasion demanded, the pen in self-defense. This is evidenced by numerous works which were written in response to certain narrow-minded pamphleteers — *miseri pedanti*, pitiful pedants, — who would have the activities of women limited to the nursery or the kitchen.[61]

A striking characteristic of these learned women was the entire absence of all priggism or pedantry. Whether lecturing on law or philosophy, or discoursing in Latin before Popes and cardinals, or taking part in discussions on art and literature with the eminent humanists of the day, they ever retained that beautiful simplicity which gives such a charm to true greatness of mind and is the best index of true scholarship and noble, symmetrical womanhood.

Nor did the rare intellectual attainments of these daughters of Italy destroy that harmony of creation which, some will have it, is sure to be jeopardised by giving women the same educational advantages as men. So far was this from being the case that there were never more loyal and helpful wives nor more devoted and stimulating mothers than there were among those women who wrote verses in the language of Sappho, or delivered public addresses in the tongue of Cicero. Still less did their serious and long-protracted studies entail any of the dangers we hear so much of nowadays. The large and healthy families of many of them prove that intellectual work, even of the highest order, is not incompatible with motherhood; and still less that it, *per se*, conduces, as is so often asserted, to race-suicide. These facts are commended to the consideration of our modern opponents of the higher education of women and to those militant conservatives and old-time reactionaries who are still averse to opening the doors of some of our older universities to women — even such universities as Oxford, several of whose colleges were founded on the revenues derived from suppressed educational institutions which had been built and used for generations for the sole behoof of women.

But distinguished as were the women of Italy for their culture and scholarship, they were yet more distinguished as patrons of learning, as leaders and inspirers of the eminent men who were the chief representatives of the Renaissance. Reference has already been made to the influence of Vittoria Colonna on Michaelangelo — "who saw with her eyes, acted by her inspiration, was lifted by her beyond the stars" — but this is only one of many similar instances that might be adduced. Indeed, to the student of the Italian Renaissance, the most interesting feature of it was, not its women doctors and professors, but those noble and accomplished ladies who made the courts of Ferrara, Mantua, Milan and Urbino the most noted intellectual centres of Europe.

The most beautiful ornaments of the first three courts were Renee, duchess of Ferrara; Isabella d'Este, marchioness of Mantua, and Beatrice d'Este, duchess of Milan. They were all women of exceptional learning and culture, and each was the centre of a galaxy of talent such as is rarely witnessed in any one place.

Among the men attracted to their courts were the most illustrious scholars, artists, poets and musicians of the Renaissance. Here they found congenial homes and breathed an atmosphere made fragrant by the appreciation shown by their charming hostesses for their power and genius. Here they found inspiration and a stimulus that spurred them on to their greatest achievements. In Ferrara, where it was said that "there were as many poets as there were frogs in the country round about," were gathered the most gifted poets of the Renaissance who had been attracted there to recite their latest masterpieces. Among them were Clement Marot, the first poet of modern France, and Ariosto, the immortal author of *Orlando Furioso*. There were the great painters, Titian and Bellini, and the illustrious poet, Torquato Tasso, whose love subsequently immortalised Renee's youngest daughter Leonora.

A similar artistic and intellectual supremacy was held by Isabelle d'Este. For portrait painters she had Titian and Leonardo da Vinci, while, as decorators of her home, she had Bellini and Perugino, whose compositions she herself arranged, even in the minutest details. So it was likewise in the gay and brilliant court of Beatrice d'Este, in Milan — a place where artists and scholars of all nationalities were always sure of a cordial welcome.

But the ideal centre of intellectual culture was the court of Urbino, the central figure of which was the learned and accomplished Elizabetta Gonzaga. This picturesque city of the eastern slope of the Apennines was then to Italy what Athens had been to Greece in the days of Pericles; and Elizabetta was to its court what Aspasia was in her own matchless salon — the magnet which attracted all the artists and men of letters of the age.

Castiglione, whose great work, *The Courtier*, was partly written as a memorial of the peerless woman who inspired it, gives us a vivid picture of "the fair ladies, with their quick intelligence and ready sympathy," discussing questions of art, literature, philosophy and Platonism, with the most eminent scholars and artists of Europe. But Castiglione confesses that he is unable to give us more than the mere outline of the picture. "To paint the polished society of Urbino," as has been well said, "we should need colours no palette contains — transparencies of the Grecian sky, the indigo of certain seas, the liquid azure of certain eyes. For more than a century the court of Urbino was regarded as the supreme exemplar. In the seventeenth century, the Hotel de Rambouillet was still striving to make itself a copy of it; unluckily such things as these are not easily copied."[62]

We are not surprised, then, at being told that "men moulded by Italian ladies" — such ladies as graced the court of Urbino — "could be distinguished among a thousand." Still less are we surprised to note the immense difference between the refined and brilliant discussions of *The Courtier* as compared with the coarse tales of the *Decameron* and *Heptameron*. And we can understand the marvellous influence which Castiglione's matchless work — inspired by the beloved Duchess Elizabetta — had upon the masters of English literature — on Shakespeare, Ben Jonson, Spenser, Marlow, Shelley.

Cardinal Bembo, who was one of the most assiduous frequenters of this famous court, in writing of Elizabetta, does not hesitate to declare: "I have seen many excellent and noble women, and have heard of some who were as illustrious for certain qualities, but in her alone among women, all virtues were united and brought together. I have never seen nor heard of any one who was her equal, and know very few who have even come near her."

It was Castiglione's experience at the court of Urbino, where he was a daily witness of the irresistible influence of Elizabetta, that made him give expression to the sentiment, "Man has for his portion physical strength and external activities; all doing must be his, all inspiration must come from woman." It was also this keen student of the mysterious workings of woman's genius and of her secret, all-pervading influence, at times and in places least suspected, who penned the notable statement — worthy of the Renaissance — "Without women nothing is possible, either in military courage, or art, or poetry, or music, or philosophy, or even religion. God is truly seen only through them."

Only a few words are necessary to tell of the learned women of the Renaissance outside of Italy. On account of its intimate connection with the Italian peninsula, Spain was the second country in Europe to experience the effects of the new intellectual movement. Among the educated Italians whom Isabella, the Catholic, had attracted to her court were the brothers Geraldini, whom she appointed as teachers of her children. Of her daughter, Juana, Juan Vives, the eminent Spanish scholar, says she was able to make impromptu speeches in Latin, while Catherine, who became the wife of Henry VIII, excited the admiration of Erasmus by the extent and accuracy of her knowledge. It was from Salamanca that Isabella summoned her own teacher of Latin, the learned Beatrix Galindo,[63] who was a professor of rhetoric in the university long before Elizabeth of England had studied the language of Virgil under Ascham.

Then there was Francisca de Lebrixa who often filled the chair of her father, who was professor of history and rhetoric in the University of Alcala, and Isabella Losa, of Cordova, who, among her other acquirements, counted a knowledge of Greek and Hebrew. To his learned daughters, Gregoria and Luisa, Antonio Perez, minister of Philip II, wrote saying: "Do not imagine, when you are writing to me, that you are addressing Cicero or some Greek author; lower your style to my level." There were also Isabella de Joya, who commented on Scotus Erigena; Catherine Ribera, the bard of love and faith; Doña Maria Pacheco de Mendoza; Bernarda Ferreyra, to whom, on account of her rare scholarship, Lopez de Vega dedicated his beautiful elegy *Phillis*; Juana Morella, who, besides having a profound knowledge of music, philosophy, divinity and jurisprudence, was the mistress of fourteen languages; Juana de la

Cruz, the famous Mexican nun whose poetry of superior merit, as well as her exceptional attainments in many branches of knowledge, won for her the epithet of the "Tenth Muse;" Luisa Sigea, who besides being a poet was a mistress of the classical and several oriental languages, including Hebrew and Syro-Chaldaic, and other learned women whom "no one was astonished to see taking by main force the first rank in the spheres of literature, philosophy and theology."

So profoundly had the Renaissance affected the women of a limited circle in England, that Erasmus could declare without exaggeration: "It is charming to see the female sex demand classical instruction. The queen is remarkably learned and her daughter writes good Latin. The home of More is truly the abode of the Muses."

The queen of whom Erasmus speaks is Catherine of Aragon, who was educated in Spain, who was a pupil of Vives, and who, besides having a thorough knowledge of Latin and Greek, was well acquainted with several modern languages. The daughters of Sir Thomas More were among the most learned women of their time and were, indeed, worthy of dwelling in "the home of the Muses."

Lady Jane Grey read Plato in the original at the age of thirteen.[64] Anne, Margaret and Jane Seymour were likewise celebrated for their knowledge of the classics, as were Anne Boleyn and Mary Stuart, who both received their education in France, and especially Queen Elizabeth, who was not only one of the most learned women of her time but was probably also the most learned queen England has ever produced. There were, however, no university professors or poets of eminence among the English women, as there were in Italy and Spain, and their creative work was practically nothing.

Since the time of Hroswitha, Gertrude, the Matildas and Hildegard, the learned woman has never been the ideal woman in Germany. When Olympia Morati was on her way from Ferrara to Heidelberg to take the chair of Greek, she found the daughters of professors and humanists devoting themselves to sewing and embroidery instead of art and literature. Anna, the eldest daughter of Melanchthon, was almost alone among the German women of the Renaissance who had a knowledge of Latin.

In France the most learned woman of her time was undoubtedly Margaret of Angoulême, queen of Navarre. So great was her knowledge and so enthusiastic was she in promoting the study of the Latin and Greek classics that Michelet, with something of exaggeration, perhaps, calls her "the amiable mother of the Renaissance in France."[65] She was noted for her devotion to the study of Scripture and theology as well as Greek and Hebrew. She always had around her, or was in correspondence with, the most distinguished scholars, poets, artists, philosophers and theologians of the age, and undoubtedly did much, as a patroness of men of letters, toward furthering the literary movement in France. She is, however, chiefly known to modern readers by her *Heptameron* — a work which reveals too clearly the tastes of her associates and the manners and customs of the time.

With the exception of Margaret of Navarre, there were but few literary women of more than ephemeral reputation during the French Renaissance. Among these Louise Labe deserves mention, as she was the most distinguished poetess in France during the sixteenth century.[66] She, like Margaret, was the centre of a coterie of men of letters; but the reunions over which she presided, as well as those of the author of the *Heptameron*, were entirely lacking in the dignity and refinement of those of the polished court of Urbino in the days of the peerless Elizabetta Gonzaga.

From what has been said respecting the rare learning of the women of the Renaissance, one might infer that women in general enjoyed special educational facilities during this period of intellectual activity. Paradoxical as it may seem, the very contrary was the case. For, as history tells us, the education of the Renaissance was essentially aristocratic. It was only for the women of the nobility and for the wives and daughters of scholars, while the great majority of the sex remained in a state of complete illiteracy.

The environment of the daughters of scholars was peculiarly favourable to their intellectual development, and learning was in a certain measure their natural heritage. They did not receive their education in schools, for there were then few or no schools for girls, but from their fathers or from the men of letters who frequented their homes. A typical home of this kind was that of the noted savant, Robert Estienne of Paris, printer to Francis I. Here the language of conversation was Latin, not only for the members of the family but also for the servants as well.[67] Under such conditions we are not surprised to be informed that the girls, as well as the boys, learned to speak Latin as well as their mother tongue. And listening, as they did, to the daily discussions on art and literature by the most learned men of a most learned age, it was inevitable that they should acquire those vast stores of knowledge on all subjects that so excite the astonishment of our less studious women of today.

With the daughters of the nobility it was the same. In their youth they had, under the paternal roof, the benefit of the instruction of the most eminent masters of the time. And as they grew up their constant intercourse with learned men and the part they took in all literary and social assemblies, which were so prominent a feature of the period, enabled them to complete their education under the most favourable auspices, and to have, before they were out of their teens, a fund of information on all subjects that could not be obtained so well, even in the best of our modern institutions of learning.

It was to these daughters of the elite — *ingenuae puellae* — that Erasmus and Vives addressed their treatises on education. They were the privileged class at whose disposition were placed all the treasures of Greek and Latin letters. It was, then, an easy matter for them to write poetry and dissertations in the languages of Horace and Plato. And it was often a necessity for them to speak Latin, for it was then the universal language of the learned — the language that was understood everywhere — in England as in Italy, in Germany as in France, in Flanders as well as in Spain and Portugal.

It was then that The Republic of Letters was a reality as never before; that the man of letters was, of a truth, "a citizen of the world;" that his country was wherever the cult of letters had priests or devotees. He was what the ballad singer was during the Middle Ages, but with more dignity and seriousness. He was the agent and representative of intellectual life, the living symbol of the unity and solidarity of the human mind. And as in time he linked the past to the present so likewise in space he bound all peoples together and belonged equally to all. Such was Erasmus of Holland, who was equally at home in France and Switzerland, in Italy and England — everywhere received with the honour accorded to princes of the blood royal. Such was Vives, of Spain, the teacher of Catherine of Aragon, of Mary, the daughter of Henry VIII — at one time professor in Louvain, at another in Oxford — always and everywhere an ardent exponent of humanism for women as well as for men. Such was Politian and such were scores of his contemporaries, who carried the torch of knowledge from castle to castle and from court to court, where maidens equally with youths enjoyed all the advantages derivable from the lessons of such distinguished teachers and such eminent leaders of culture.

For it was a peculiarity of the scholar of the Renaissance that he was a great traveller — seeking knowledge wherever it was to be found — and carrying it with him whithersoever he went. He journeyed from university to university, everywhere exchanging views with his intellectual compeers, and everywhere diffusing the knowledge he had so laboriously acquired. The consequence was a wonderful uniformity of education among the higher classes — among women as well as among men — something that was never known before. Through the generally diffused knowledge of Latin, the common literary medium of communication, all the nations of Europe, even those at war with one another, were brought together in an intellectual brotherhood and in a way which gave scholarship a power and a prestige that accrued to the benefit of women and men alike.

But the educational advantages enjoyed by the women of the Renaissance were not for the bourgeoisie — not for the daughters of peasants, tradesmen and artisans. They were solely, as has been stated, for the benefit of the children of princes or of scholars — of those only who could claim either nobility of birth or nobility of genius.[68] Even the most zealous of the humanists would have been surprised if they had been asked to diffuse a portion of their light among the women of the masses. For education, as they viewed it, was something solely for the elect — for ladies of the court and not for women of a lower condition. So far as the rest of womankind was concerned, their occupation was limited, according to a Breton saying, to looking after altar, hearth, and children — "*La femme se doit garder l'autel, le feu, les enfants.*"

It was about this time, too, that men began, especially in France and Germany, to revive the anti-feminist crusade which had so retarded the literary movement among the women of ancient Greece and Rome. They refused to hear women and intellect spoken of together. The Germans recognised no intelligence in them apart from domestic duties, and seemed to belong to that strange race, that has not yet died out, which believes woman to be "afflicted with the radical incapacity to acquire an individual idea." "What the Italians called intelligence a German would call tittle-tattle, trickery, the spirit of opposition. They rejected such gratifications and had no intention of allowing Delilah to shear them."[69]

In the estimation of Luther, the intellectual aspirations of women were not only an absurdity, but were also a positive peril. "Take them," he says, "from their housewifery and they are good for nothing." He treated the humanist Vives, preceptor of Mary Tudor, as "a dangerous spirit," because the learned Spaniard was an ardent advocate of the higher education of women. As to abstract and severe studies they were for girls, according to one of Luther's contemporaries, but "vain and futile quackeries." For an accomplished woman to quote the Fathers or the ancient classical writers was to provoke ridicule, because to do so was considered an indication of pedantry or affectation. Montaigne gave expression to the age-old prejudice against woman by refusing to regard her as anything but a pretty animal, while Rabelais, the coryphaeus of the French Renaissance, declared that "Nature in creating woman lost the good sense which she had displayed in the creation of all other things."

Such being the views of the great leaders of thought and formers of public opinion respecting the mental inferiority of woman — views which, outside of Italy, had, with few exceptions, the cordial approval of the supercilious, cockahoop male — is it necessary to add that the Renaissance did nothing for popular education? The masses of women, especially after the suppression of the convent schools in England and Germany, were, in many parts of Europe, and notably in the two countries mentioned, in a worse condition than they were during the Dark Ages.[70]

## Woman And Education Between The Renaissance And The Twentieth Century

The period following the Renaissance was not a brilliant one for woman, especially outside of Italy. For in this favoured land, even after the decadence in literature that followed the glorious cinquecento, intellectual life opposed so effective a barrier to the forces of extinction which were at work in other parts of Europe, notably Germany and England, that there were still in every part of the peninsula from the fertile plains of Lombardy to the sunny Ionian sea, learned and cultured women who were eager to emulate the achievements of their illustrious sisters of Italy's golden age of art, and letters. We do not, it is true, find among them a Properzia de Rossi, a Veronica Gambara, or a Vittoria Colonna; but we find many earnest and enthusiastic students in every department of knowledge.

That which most impresses the student of education during this period of Italian history is not the splendour of art and letters in court and castle, which so dazzled Europe during the time of Renee of Ferrara and Elizabetta Gonzaga of Urbino. We find, it is true, a goodly number of women who won distinction as poets and artists; but it is rather those who were devoted to more serious studies that arrest our attention — women who attained eminence in physical and natural science, in mathematics, in the classical and oriental languages, in philosophy, law and theology. Space precludes the mention of more than a few of these, but these few may be accepted as typical of many others almost equally distinguished.

Chief among those of whom their countrymen are specially proud are Rosanna Somaglia Landi, of Milan, linguist and translator of Anacreon; Maria Selvaggia Borghini, of Pisa, translator of the works of Tertullian; Eleonora Barbapiccola, of Salerno, who translated into Italian the *Principa Philosophiae* of Descartes; Maria Angela Arginghelli, of Naples, who was famed for her profound knowledge of physics and the higher mathematics and who gave an Italian version of Stephen Hales' *Vegetable Statics*. Then there was Clelia Grillo Borromeo, of Genoa, who was so distinguished in science, mathematics, mechanics and languages that a medal was struck in her honour bearing the inscription, *Gloria Genuensium* — glory of the Genoese; and the still more famous Elena Cornaro Piscopia, of Venice, who was truly a prodigy of learning as well as a paragon of virtue. In addition to a knowledge of many modern, classical and oriental tongues, she exhibited remarkable proficiency in astronomy, mathematics, music, philosophy and theology. After a course of study in the University of Padua and after the usual examination and discourse in classic Latin on some of the questions of Aristotelian philosophy, she had the doctorate of philosophy conferred on her in the cathedral of Padua, in the presence of thousands of learned men and applauding students from all parts of Europe. But not content with conferring on this extraordinary woman the ring, wreath of laurel and the ermine mozetta — the usual insignia of the doctorate — the University, as a special mark of distinction, had a medal coined in honour of the illustrious graduate bearing her effigy, with the words, as the decree of the University expressed it, *ad perpetuam rei memoriam*. That there was nothing superficial about this young woman's knowledge of languages, it suffices to state that she was able to speak Latin and Greek as fluently as her own Italian, and that so profound was her knowledge of divinity that there were many distinguished ecclesiastics in both Italy and France who favoured conferring on her the doctorate in theology.

Among other young women who obtained the doctorate in various universities were Maddalena Canedi-Noe and Maria Vittoria Dosi who, after the usual course of study in the university of Bologna, obtained the degree of doctor of civil law, and Maria

Pellegrina Amoretti, who received the degree of doctor in both canon and civil law in the University of Pavia and with it the doctor's cap — *berreto dottorale*. But more remarkable for learning than any of these university graduates was Maria Gaetana Agnesi, one of the most extraordinary women scholars of all time. On account of her wonderful knowledge of languages she was called "The Oracle of Seven Tongues." This, however, is not her chief title to fame. It is rather her marvellous achievements in the domain of the higher mathematics. After the appearance of her most noted work, *Instituzioni Analytiche*, she would at once have been elected a member of the French Academy of Sciences had not the laws of this learned body precluded the admission of women.[71] That great Maecenas of learning, Benedict XIV, showed his appreciation of Maria Gaetana's exceptional attainments by appointing her — *motu proprio* — to the chair of higher mathematics in the University of Bologna. A similar honour had, in the preceding century, been conferred on Marta Marchina, of Naples, when, on account of her rare knowledge of letters, philosophy and theology, she was offered a chair in the Sapienza, in Rome, an honour which her modesty and love of retirement caused her to decline.

We have seen that women professors achieved distinction in the Italian universities even as early as the closing centuries of the Middle Ages. The same was true during the Renaissance, and it has been equally true during the period that has elapsed since the cinquecento.

Among the most eminent of those who taught in the universities were Laura Bassi, who had the chair of physics in the University of Bologna, and Clotilde Tambroni, professor of the Greek language and literature in the same institution of learning. So thorough was her knowledge of the language of Plato that it was the opinion of her contemporaries that there were then only three persons in Europe who equalled her in her mastery of this classic tongue. It was this distinguished Hellenist who graciously delivered the address when one of her countrywomen, Maria dalle Donne, received her doctorate in medicine and surgery. After her graduation Dr dalle Donne was given charge of a school for midwives in which she rendered the greatest service to her sex. Even the chair of anatomy in the University of Bologna was held by a woman, Anna Morandi-Menzolini, and her work was of the highest order. The same position was held by another woman, Maria Petraccini-Terretti, in the University of Ferrara.

What a contrast between the attitude of the universities of Italy and those of other parts of the world toward women as students and professors! For a thousand years the doors of the Italian universities have been open to women, as well as to men; and for a thousand years women, as well as men, have received their degrees from these noble and liberal institutions, and occupied the most important positions in their gift, and that, too, with the approval and encouragement of both spiritual and temporal rulers. For these wise and broad-minded men did not regard it unwomanly for Laura Bassi to teach physics, for Clotilde Tambroni to teach Greek, for Dorotea Bucca to teach medicine, for Maria Gaetana to teach differential and integral calculus, for Anna Morandi to teach anatomy, for Novella d'Andrea to teach canon law, or even, if we may believe Denifle, one of the best of authorities, for the daughters of a Paris professor to teach theology.[72] Yes, what a contrast, indeed, between the Universities of Bologna and Padua, with their long and honoured list of women graduates and professors, and the Universities of Cambridge and Oxford from which women have always been and are still excluded, both as students and professors.

Contrast, also, the honours shown to women as students and professors of medicine in Salerno, in the thirteenth century, with the riots excited among the chivalrous male students of the University of Edinburgh, when, less than a half century ago, seven young women applied for the privilege of attending the courses of lectures on medicine and surgery in that institution. And contrast the sympathy and encouragement of Italy with the almost brutal opposition which women in our own country encountered when, but a few decades ago, they applied for admittance to the medical schools of New York and Philadelphia. The difference between the

Italian and the Anglo-Saxon attitude toward women in the all-important matters in question requires no comment.[73]

One reason for the great difference between the women of Italy and those of other parts of Europe in the matter of higher education during the period we have been considering was the old Roman spirit of independence of the former and their always insisting on what they regarded as their natural and indefeasible rights. Following the example of the matrons of ancient Rome, they insisted on being treated as the equals of men, and, as a consequence, they demanded in the intellectual order all the advantages that were accorded to men. They would never admit their mental inferiority to man, and woe betide the luckless wight who even insinuated such inferiority. The shafts of satire and ridicule were at once directed against him by a score of women who were able to use the pen as well as, if not better than, himself. Sometimes, however, such an one was taken seriously, and then the result was a book by some clever woman to prove that there was no difference in the intellectual power of the two sexes — that, if there was a difference, it was in favour of the gentler sex. There is quite a large number of such works in Italian; and it must be said that the women always met the arguments of their adversaries in a manner that does them the greatest credit.

It was probably because of their insistence on the equality of the sexes, as well as because of their achievements in every department of mental activity, that the educated women of Italy enjoyed so many privileges denied their sisters in other parts of Europe. Thus, in addition to being treated as the equals of men in the universities, they met them on an equal footing in the art, literary and scientific societies and academies, in the proceedings of which they always exhibited an active and enthusiastic interest. In these reunions the women gained strength of mind and independence of character from the men, while the men imbibed refinement and gentleness from the women. Compare this condition with the systematic exclusion of women from similar societies in other countries — even in this twentieth century of ours — and one of the not least potent reasons for the intellectual supremacy of the women of Italy will be apparent.

Next after Italy, France was the country in which, during the post-Renaissance period, women enjoyed the greatest advantages of mental development. But we look in vain, even during the age of Louis XIV, for that flowering of the female intellect that, at the same period, rendered the daughters of Italy so famous. It is true that there was a certain number of learned women in France during the seventeenth century, and notably during the golden age of Louis XIV, for during this period the traditions of the Renaissance were perpetuated and there was still a lingering love of letters, at least among certain classes of the aristocracy.

Prominent among those who attracted attention for their learning were Gilberte and Jaqueline Pascal, of the celebrated convent of Port Royal; Marie-Eleanore de Rohan and Gabrielle de Rochechouart, both, like the Pascal sisters, inmates of the cloister; Marie Cramoisy, wife of the first director of the royal printing office, and Mlle de Luynes, a friend of Pascal. All these counted among their attainments a writing knowledge of Latin, but were far from being able, like the Italian women above mentioned, to speak it with the same fluency as they did their mother tongue.

In addition to the learned French women just named, there was Elizabeth de Rochechouart, a niece of Mme. de Montespan, who was able to read Plato in Greek, and Anne de Rohan, Princess of Guemene, who surprised her countrymen by studying Hebrew. Then there were Mme. de Grignan, Marie Dupre, Louise Serment, Anne de La Vigne, who, like the Princess Palatine, Elizabeth, and Christine of Sweden, were ardent disciples of Descartes, and took the lead among the *femmes philosophes* of their time.

But for profound and varied scholarship Mme Dacier, the daughter of the erudite Tanquil Le Fevre, was the most famous of all the women of her time in France. Possessed of rare power of eloquence and beauty of style, together with an extraordinary capacity for criticism, there was not a man in Europe who did not respect her judgement in matters of literature and culture. But that for which she was specially celebrated was her exceptional knowledge of Latin and Greek. She not only translated the *Iliad* and the *Odyssey* but also several others of the ancient classics. None of her contemporaries had a more thorough mastery of the tongues of Homer and Virgil, nor did any of her countrymen contribute more than she toward the advancement of the knowledge of the literature of ancient Greece and Rome. So highly prised was her version of the Iliad that it was translated by Ozell into English. Her version of Plato's *Phaedo* was also translated into English and published by a New York bookseller more than a century after her death. The scholarly Menagius, in his *Historia Mulierum Philosopharum*, did not hesitate to pronounce her the most learned woman of all time — *Feminarum quot sunt, quot fuere doctissima.*[74]

To Mme. de Maintenon, the morganatic wife of the Great Monarch, is due the Institut de Saint-Cyr, the first state school for girls founded in France. It was, however, solely for the daughters of the nobility. And, although it was from the first under the direction of the foundress, a woman who was before all else a teacher as well as one of the most enlightened women of the most literary and philosophic age France ever knew — the age when the French language was perfected, the age of the Academy, of Boileau, Moliere, Racine, Bossuet, Descartes — the studies prescribed in this institution, which was under the special patronage of the king, were of the most elementary character. They comprised reading, writing, arithmetic, grammar, music, drawing, dancing, and the elements of history, mythology and geography. As to history, Mme de Maintenon was satisfied if the pupils of Saint-Cyr knew enough not to confound the kings of France with those of other nations, and were able to avoid mistaking a Roman emperor for the Emperor of China or Japan; or the King of Spain or England for the King of Persia or Siam. And yet, restricted as it was, her programme of studies was more complete than that of any other girls' school in the kingdom. One of her reasons for not insisting on a more thorough course was that "women never know but by halves, and the little that they do know usually makes them proud, haughty and talkative and disgusted with solid things."[75]

In Saint-Cyr, the best girls' school in the kingdom, there was not a word about the first principles of philosophy, nor about the physical and natural sciences recommended by Fenelon. The elements just referred to, combined with a goodly amount of esprit — *bien de l'esprit* — were considered quite sufficient to prepare the future wives of the nobility for all the duties they would be called upon to perform.

Mme de Maintenon had probably been unconsciously influenced by what she had seen at the court of her liege lord, where the greater part of the women were extremely ignorant. Even Mme de Montespan, the king's favourite, and for years the leading figure at the court, was no exception. So ignorant was she that she was not even able to spell the simplest and most common words.[76]

And so it was with the most illustrious ladies of France. Many of them were so devoid of instruction that they were unable either to read or to write. Even the teachers in Saint-Cyr were so deficient in the simplest rudiments of an education that Mme. de Maintenon found it necessary to correct their letters, in order to teach them the most essential rules of epistolary correspondence. In reality, the women of the age of Louis XIV did not trouble themselves about an education as we understand it. Endowed with esprit, with a natural and acquired taste for things intellectual, they were satisfied with such knowledge as they could glean from reading or conversation, and with comparatively few exceptions, showed no disposition to devote long years to study in school, much less in a university, as did their sisters to the south of the Alps.

The foundress of Saint-Cyr had likewise been influenced by her environment as well as by the court — an environment which was becoming daily more and more unfavourable to the education, especially anything approaching the higher education, of women. A young woman's education was considered complete when she was able to read, write, dance and play some musical instrument. Anything more was deemed superfluous and deserving of censure and ridicule rather than praise.

It was at this time that Moliere's two celebrated plays, *Les Femmes Savantes* and *Les Precieuses Ridicules,* were given to the world. These well-known productions, replete with the author's brightest flashes of wit, and abounding in his most effective shafts of satire, produced at once an immense sensation. As soon as published, they were in the hands of everybody. Those who were opposed to the education of women — and the number was daily increasing — had recourse to them as to arsenals which supplied them with just the arms they had so long needed to decide in their favour the long warfare which they had been conducting against the gentler sex. The views of the bourgeois Chrysale as expressed to his sister, Belise, were so in harmony with their own that they loved on every occasion to repeat with him:

> No,
> It isn't decent, and for many reasons,
> That womankind should study and know too much.
> To teach her children what is right and wrong,
> Manage her household, oversee her servants,
> And keep expenses within bounds, should be
> Her only study and philosophy.
> Our fathers, on this point, showed great good sense;
> They said a woman always knows enough
> If but her understanding reaches
> To telling, one from t'other, coat and breeches.
> Their wives, who couldn't read, led honest lives,
> Their households were their only learned theme,
> And all their books were thimble, thread and needles.
> With which they made their daughters' wedding outfits.
> But now our women scorn to live like that;
> They want to write and all be authoresses.
> They think no knowledge is too deep for them.[77]

Moliere's intention in writing these justly famous comedies was not, as is so often asserted, to ridicule women of learning, but only those superficial pedants who affected knowledge or loved to make a display of the little knowledge they happened to possess. The result, however, was quite different from what had been intended, for the poet's pleasantries were taken so seriously, that even women of real learning, in order to avoid ridicule, were condemned to absolute silence. The comic dramatist, Destouches, expressed the prevailing opinion when he wrote:

> Une femme savante
> Doit cacher son savoir, ou c'est une imprudente.[78]

Few French women thereafter had the courage to defend their sex, as did their sisters in Italy, and the result was that, with a few exceptions, like Mme du Chatelet, Sophie Germain, and Mme Lepaute, there were no more learned women in France for fully two centuries.

Never did satire and ridicule accomplish more, except probably in the case of *Don Quixote* — that masterly creation of Cervantes which dealt the death-blow to knight-errantry — than did *Les Femmes Savantes* and *Les Precieuses Ridicules.* The learned woman became as much an object of derision in France as was the knight-errant in Spain.

It was not, however, in the nature of the French woman, with all her vivacity and energy, to be suppressed entirely or to be relegated for long to the background in things of the mind. But, not then daring to face the ridicule which was inevitable, if she devoted herself to science or philosophy, she sought a substitute for her intellectual activity in the salon.

The first salon was established by an Italian woman, the Marquise de Rambouillet, in 1617, and was modeled after the famous reunions held at the court of Urbino under Elizabetta Gonzaga, a century before. Although it never exhibited the splendour of its Italian prototype, the Hotel de Rambouillet was for more than fifty years the most important literary centre of the kind in France. Here, owing to the tact, esprit, and magnetic personality of Mme. de Rambouillet, were gathered the most distinguished men and women of the time. Among them were poets, philosophers, statesmen, ecclesiastics and ladies of rank, whose names still dazzle us by their brilliancy. Bossuet, Moliere, La Fontaine, Corneille and the great Conde were there; so were Flechier, Balzac, Voiture, Saint-Evremont, Descartes and La Rochefoucauld; and so, too, were Mme. de Sevigne, the Duchess of Montpensier, Madeleine de Scudery, La Comtesse de La Fayette, Charlotte de Montmorency, and Cardinal Richelieu who got from this noted salon the idea which led to his greatest foundation — the French Academy.

It was Mme de Rambouillet who, through her reunions in her exquisite *Chambre Bleue*, for the first time brought together elements that were previously considered as belonging to different castes. It was she, also, who created modern society with its purely intellectual hierarchy, by having the representatives of the nobility meet men of science and letters on an equal footing. It seems to us now the most natural thing in the world for a great savant, a great poet, or a great philosopher, to be received in the same salon with the Duchess of Montpensier — *La Grande Mademoiselle* — but it was far from being so when the brilliant young Italian matron — for she was a daughter of the noble Roman family of the Savelli — began her epoch-making work in the Hotel de Rambouillet, where, after overcoming countless difficulties and prejudices, she eventually succeeded in bringing together, and in enlisting in a common cause, the nobility of birth and the nobility of intellect, and introducing into the exclusive set of Paris the same kind of social coteries that had so long been popular in Urbino and Ferrara.

The Hotel de Rambouillet was the exemplar of that long series of salons which, for two centuries, were the favourite trysting-places of the talent, the wit, the beauty of Europe, and which exerted such a potent influence on society and on the progress of science and literature. The mistress of the salon was supreme, and she maintained her supremacy by her tact, sympathy, intelligence and mental alertness, rather than by learning and superior mental power.

Indeed, it is a singular fact that very few of the *salonieres* were learned women. The most gifted and the most learned of them were Mlle Lespinasse, Mme de Staël, and Mme Swetchine. Mme Geoffrin, who was of bourgeois origin, was so devoid of education that Voltaire said she was unable to write two lines correctly. And yet, despite her educational limitations, she became, by her own unaided efforts, the queen of intellectual Europe.

And, if we may judge by their portraits, most of the great leaders of salons were homely, if not positively ugly, and many of them were advanced in years. Thus, Mme du Deffand — the female Voltaire — was sixty-eight years old and blind when her friendship with Horace Walpole, one of the wittiest Englishmen who ever lived, began — a friendship that endured until her death at the age of eighty-three. The face of Mlle de Lespinasse was disfigured by small-pox and her eyesight was impaired; and yet, without rank, wealth or beauty, she was the pivot around which circled the talent and fashion of Paris, and whose personal magnetism was so great that the state, the church, the court, as well as foreign countries, had their most distinguished representatives in her salon.

Here she received and entertained her friends every evening from five until nine o'clock. "It was," writes La Harpe, "almost a title to consideration to be received into this society." So great was the influence exerted by Mlle de Lespinasse that she bent savants to her will by the sheer force of genius. Her salon became known as "the ante-chamber of the French Academy;" for it was asserted that half the academicians of her time owed their fauteuils to her active canvass in their behalf. And so successful was she in opening the lips and minds of her habitues, whether an historian like Hume, a philosopher like Condillac, a statesman like Turgot, a mathematician like d'Alembert, a litterateur like Marmontel or an encyclopaedist like Condorcet, that it was said of her that she made "marble feel and matter think."

She was a veritable enchantress of the great and the learned of her time. She did not, however, wield her magic wand through her learning, or the accident of birth, or the physical attractions of person, but solely by reason of her wonderful vivacity, charm of mind, and exquisite tact, which consisted, as those who knew her well tell us, "in the art of saying to each that which suits him," and in "making the best of the minds of others, of interesting them, and of bringing them into play without any appearance of constraint or effort." This rare faculty it was which secured for her a supremacy in the world of thought and action that has been accorded to but few women in the world's history. Vibrant with emotion and passion, she reminds one of the gifted but hapless Heloise. Marmontel, who had such a high opinion of her judgement that he submitted his works for her criticism, as Moliere had submitted his to Ninon de Lenclos, describes her as "the keenest intelligence, the most ardent soul, the most inflammable imagination that has existed since Sappho."

But aside from what she achieved indirectly through the habitues of her salon, what has this supremely clever woman left to the world? Only a few love letters to a heartless coxcomb.

And what have the other noted salonieres from the time of the Marquise de Rambouillet to that of Mme. Swetchine — full two centuries — bequeathed to us that is worth preserving? With the exception of the works of Mme de Staël, whom Lord Jeffrey declared to be "the greatest female writer in any age or country," we have little more than certain *Memoires* and *Correspondances* whose chief claims to fame rest on the vivid pictures which they present of the manners and customs of the time and of the celebrities who were regarded as the chief ornaments of the salons which they severally frequented. Most of these works were posthumous; for few women, after Moliere's merciless scoring of learned women, had the courage to appear in print. Even Mme de Scudery, one of the most gifted and prolific writers of the period, gave her first novel to the world under her brother's name. And so tabooed was female authorship that Mme de La Fayette, one of the most brilliant of the *precieuses,* disclaimed all knowledge of her *Princesse de Cleves,* while her masterpiece, *Histoire d'Henriette d'Angleterre,* was not published until after her death.

The truth is that the period of the salon was for the most part a period of contrasts and contradictions. At first the better educated *salonieres* were chiefly interested in belles-lettres. Then they devoted themselves more to science and philosophy, and finally, during the years immediately preceding the Revolution, they found their greatest pleasure in politics. As for the men, while professing to adore women, they had little esteem for them, and still less respect. Often, it is true, the women who frequented the salons were deserving neither of respect nor of esteem.

Sydney Smith spoke of those under the old regime as "women of brilliant talents who violated all the common duties of life and gave very pleasant little suppers." It was certainly true of many of them — even of some of the most distinguished — such, for instance, as Mme d'Epinay, Mme du Deffand, Ninon de Lenclos and Mme Tencin, the mother of D'Alembert. There was little in their manner of life to distinguish them from the *hetaerae* of ancient Athens, and it was probably owing to this fact, as well as their wit and brilliancy, that many of them attained such pre-eminence as social leaders. The statesmen, philosophers, men of science and letters of France, like those of Greece more than two thousand years before, wanted distraction and amusement. That the mistresses of the salons should be women of learning was of little moment. The all important thing for their habitues was that they should be good entertainers — that they should be witty, tactful and sympathetic — and, if ignorant, that they should be brilliantly ignorant, and, at the same time, enchantingly frank and naïve.

Strange as it may appear there was as much hostility to learned women at the close of the eighteenth century as there was in the time of Louis XIV. And the remarkable fact is that the strongest opponents of women's education were found among the most prominent writers and scholars of the day — men who, like their predecessors of old, based their opposition on the assumed mental inferiority of woman. Thus, to Rousseau, woman was at best but "an imperfect man," and, in many respects, little more than "a grown-up child." Search after abstract and speculative truths, principles and axioms in science, "everything that tends to generalise ideas is outside of her competence." That means that women are to be excluded from the study of mathematics and the physical sciences, because they are incapable of generalisation, abstraction, and the mental concentration that these subjects demand. Even the masterpieces of literature, according to him, are beyond their comprehension. In a word, feminine studies, Rousseau will have it, should relate exclusively to practical and domestic matters and he endorses the words of Moliere that

> It is not seemly, and for many reasons,
> That a woman should study and know so many things.

Diderot, Montesquieu, Voltaire and the Encyclopaedists share the views of Rousseau. Diderot declares that serious studies do not comport with woman's sex, while Montesquieu would limit female education to mere accomplishments.

But this is not all. Antagonistic as these men were to the education of the daughters of the nobility and the well-to-do, they were entirely opposed to the education of the children of the poor. "The good of society," it was averred, "demands that the instruction of the people extend not beyond their occupations." "The poor," declares Rousseau, "have no need of instruction," and Voltaire and the Encyclopaedists say, "Amen."[79]

Very little need be said about the education of women in Germany during the period we have been considering. When there was any at all, it was of the most rudimentary character, while as to books, they were limited to the kind recommended by Byron for the women of modern Greece — "books of piety and cookery." The attitude of the Germans generally toward female education, for centuries past, was clearly defined by the Kaiser Wilhelm II, when, a few years ago, he publicly stated: "I agree with my wife. She says women have no business to interfere with anything outside of the four K's, that is, *Kinder, Kirche, Küche, Kleider* — children, church, kitchen, clothes."

There was, however, during the period we are now considering, one remarkable example of a learned woman of Teutonic origin. This was the famous Anna Maria van Schurman, who was one of the most gifted women that ever lived. She was, probably, as near to being a universal genius as any one of her sex of whom we have knowledge. Artist, musician, poet, philosopher, theologian, linguist, she was the admiration of the scholars of the world and the pride of the Low Countries — the land of her birth. She lived when Holland was in the van of human progress and amidst of the splendours of the Dutch Renaissance. She was the friend and correspondent of the most distinguished scholars and most noted celebrities of her time. Among these were Voet, Spanheim, Descartes, Gassendi, Constantine Huyghens, Princess Elizabeth of Bohemia, Queen Christina of Sweden, and Cardinal Richelieu. To go to the Netherlands, it was then said, without seeing Anna van Schurman, was like going to Paris without seeing the king. She was hailed as "The Tenth Muse," "The Sappho of Holland," "The Oracle of Art," "The Star of Utrecht."

That, however, which gave the greatest renown to the "Learned Maid," as Anna was called, was her extraordinary knowledge of languages. For, besides being proficient in the chief modern tongues of Europe, she was well acquainted with Latin, Greek, Hebrew, Syro-Chaldaic and Ethiopic. The oriental languages she studied as an aid to the better understanding of Holy Scripture.

She was the author of several works, among which was an Ethiopic grammar which was acclaimed by the professors of the Dutch universities as a marvellous achievement. Her best known volume is designated *Opuscula*. It was brought out by the Elzevirs in Leyden and went through several editions. It is composed of letters and short treatises in French, Latin, Greek and Hebrew — in verse as well as prose.

Of more value, if less striking, than the productions named were the "Learned Maid's" writings in favour of the intellectual enfranchisement of her own sex. In a letter to Dr Rivet, Professor of Theology in Leyden, she declares:

> My deep regard for learning, my conviction that equal justice is the right of all, impel me to protest against the theory which would allow only a minority of my sex to attain to what is in the opinion of all men most worth having. For,

> since wisdom is admitted to be the crown of human achievement, and is within every man's right to aim at in proportion to his opportunities, I cannot see why a young girl, in whom we admit a desire of self-improvement, should not be encouraged to acquire the best that life affords.

To those who objected that the distaff and the needle were sufficient to occupy women's minds, Anna Maria made answer that the words of Plutarch — "It becomes a perfect man to know what is to be known and to do what is to be done" — applied with equal truth to a perfect woman.[80]

In England, until the latter part of the nineteenth century, the educational status of women was but little better than in Germany. During the Stuart period schools for girls were so scarce that most of those who received any education at all obtained it at home under private tutors. Even then it rarely embraced more than reading, writing, needlework, singing, dancing and playing on the lute or virginal.[81]

As to the higher studies for women, Lady Mary Wortley Montagu writes as follows: "My sex is usually forbid studies of this nature and folly reckoned so much our proper sphere that we are sooner pardoned any excesses of that than the least pretensions to reading or good sense. We are permitted no books but such as tend to the weakening or effeminating of the mind. Our natural defects are in every way indulged, and it is looked upon as in a degree criminal to improve our reason or fancy we have any.... There is hardly a creature in the world more despicable or more liable to universal ridicule than that of a learned woman: these words imply, according to the received sense, a tattling, impertinent, vain and conceited creature."[82]

Higher studies for their daughters were regarded by the generality of men, the same writer tells us, "as great a profanation as the clergy would do if the laity would presume to exercise the functions of the priesthood."

Referring to the handicaps suffered by the women of England

in the pursuit of knowledge, the same writer declares: "We are educated in the grossest ignorance, and no art is omitted to stifle our natural reason; if some few get above their nurses' instructions, our knowledge must be concealed and be as useless to the world as gold in the mine."

Lord Chesterfield, in *His Letters to His Son*, expresses the opinion of his contemporaries when he writes on the same subject as follows: "Women are only children of a larger growth; they have an entertaining tattle, sometimes wit; but, for solid reasoning, good sense, I never in my life knew one who had it, or who reasoned or acted consequentially for twenty-four hours together.... A man of sense only trifles with them, plays with them, humours and flatters them as he does a sprightly forward child; but he neither consults them about nor trusts them with serious matters, though he often makes them believe he does both, which is the thing in the world which they are proud of; for they love mightily to be dabbling in business, which, by the way, they always spoil, and, being distrustful that men in general look upon them in a trifling light, they almost adore that man who talks to them seriously and seems to consult and trust them."[83]

And this was written by that "mirror of politeness and chivalry" whose name has for two centuries been synonymous with that of a perfect gentleman! And Lady Montagu was compelled to pen her caustic and pathetic plaints during the age of Pope, Steele, Addison, Swift,[84] Johnson, Dryden and Goldsmith — the most brilliant pleiad of literary men that England had known since the days of Shakespeare.

So unnatural for women were literary and scientific pursuits regarded by all classes that the few who attained any eminence in them were classed as abnormal creatures who deserved no more consideration than did the *Precieuses* across the Channel. And so great was the power of public sentiment against women writers that Fanny Burney was afraid to acknowledge the authorship of *Evelina*. Even in Jane Austen's days, the feeling that a woman, in writing a book, was overstepping the limitations of her sex was so pronounced that she never actually avowed the authorship of those charming works which have been the delight of three generations of readers. It was this same sentiment that caused the Brontë sisters and George Eliot, as well as many other notable women, to write under pseudonyms. They feared to disclose their sex lest their works, if known as the productions of women, should be *ipso facto* branded as of inferior merit.

During the period in question women fared no better in the United States than in England. They were subject to the same educational debarment and were the victims of the same snobbery and intolerance. The Pilgrim Fathers and their descendants for many generations made no secret of their belief in the mental inferiority of woman, and applied to her the gospel of liberty contained in the following words of Eve to Adam as given in *Paradise Lost*:

> My author and dispenser, what thou bidst
> Unargued I obey; so God ordains;
> God is thy law, thou mine: to know no more
> Is woman's happiest knowledge and her praise.

To the Puritan of New England, as to the Puritan Milton, the relative attainments of woman and man were tersely expressed in Tennyson's couplet:

> She knows but matters of the house,
> And he, he knows a thousand things.

To us one of the most astounding facts in the educational history of New England is the long time during which girls were without free school opportunities. Thus, although schools had been established within twenty years after the Pilgrims landed at Plymouth Rock, it was not until a century and a half later that their doors were opened to girls. The public schools of Boston were established in 1642, but were not opened for girls until 1789; and then only for instruction in spelling, reading and composition, and that but one half of the year. There was no high school in Boston, the vaunted Athens of America, until 1852.

Harvard College was founded in 1636 for the education of "ye English and Indian youth of this country in knowledge and godlyness," but in this institution no provision was made for women and its doors are still closed to them.

"The prevailing notion of the purpose of education," declares Charles Francis Adams, in speaking of Harvard College, "was attended with one remarkable consequence — the cultivation of the female mind was regarded with utter indifference; as Mrs. Abigail Adams says in one of her letters, 'it was fashionable to ridicule learning.' "[85]

It was not until 1865 that Matthew Vassar, "recognising in women the same intellectual constitution as in man," founded the first woman's college in the United States. This was soon followed by similar institutions in various parts of this country and Europe. In less than ten years thereafter Girton and Newnham colleges were founded at Cambridge, England, in order that women might be enabled to enter upon a regular university career.

In all the universities of England, Scotland and Ireland, except Oxford, Cambridge[86] and Trinity College, Dublin, women are now admitted to all departments, pass the same examinations as the men and receive the same academic degrees. Germany, whose institutions for the higher education of men have so long been justly famous, was exceedingly slow to open its universities to women, and then only after the most stubborn opposition of those who still maintained that the studies of women should be limited to the three R's and their occupations confined to the four K's. But even in this conservative country the cause of woman has at length triumphed, and she now enjoys educational advantages that a few decades ago were deemed forever impossible.

And so it is in every civilised country. Woman's long struggle for complete intellectual freedom is almost ended, and certain victory is already in sight. In spite of the sarcasm and ridicule of satirists and comic poets, in spite of the antipathy of philosophers and the antagonism of legislators who persisted in treating women as inferior beings, they are finally in view of the goal toward which they have through so many long ages been bending their best efforts. Moreover, so effective and so concentrated has been their work during recent years that they have accomplished more toward securing complete intellectual enfranchisement than during the previous thirty centuries.

From the former home of the Vikings to the romantic land of the Cid, from the capital of Holy Russia to the fair metropolis of the Golden Gate, women are now welcomed to the very institutions from which but a few years ago they were so systematically excluded. They attend the same courses as men, pass the same examinations and receive the same degrees and honours. Their sex is no longer a bar to positions and employment that only a generation ago were considered proper only for the proud and imperious male. They have proved beyond cavil that genius knows not sex, and that, given a fair opportunity, they are competent to achieve success in every department of human effort.

Thus, to speak only of Europe, there are today women professors in the universities of Norway, Sweden, Switzerland, France, Greece and Russia, as there have been in Italy since the closing years of the Dark Ages. They lecture on science, literature, law and medicine, and in a manner to extort the admiration of their erstwhile antagonists. In Germany and Hungary there are women chemists and architects, while it is a matter of record that the best construction work done on the trans-Siberian railroad was that in charge of a woman engineer.

As an illustration of the marvellous change which has been brought about during the last three-quarters of a century in the educational status of woman, I can do no better than transcribe a few passages from a work by Sir Walter Besant describing the transformation of woman during the reign of Queen Victoria; for it applies to all civilised countries as well as to England.

"The young lady of 1837 has been to a fashionable school; she has learned accomplishments, deportment and dress. She is full of sentiment; there was an amazing amount of sentiment in the air about that time; she loves to talk and read about gallant knights, crusaders and troubadors; she gently touches the guitar; her sentiment, or her little affectation, has touched her with a graceful melancholy, a becoming stoop, a sweet pensiveness. She loves the aristocracy, even although her home is in that part of London called Bloomsbury, whither the belted earl cometh not, even though her papa goes into the City; she reads a deal of poetry, especially those poems which deal with the affections, of which there are many at this time. On Sunday she goes to church religiously and pensively, followed by a footman carrying her prayer book and a long stick; she can play on the guitar and the piano a few easy pieces which she has learned. She knows a few words of French, which she produces at frequent intervals; as to history, geography, science, the condition of the people, her mind is an entire blank; she knows nothing of these things. Her conversation is commonplace, as her ideas are limited; she can not reason on any subject whatever because of her ignorance; or, as she herself would say, because she is a woman. In her presence, and indeed in the presence of ladies generally, men talk trivialities. There was indeed a general belief that women were creatures incapable of argument, or of reason, or of connected thought. It was no use arguing about the matter. The Lord had made them so. Women, said the philosophers, can not understand logic; they see things, if they do see them at all, by instinctive perception. This theory accounted for everything, for those cases when women undoubtedly did 'see things.' Also it fully justified people in withholding from women any kind of education worthy the name. A quite needless expense, you understand."

Her amusements, we are told, were "those of an amateur — a few pieces on the guitar and the piano and some slight power of sketching or flower painting in water-colours." The literature she read "endeavoured to mould woman on the theory of recognised intellectual inferiority to man. She was considered beneath him in intellect as in physical strength; she was exhorted to defer to man; to acknowledge his superiority; not to show herself anxious to combat his opinions....

"This system of artificial restraints certainly produced faithful wives, gentle mothers, loving sisters, able housewives. God forbid that we should say otherwise, but it is certain that the intellectual attainments of women were then what we should call contemptible, and the range of subjects of which they knew nothing was absurdly narrow and limited. I detect the woman of 1840 in the character of Mrs. Clive Newcome, and, indeed, in Mrs. George Osborne, and in other familiar characters of Thackeray."

Then Sir Walter, turning to the young Englishwoman of 1897, thus describes her:

> She is educated. Whatsoever things are taught to the young man are taught to the young woman; the keys of knowledge are given to her; she gathers of the famous tree; if she wants to explore the wickedness of the world she can do so, for it is all in the books. The secrets of nature are not closed to her; she can learn the structure of the body if she wishes. The secrets of science are all open to her if she cares to study them. At school, at college, she studies just as the young man studies, but harder and with greater concentration. She has proved her ability in the Honours Tripos of every branch; she has beaten the senior wrangler in mathematics; she has taken a 'first-class' in classics, in history, in science, in languages. She has proved, not that she is a man's equal in intellect, though she claims so much, because she has not yet advanced any branch of learning, of science, one single step, but she has proved her capacity to take her place beside the young men who are the flower of their generation — the young men who stand in the first class of honours when they take their degree....
>
> Personal independence — that is the keynote of the situation. Mothers no longer attempt the old control over their daughters; they would find it impossible. The girls go off by themselves on their bicycles; they go about as they please; they neither compromise themselves nor get talked about; for the first time in man's history it is regarded as a right

> and proper thing to trust a girl as a boy insists upon being trusted. Out of this personal freedom will come, I dare say, a change in the old feelings of young man to maiden. He will not see in her a frail, tender plant which must be protected from cold winds; she can protect herself perfectly well. He will not see in her any longer a creature of sweet emotions and pure aspirations, coupled with a complete ignorance of the world, because she already knows all that she wants to know....
>
> Perhaps the greatest change is that woman now does thoroughly what before she only did as an amateur.[87]

Yes, the world is beginning at last to realise the truth of the proposition which the learned Maria Gaetana Agnesi so eloquently defended nearly two centuries ago — to wit, that nature has endowed the female mind with a capacity for all knowledge, and that, in depriving women of an opportunity of acquiring knowledge, men work against the best interests of the public weal.[88]

We are at the long last near that millennium which Emerson had in mind when, in 1822, he predicted "a time when higher institutions for the education of young women would be as needful as colleges for young men" — that millennium for which women have hoped and striven ever since Sappho sang and Aspasia inspired the brightest, the noblest minds of Greece.

## Notes & References

1) Demosthenes *In Neaeram, 122. Ôáò ìåí ãáñ ?ôáéñáò ?äïíçò ?íåê' å÷ïìåí, ôáò äå ðáëëáêáò ôçò êáè' ?ìåñáí èåñáðåéò ôïõ óùìáôïò, ôáò äå ãõíáéêáò ôïõ ðáéäïðïéåéóèáé ãíçóéùò êáé ôùí åíäïí æõëáêá ðéóôçí å÷åéí.*

As indicative of the comparative value of men and women, as members of society, in the estimation of the Greeks, Euripides makes Iphigenia give utterance to the following sentiment :

More than a thousand women is one man
Worthy to see the light of life.

2) *Ôçò ôå ãáñ, ?ðáñ÷ïõóçò æõóåùò ìç ÷åéñïóé ãåíåóèáé ?ìéí ìåãáç ç äïîá'êáé ?ò áí åð' åëá÷éóôïí áñåôçò ðåñé ç øïãïõ åí áñóåóé êëåïò ç.*

*Thucidides, History of the Peloponnesian War,* II, 45.

"Phidias," Plutarch tells us in his *Conjugal Precepts*, "made the statue of Venus at Elis with one foot on the shell of a tortoise, to signify two great duties of a virtuous woman, which are to keep at home and be silent. For she is only to speak to her husband or by her husband."

3) Ariosto, referring to the undying fame of Sappho and Corinna, expresses himself in words as beautiful as they are true, as witness the following couplet :

Saffo e Corinna, perche furon dotte, Splendono illustri, e mai non veggon notte.

— Orlando Furioso, Canto XX, strophe I.

4) The nine "Terrestrial Muses" were Sappho, Erinna, Myrus, Myrtis, Corinna, Telesilla, Praxilla, Nossis and Anyta.

The Greek poet Antipater embodies the names of the "Terrestrial Nine" in an epigram which is well rendered in the appended Latin translation :

Has divinis linguis Helicon nutrivit mulieres
Hymnis, et Macedon Pierias scopulus,
Prexillam, Myro, Anytae os, foeminam Homerum,
Lesbidum Sappho ornamentum capillatarum.
Erinnam, Telesillam nobilem, teque Corinna,
Strenuum Palladis scutum quae cecinit.
Nossidem muliebri lingua, et dulsisonam Myrtin,
Omnes immortalium operatrices librorum.
Novem quidem Musas magnum coelum, novem vero illas
Terra genuit hominibus, immortalem laetitiam.

5) Cf *Poetriarum octo, Erinnae, Myrus, Mytidis, Corinnae, Telesillae, Praxillae, Nossidis, Anytae fragmenta et elogia*, by J C Wolf Hamburg, 1734. See also the charming memoir *Sappho* by H T Wharton, London, 1898, and *Griechische Dicterinnen*, by J C Poestion, Vienna, 1876.

6) See *Mulierum Gracarum qua oratione prosa usae sunt fragmenta et elogia Graece et Latine*, by J C Wolf, London, 1739, *Historia Mulierum Philosopharum*, scriptore Aegidio Menagio; Lugduni, 1690, *Griechische Philosophinnen*, by J C Poestion, Norden, 1885, and *Le Donne alle Scuole dei Filosofi Greci* in *Saggi e Note Critiche*, by A Chiappelli, Bologna, 1895.

7) *Woman: Her Position and Influence in Ancient Greece and Rome and Among the Early Christians*, pp 58 and 59, by James Donaldson, London, 1907.

8) There were several hetaerae named Lais. One of them, apparently a native of Corinth, was celebrated throughout Greece as the most beautiful woman of her age.

9) For information respecting the hetaerae the reader is referred to the *Letters* of Alciphron, to Lucian's *Dialogues* on courtesans, and more particularly to the *Deipnosophists* of Athenaeus, Ch XIII. See also *The Lives and Opinions of the Ancient Philosophers*, by Diogenes Laertius, Bohn Edition, London.

10) Donaldson, *op cit*, pp 61 and 62.

Adolph Schmidt, one of the late biographers of Aspasia, accepts these statements as true and credits to Aspasia the making of both Pericles and Socrates. His views are also shared by other modern writers who have made a special study of the subject.

According to some writers an indirect allusion to Aspasia's intellectual superiority is found in the *Medea* of Euripedes in the following verses of the women's chorus:

In subtle questions I full many a time
Have heretofore engaged, and this great point
Debated, whether woman should extend
Her search into abstruse and hidden truths.
But we too have a Muse, who with our sex
Associates to expound the mystic lore
Of wisdom, though she dwell not with us all.

11) It is proper to add that certain modern writers will not admit that Aspasia was ever a hetaera in the sense of being a courtesan. After Pericles had divorced his first wife, he lived with Aspasia as his second wife, to whom he was devoted and faithful until death. According to Greek law, which forbade Athenian citizens to marry foreign women, he could not be her legal husband; but, there can be no doubt that he always treated her with all the respect and affection due to a wife. His dying words: "Athens entrusted her greatness and Aspasia her happiness to me," clearly evince her nobility of character and the place she must ever have occupied in the great statesman's heart.

The most important notices in ancient writings, respecting Aspasia, are found in Plutarch's *Pericles*, Xenophon's *Memorabilia* of Socrates and Plato's *Menexenus*. Among the most valuable of modern works on the same subject is *Aspasie de Milet*, by L Becq de Fouquieres, Paris, 1872. Cf also *Aspasie et le Siecle de Pericles*, Paris, 1862; *Histoire des Deux Aspasies*, by Le Comte de Bievre, Paris, 1736, and A Schmidt's *Sur l'Age de Pericles*, 1877-79.

12) Under the term music, Plato, like his contemporaries, included reading, writing, literature, mathematics, astronomy and harmony. It was opposed to gymnastic as mental to bodily training. Both music and gymnastic, however, were intended for the benefit of the soul.

13) *The Dialogues of Plato, Laws*, VII, 805, Jowett's translation, New York, 1892.

14) *Op cit, The Republic*, V, 451 *et seq* and 466.

15) It was the boast of the Emperor Augustus that all his clothes were woven by his wife, sister or daughter. Suetonius, in his *Lives of the Twelve Caesars*, informs us that this great master of the world *filiam et neptes ita instituit ut etiam lanificio assuefaceret.*

16) This type of the old Roman schoolmaster is alluded to in the following well known verses of Martial:

Quid tibi nobiscum est, ludi scelerate magister,
Invisum pueris virginibusque caput?
Nondum cristati rupere silentia Galli
Murmure jam saevo verberibusque tonas.
— Lib IX, 79.

which have been rendered as follows:

Despiteful pedant, why dost me pursue,
Thou head detested by the younger crew?
Before the cock proclaims the day is near
Thy direful threats and lashes stun my ear.

Martial elsewhere refers to "Ferulaeque tristes, sceptra pedagogorum" — melancholy rods, sceptres of pedagogues — and it appears from one of Juvenal's satires that "to withdraw the hand from the rod" was a phrase meaning "to leave school."

17) *Woman through the Ages*, Vol I, pp 110-11, by Emil Reich, London, 1908.

Schoolhouses among the Romans, as well as among the Greeks, were quite different from our modern, well-equipped buildings. Usually, at least, in earlier times, instruction was given in the open air, in some quiet street corner or in *tabernae* — sheds or lean-tos— as in certain Mohametan countries today. Horace refers to this in *Epistola* XX, Lib I, when he writes:

Ut pueros elementa docentem
Occupet extremis in vicis balba senectus.

In such schools the pupils sat on the floor or the bare ground, or, if the lessons were given on the street, they sat on the stones. There were no desks, or, if there were any benches, they had no backs. The pupils were, therefore, perforce obliged to write on their knees. Cf *Historical Survey of Pre-Christian Education*, pp 278 and 346, by S S Laurie, London, 1900.

18) Cf his *Tiberius Gracchus*. Cicero says of them, "Non tam in gremio educatos quam sermone matris."

19) Ibidem, *Life of Pompey*.

20) *De Oratore*, Lib III, Cap XII.

21) "Potiorem iam apud exercitus Agrippinam quam legatos, quam duces; compressam a muliere seditionem, cui nomen principis obsistere non quiverit." *Annales*, Lib I, Cap 69.

22) *Oeconomicus*, VII, 5-6.

23) *Epistolae*, Lib I, 16.

24) Sit mihi verna satur, sit non doctissima conjux. *Epigrammata*, Lib II, 90.

Martial's taste in this respect was the same as that of Heine, who said of the woman he loved: "She has never read a line of my writings and does not even know what a poet is," and the same as that of Rousseau, who declared that his last flame, Therese Lavasseur, could not tell the time of day.

25) Satire VI, 434-40.

26) *Joannis Stobaei Florilegium*, Vol IV, p 212, Teubner's edition, 1857.

27) The following is the epitaph as written by St Jerome, "the Christian Cicero":

Scipio quam genuit, Pauli fudere parentes,
Gracchorum soboles, Agamemnonis inclyta proles,
Hoc jacet in tumulo, Paulam dixere priores,
Euxtochii genetrix, Romani prima senatus,
Pauperiem Christi et Bethlehemitica rura secuta est.

28) In his preface to the *Commentary on Sophonius*.

29) For an exhaustive account of the lives and achievements of St Jerome and his noble friends, Paula and Eustochium, the reader is referred to *L'Histoire de Sainte Paule*, by F Lagrange, Paris, 1870, and *Saint Jerome, La Societe Chretienne à Rome et l'Émigration Romaine en Terre Sainte*, by A Thierry, Paris, 1867. Cf also *Woman's Work in Bible Study and Translation*, by A H Johns in *The Catholic World*, New York, June 1912.

30) See *Histoire de Sainte Radegonde, Reine de France*, in Ch XX, by Em Briand, Paris, 1897.

31) *Historia Ecclesiastica Gentis Anglorum*, Lib IV, Cap 23.

32) *The Monks of the West*, Book XI, Ch II.

33) Vol I, pp 46 and 49, New York, 1871.

34) *Op cit*, Book XI, Ch II.

It will interest the reader to know that Caedmon has a place among the saints in the *Acta Sanctorum* of the Bollandists. See the special article on him in Vol II, p 552, under the caption of "*De S Cedmono, cantore theodidacto*."

35) *Woman Under Monasticism*, Ch IV, §2, by Lina Eckenstein, Cambridge, 1896. In this chapter is an interesting account of the

Anglo-Saxon nuns who were among the correspondents of Boniface.

36) The reader will recall Chaucer's account in the *Canterbury Tales* of the wife of the well-to-do miller of Trumpyngton:
A wyf he hadde y-comen of noble kyn;
She was y-fostred in a nonnerye.
There dorste no wight clepen hir but 'Dame;'
What for hire kynnrede and hir nortelrie,
That she had lerned in the nonnerie."
— *Reeve's Tale.*

37) Pp 78-79, London, 1897.

38) *History of European Morals*, Vol II, p 369, New York, 1905.

39) *Henry VIII and the English Monasteries*, London, 1895.

40) *The English Historical Review*, July 1888.
Another recent writer affirms without hesitation that "Hroswitha has earned a place apart in the Pantheon of women poets and writers. She alone in those troublous times of the tenth century recalls to our minds the existence of dramatic art; her name, indeed, deserves to be rescued from oblivion and to become a household word." *Fortnightly Review*, p 450, March 1896.

41) *Histoire de l'Éducation de Femmes en France*, Vol I, p 72 *et seq* by Paul Rousselot, Paris, 1883.
A certain jurisconsult of the thirteenth century, one Pierre de Navarre, expressed the sentiment of many of his contemporaries when he wrote the following paragraph:
Toutes fames doivent savoir filer et coudre; car la pauvre en aura mestier et la riche conoistra mieux l'oeuvre des autres. A fame ne doit-on apprendre lettre ni escrire, si ce n'est especiaument pour estre nonain, car par lire et escrire, de fame sont maint mal avenu.

42) *Opera Omnia S Hildegardis*, Vol 197, Col 48 of Migne's *Patrologiae Cursus Completus*. Cf also *Nova S Hildegardis Opera*, editor: Cardinalis Pitra, Paris, 1882, and *Das Leben und Wirken der Heiligen Hildegardis*, von J P Schmelzeis, Freiburg im Breisgau, 1878.

43) It was Peter Lombard, whose *Sentences* "became the very canon of orthodoxy for all succeeding ages," who, in marked contrast with those of ancient and modern times that regarded woman as the inferior or slave of man, asserted her equality with him in a sentence that should be written in letters of gold. "Woman," he declares, *Sententiarum*, Lib II, Disp 18, "was not taken from the head of man, for she was not intended to be his ruler, nor from his foot, for she was not intended to be his slave, but from his side, for she was intended to be his companion and comfort."

In this view the great Schoolman but follows the teachings of St Augustine. For in his commentary, *De Genesi ad Litteram*, Lib 9, Cap 13, the learned bishop of Hippo writes: "Quia igitur viro nec domina nec ancilla parabatur, sed socia, nec de capite, nec de pedibus, sed de latere fuerat producenda, ut juxta se producendam cognosceret, quam de suo latere sumptam didecisset." Again the same illustrious doctor declares that woman was formed from man's side in order that it might be manifest that she was created to be united with him in love — in consortium creabatur dilectionis.

44) Cf *Hortus Deliciarum*, by Herrad de Lansberg, folio with one hundred and ten plates, Strasburg, 1901, and *Herrade de Landsberg*, by Charles Schmidt, Strasburg.

The erudite academician, Charles Jourdain, says of Herrad's great work "L'encyclopaedie qu'on lui doit, *l'Hortus Deliciarum*, embrasse toutes les parties des connaissances humaines, depuis la science divine jusqu'à l'agriculture et la metrologie, et on s'etonne à bon droit qu'un tel ouvrage, qui supposait une erudition si variee et si methodique, soit sorti d'une plume feminine. Quelle impression produirait aujourd'hui l'annonce d'une encyclopedie qui aurait pour auteur une simple, religieuse? Parlerons-nous des femmes du monde? Il n'existe d'elles, au XX$^{e}$ siecle, non plus que dans les siecles precedents aucun ouvrage comparable à *l'Hortus Deliciarum*." *Excursions Historiques et Philosophiques*, p 480, Paris, 1888.

45) See *Revelationes Mechtildianae ac Gertrudianae*, editor: Oudin, for the Benedictines at Solesmes, 1875.

46) In her scholarly work on *Woman Under Monasticism*, p 479, Lina Eckenstein writes as follows regarding the studies pursued in the convents of the Middle Ages:

The contributions of nuns to literature, as well as incidental remarks, show that the curriculum of study in the nunnery was as liberal as that accepted by the monks, and embraced all available writing whether by Christian or profane authors. While Scripture and the writing of the Fathers of the Church at all times formed the groundwork of monastic studies, Cicero at this period was read by the side of Boethus, Virgil by the side of Martianus Capella, Terence by the side of Isidore of Seville. From remarks made by Hroswitha we see that the coarseness of the Latin dramatists made no reason for their being forbidden to nuns, though she would have seen it otherwise; and, Herrad was so far impressed by the wisdom of the heathen philosophers of antiquity that she pronounced this wisdom to be the 'product of the Holy Spirit also.' Throughout the literary world, as represented by convents, the use of Latin was general, and made possible the even spread of culture in districts that were widely remote from each other and practically without intercourse.

47) *The Lady*, p 71, by Emily James Putnam, New York, 1910.

48) Eckenstein, *op cit*, p 478.

49) Ut, Sup, 479-80.

50) See *Womankind in Western Europe*, p 288 *et seq*, by Thomas Wright, London, 1869.

51) "Pertinere videtur ad haec tempora Betisia Gozzadini non minus generis claritate quam eloquentia ac legum professione illustris.... Betisiam Ghirardaccius et nostri ab eo deinceps scriptores eximiis laudibus certatim extulerunt." *De Claris Archigymnasii Bononiensis Professoribus a Saeculo XI usque ad Saeculum XIV*, Vol I, p 171, Bologna, 1888-96.

52) L'École de Salerne, p 18, by C Meaux, Paris, 1880. Among the most noted of these women was Trotula, who, about the middle of the eleventh century, wrote on the diseases of women as well as on other medical subjects. Compare the attitude of the school of Salerno towards women with that of the University of London, eight hundred years later. When, in the latter half of the nineteenth century, women applied to this university for degrees in medicine, they were informed, as H Rashdall writes in *The Universities of Europe in the Middle Ages*, Vol II, Part II, p 712, Oxford, 1895, that "the University of London, although it had been empowered by Royal Charter to do all things that could be done by any University, was legally advised that it could not grant degrees to women without a fresh Charter, because no University had ever granted such degrees." Cf also Haeser's *Lehrbuch der Geschichte der Medicin*, Vol I, p 645, *et seq*, Jena, 1875. Verily, the so-called dark ages have risen up to condemn our vaunted age of enlightenment!

53) *Die Entstehung der Universitäten des Mittelalters bis 1400*, Vol I, p 233, Berlin, 1885, von P Heinrick Denifle, assistant archivist of the Vatican Library, and *Histoire Literaire de la France, Commence par des Religieux Benedictins de S Maur et Continue par des Membres de l'Institut*, Vol IX, 281, Paris, 1733-1906.

54) "Une de ces nuits lumineuses ou les dernieres clartes du soir se prolongent jusqu'aux premieres blancheurs du matin." *Documents Inedits*, p 78, Paris, 1850.

55) *The Universities of Europe in the Middle Ages*, Vol I, p 31, Oxford, 1895.

56) *A Short History of the Renaissance in Italy*, p 277, London, 1893.

57) Cecelia Gonzaga, a pupil of the celebrated humanist, Vittorino da

Feltre, read the Gospels in Greek when she was only seven years old. Isotta and Ginevra Nogorola, pupils of the humanist, Guarino Verronese, likewise distinguished themselves at an early age by their rare knowledge of Latin and Greek. In later years all three enjoyed great celebrity for their learning, and were, like Battista di Montefeltro, women of genuine humanist sympathies. Cecelia Gonzaga's scholarship was in no wise inferior to that of her learned brothers, who were among the most noted students of the famous Casa Zoyosa in Mantua, where Vittorino da Feltre achieved such distinction as an educator in the early part of the Italian Renaissance. The learned Italian writer, Sabbadini, beautifully expressed the relation of women to Humanism, when he declares, in his *Vida di Guarino, "L'Humanismo si sposa alla gentilezza feminile"* — humanism weds feminine gentility.

58) Among them are the pictures of Caterina Vigri, which are preserved in the Pinacoteca of Bologna and in the Academia of Venice.

59) No less an authority than the illustrious sculptor, Canova, declared that her early death was one of the greatest losses ever suffered by Italian art.

60) It was also said of the Venetian artist, Irene di Spilimbergo, that her pictures were of such excellence that they were frequently mistaken for those of her illustrious master, Titian.

61) Among these works may be mentioned *Il Merito delle Donne*, by Modesta Pozzo di Zorgi, Venice, 1600; *La Nobilità e l'Excellenza delle Donne*, by Lucrezia Marinelli, Venice, 1601; *De Ingenii Muliebris ad Doctrinam et Meliores Litteras Aptitudine*, by Anna van Schurman, Leyden, 1641; *Les Dames Illustres*, by Jaquette Guillame, Paris, 1665, and *L'Egalite des Hommes et des Femmes*, by Marie le Jars de Gournay, Paris, 1622. The last named work was by the celebrated *fille d'alliance* — adopted daughter — of Montaigne. It is to her that we owe the *textus receptus* of the *Essais* of the illustrious litterateur.

62) *The Women of the Renaissance*, p 290, by R de Maulde la Claviere, New York, 1901.

63) Called *La Latina*, because of her thorough knowledge of the Latin language.

64) The famous Hellenist, Roger Ascham, tells of his astonishment on finding Lady Jane Grey, when she was only fourteen years of age, reading Plato's *Phaedo* in Greek, when all the other members of the family were amusing themselves in the park. On his inquiry why she did not join the others in their pastime, she smilingly replied: "I wit all their sport in the park is but a shadow to that pleasure I find

in Plato. Alas, good folk, they never knew what true pleasure meant."

65) To the poet Ronsard, she was a woman beyond compare, as is evinced by the following lines of a pastoral ode addressed to her:
La Royne Marguerite,
La plus belle fleur d'elite
Qu'onques la terre enfanta.

66) Cf *Oeuvres de Lovise Labe*, nouvelle edition, emprimee en caracteres dits de civilite, Paris, 1871.

67) The French poet, Jean Dorat, who was then professor of Latin in the College de France, expresses this fact in the following strophe:
Nempe uxor, ancillae, clientes, liberi,
Non segnis examen domus,
Quo Plautus ore, quo Terentius, solent
Quotidiane loqui.

68) A prominent writer of the time, Jean Bouchet, expressed the prevailing opinion regarding the education of the women of the masses in the following quaint sentence: "Je suis bien d'opinion que les femmes de bas estat, et qui sont contrainctes vaquer aux choses familieres et domestiques, ne doivent vaquer aux lettres, parce que c'est chose repugnante à rusticite; mais, les roynes, princesses et aultres dames qui ne se doib vent pour reverence de leur estat, appliquer à mesnage." Cf Rousellot's *Histoire de l'Éducation des Femmes en France*, Vol I, p 109, Paris, 1883.
His ideal of a woman of the peasant type was apparently Joan of Arc, who, according to her own declaration, did not know a from b— "*elle declarait ne savoir ni a ni b*."

69) Claviere, *op cit*, p 415.

70) The noted English divine, Thomas Fuller, chaplain to Charles II, recognised the irreparable loss to women occasioned by the destruction of the nunneries by the Reformers. "There were," he tells us in his quaint language, "good she schools wherein the girls and maids of the neyghbourhood were taught to read and work..... Yea, give me leave to say, if such feminine foundations had still continued..... haply the weaker sex, besides the avoiding modern inconveniences, might be heyghtened to a higher perfection than hitherto hath been attained." *Church History*, Vol III, p 336, 1845.

71) M. Thureau Dangin, the perpetual secretary of the French Academy, wrote, "La tradition ne veut pas d'academiciennes."

72) Carlyle, in a lecture on Dante, and the *Divina Commedia*, declares that "Italy has produced a greater number of great men than any other nation, men distinguished in art, thinking, conduct, and everywhere in the departments of intellect." He could with equal

truth have said that Italy has produced more great women than any other nation.

73) *Medical Women*, p 63, *et seq*, by Sophia Jex-Blake, Edinburgh, 1886, and *Pioneer Work in Opening the Medical Profession to Women*, Ch III, by Elisabeth Blackwell, London, 1895.

74) Mme Dacier was a remarkable exception chiefly because she was the daughter and pupil of one Hellenist before becoming the wife of another.

75) *Lettres et Entretiens sur l'Éducation de Filles*, Vol I, pp 225-31. Compare this superficial course of study at Saint-Cyr with the elabourate course mapped out by Lionardo d'Arezzo in a letter addressed to the illustrious lady, Baptista Malatesta. In the broad programme of education for women recommended by this eminent man of letters, "poet, orator, historian, and the rest, all must be studied, each must contribute a share. Our learning thus becomes full, ready, varied, elegant, available for action or for discourse on all subjects."

Lionardo's curriculum of studies for women was quite as comprehensive as that required for men, "with perhaps a little less stress upon rhetoric and more upon religion. There was no assumption that a lower standard of attainment is inevitably a consequence of smaller capacity."

Nor was this thorough study of letters by the women of Italy "unfavourably regarded by social opinion;" neither did it introduce "a new standard of womanly activity. Women, indeed, at this epoch, seem to have preserved their moral and intellectual balance under the stress of the new enthusiasm better than men. The learned ladies were, in actual life, good wives and mothers, domestic and virtuous women of strong judgement and not seldom of marked capacity in affairs." Cf *Vittorino da Feltre and Other Humanist Educators*, pp 122, 132, 197, by W H Woodward, Cambridge, 1905.

76) Thus, in a letter of hers to Mme. de Lauzun occurs a sentence like the following: "Il lia sy lontant que je n'ay antandu parler de vous." The duchess of Monpensier, daughter of Gaston d'Orleans, in a letter to her father exhibits a similar ignorance of her own language, when she writes: "J'ai cru que Votre Altesse seret bien ese de savoir sete istoire." Quoted by Rousselot in his *Histoire de l'Éducation des Femmes en France*, Vol I, p 287.

77) *Les Femmes Savantes*, Act II, Scene 7.

78) Destouches, in his *L'Homme singulier*, makes one of his female characters, who loves study, speak in the following pathetic fashion:
A learned woman ought — so I surmise —

Conceal her knowledge, or she'll be unwise.
If pedantry a mental blemish be
At all times outlawed by society,
If 'gainst a pedant all the world inveighs,
Shall pass unchecked in woman pedant's ways?
I hold it sure, condemned my sex is quite
To trifling nothings as its sole birthright;
Ridiculous 'tis thought outside its 'sphere;'
The learned woman dare not such appear;
Nay, she must even cloak her brilliancy
So envy leave in peace stupidity;
Must keep the level of the common kind,
To subjects commonplace devote her mind,
And treating these she must be like the rest.
Lo, in such garb refinement must be dressed:
That knowledge shall not make her seem unwise,
She must herself in foolishness disguise."
—Act III, Scene 7.

79) No one, however, went so far in his opposition to the education of women as the notorious Silvain *Marechal*, the author of *Projet d'une Loi portant Defense d'Apprendre à Lire aux Femmes*, who would have a law passed forbidding women to learn to read. He maintained that a knowledge of science and letters interfered with their being good housekeepers. "Reason," he avers, "does not approve of women studying chemistry. Women who are unable to read make the best soup I would rather," he declares in the words of Balzac, "have a wife with a beard than a wife who is educated." See pp 40, 50 and 51, of the edition of this strange work, published at Brussels, 1847.

80) In her *Problema Practicum*, addressed to Dr Rivet, Anna van Schurman states and develops in true syllogistic form a series of propositions in defence of her thesis in favour of the higher education of women. Two of these propositions are here given as illustrative of her points of view:

I Cui natura inest scientiarum artiumque desiderium, ei conveniunt scientiae et artes. Atque feminae natura inest scientiarum artiumque desiderium. Ergo.

II Quidquid intellectum hominis perficit et exornat, id femmae Christianae convenit. Atqui scientiae et artes intellectum hominis perficiunt et exornant. Ergo. See *Nobiliss. Virginis Annae Schurman Opuscula*, pp 35 and 41, Leyden, 1656, and her *De Ingenii Muliebris ad Doctrinam et Meliores Literas Aptitudine*, Leyden, 1641. Cf also *Anna van Schurman*, Ch IV, by Una Birch, London, 1909.

81) A writer of the seventeenth century gives the following as the popular programme of female study: "To learn alle pointes of good housewifery, spinning of linen, the ordering of dairies, to see to the salting of meate, brewing, bakery, and to understand the common prices of all houshold provisions. To keepe account of all things, to know the condition of the poultry — for it misbecomes no woman to be a hen-wife. To know how to order your clothes and with frugality to mend them and to buy but what is necessary with ready money. To love to keep at home." How like the German four K's and the words on the sarcophagus of a Roman matron — *lanifica, frugi, domiseda* — a diligent plyer of the distaff, thrifty and a stay-at-home.

82) *The Letters and Works of Lady Mary Wortley Montagu*, Vol II, p 5, Bohn Edition, 1887.

83) Letter XLIX, London, Sept 5, OS, 1748.

Walpole, writing in 1773, makes the following curious declaration: "I made a discovery — Lady Nuneham is a poetess, and writes with great ease and sense some poetry, but is as afraid of the character, as if it was a sin to make verses." And Lord Granville tells us of an eminent statesman and man of letters who, in the early part of the last century, was so troubled on discovering in his daughter a talent for poetry that he "appealed to her affection for him, and made a request to her never to write verses again. He was not afraid of her becoming a good poetess, but he was afraid of the disadvantages which were likely to be suffered by her, if she were supposed to be a lady of literary attainments."

84) It was Swift who had such a low opinion of woman's intellect that in writing to one of his fair correspondents he told her that she could "never arrive in point of learning to the perfection of a schoolboy." Lady Pennington, strange to say, seems to have shared his views, for in a manual of advice to young ladies, she declares: "A sensible woman will soon be convinced that all the learning the utmost application can make her master of will be in many points inferior to that of the schoolboy." "At the time the Tatler first appeared in the female world any acquaintance with books was distinguished only to be censured," and it was then considered "more important for a woman to dance a minuet well than to know a foreign language."

85) The wife of President John Adams, descended from the most illustrious colonial families, writing in 1817, regarding the educational opportunities of the girls of her time and rank, expressed herself as follows:

Female education in the best families went no farther than writing and arithmetic, and, in some few and rare instances, music and dancing." According to her grandson, Charles Francis Adams, "The only chance for much intellectual improvement in the female sex was to be found in the families of the educated class, and in occasional intercourse with the learned of the day. Whatever of useful instruction was secured in the practical conduct of life came from maternal lips; and, what of farther mental development depended more upon the eagerness with which the casual teachings of daily conversation were treasured up than upon any labour expended purposely to promote it.

— *Familiar Letters of John Adams and His Wife, Abigail Adams, During the Revolution, With a Memoir of Mrs Adams*, by Charles Francis Adams, pp X and XI, New York, 1876.

86) When the students of Girton and Newnham in 1897, after passing the Cambridge examinations — many of them with the highest honours — applied for degrees, "the undergraduate world was stirred to a fine frenzy of wrath against all womankind," and an astonished world saw re-enacted scenes scarcely less disgraceful than those which characterised the riotous demonstrations which, seventeen years before, had greeted seven young women at the portals of the University of Edinburgh.

87) *The Queen's Reign*, Ch V, London, 1897.

88) Proposition third, of her *Propositiones Philosophicae*, Milan, 1738, reads as follows:

Optime etiam de universa Philosophia infirmiorem sexum meruisse nullus infirmabitur; nam praeter septuaginta fere eruditissimas, Mulieres, quas recenset Menagius, complures alias quovis tempore floruisse novimus, quae in philosophicis disciplinis maximam ingenii laudem sunt assecutae. Ad omnem igitur doctrinam, eruditionemque etiam muliebres animos Natura comparavit: quare paulo injuriosius cum feminis agunt qui eis bonarum artium cultu omnino interdicunt, eo vel maxime, quod haec illarum studia privatis, publicisque rebus non modo haud noxia futura sint verum etiam perutilia.

This admirable work, with its one hundred and ninety-one propositions, is commended to those who may have any doubt regarding the learning or capacity of the Italian women who have been referred to in the preceding pages.

# Woman's Capacity for Scientific Pursuits

IN A CURIOUS old black-letter volume entitled *The Boke of the Cyte of Ladyes,* published in England in 1521 by Henry Pepwell, occurs the following passage: "I mervayle gretely of the opynyon of some men that say they wolde in no wyse that theyr daughters or wyves or kynnes-women sholde lerne scyences, and that it sholde apayre theyr condycyons. This thing is not to say ne to sustayne. That the woman apayreth by conynge it is not well to beleve. As the proverb saythe, 'that nature gyveth may not be taken away.' "

The book from which this remarkable quotation is taken is a translation of Christine de Pisan's *La Cite des Dames,* which was written early in the fifteenth century. It is a capital defence against the slanderers of the gentler sex and an armoury of arguments for all time against those men who declare that "women are fit for nothing but to bear children and spin." It shows conclusively that conynge — knowledge — far from tending to injure women's character — apayre theyr condycyons — as was asserted by Christine's antagonists, contributes, on the contrary, to elevate and ennoble them and to render them better mothers and more useful members of society.

Notwithstanding that it was written five hundred years ago, and notwithstanding its "antiquated allegorical dress and its quaint pre-Renaissance notions of history," it is in many of its aspects a surprisingly modern production. The line of argument adopted by the writer is virtually the same as that which is adopted today in the discussion of the same questions which are so ably treated in this long-forgotten book[1] and show that Christine de Pisan was in every way a worthy champion of her sex.

No woman of her time was more competent to discuss the capacity of her sex for science as well as for other intellectual pursuits than was this learned daughter of Italy. She was not only a woman of profound and varied knowledge, but was also, as stated in the preceding chapter, the first woman to earn her living by her pen. Besides writing *The City of Ladies* and more verses — mostly ballads and virelays — than are contained in the *Divina Commedia*, she was also the author of many other works on the most diverse subjects. She is best known to historians as the author of *Livre des Fais et Bonnes Meurs du sage Roy Charles V*, which is a graphic account of the court and policy of this monarch, and of the *Livre des Faits d'Armes et de Chevalerie*. The latter work is not, as might be imagined from its title, a collection of tales of chivalry, but, incredible as it may seem, a profound and systematic treatise on military tactics and international law. It deals with "many topics of the highest policy, from the manners of a good general and the minutiae of siege operations to the wager of battle, safe-conducts and letters of marque," and was deemed so important by Henry VII that at his expressed desire it was translated into English and published by Caxton under the title of *The Boke of Fayettes of Armes and Chyvalrye*. Even so late as the time of Henry VIII it was regarded as an authoritative manual on the topics treated.

So great, indeed, was the extent and variety of Christine's attainments, so thoroughly had she studied the Latin and Greek authors, sacred and profane, and so profound was her knowledge of all the subjects which she dealt with in her numerous books that "one cannot but feel a certain astonishment when one finds in a woman in the fourteenth century an erudition such as is hardly possessed by the most laborious of men."

When we read the eloquent plea which this learned woman of five centuries ago makes in behalf of her sex, when we note the examples she quotes of women "illumined of great sciences," and consider the arguments by which she demonstrated the capacity of women for all scientific pursuits, we can easily fancy that we are reading the brief of some modern exponent of the woman's rights movement and are almost disposed to believe that La Bruyiere was right when he declared, *Les anciens ont tout dit.* For so cogent is Christine's reasoning and so thoroughly does she traverse her subject from every point of view that she has left later writers little to add to the controversy except matters of detail which were not available in her time.

In spite, however, of Christine's *Cyte of Ladyes,* "in which," according to our medieval paragon, "women, hitherto scattered and defenceless, were forever to find refuge against all their slanderers," in spite of the fact that the foundations of this city were laid by Reason, that its walls and cloisters were built on Righteousness, and its battlements and high towers on Justice, in spite of the fact that the material entering into its construction was "stronger and more durable than any marble," and that it was, as our author declares, "a city right fair, without fear and of perpetual during to the world — a city that should never be brought to nought," Christine's work was soon lost sight of, and the right of women to the same intellectual advantages as men was as strongly denied as it had been before she had so valiantly championed their cause, and denied, too, on the assumed ground of their innate incapacity.

It mattered not that during the succeeding centuries other women took up the cause for which the author of *La Cite des Dames* had so nobly battled; it mattered not that countless women in every civilised country of the globe distinguished themselves by their achievements in every department of science and gave evidence of talent and genius of the highest order; it mattered not that chivalrous representatives of the sterner sex, like John Stuart Mill, came forward to plead the case of that half of humanity which had so long been held in cruel subjection. The attitude of the world toward the intellectually disfranchised sex remained unchanged almost until our own time.

But, although women now enjoy advantages in the pursuit of

science which were undreamed of only a generation ago, the age-old prejudices respecting woman's mental powers and her capacity for the more abstract branches of science still prevail. It is useless to cite instances of women who have attained eminence in astronomy, mathematics, archaeology, or in any other science whatever. Such instances, we are assured, are only exceptions and prove nothing. Men like Lombroso are willing to admit the existence of an occasional woman of talent, but they deny the existence of genius in one who is truly a womanly woman.[2] For, with Goncourt, they flippantly assert, *Il n'y a pas de femmes de genie: lorsqu'elles sont des genies, elles sont des hommes*— there are no women of genius; when they have genius they are men.

The reasons that now influence men for affirming the intellectual disparity of the sexes are, it must be observed, quite different from what they were in the time of Christine de Pisan — quite different from what they were half a century ago. Our forebears, in their endless disputations regarding woman's mental inferiority, based their arguments on *a priori* deductions, or on metaphysical considerations which proved nothing and which were often irrelevant, if not absurd.

Thus the Aristotelians, accepting as true the doctrine of the four elements as well as the superimposed doctrine of the four elemental qualities, sought to explain the properties of all compound bodies by these primal qualities. In this way they explained the various virtues of drugs and medicines. And by the same process of reasoning they explained the assumed difference between male and female brains. They assumed, to begin with, that there was a difference between the intellectual capacities of men and women. They then assumed that this difference in capacity was due to the difference in character and texture of the female as compared with those of the male brain. They next further assumed that the doctrines of the four elements and of the four elemental qualities were established beyond question, and then assumed again that the reason of woman's inferior capacity was due to the fact that her brain was moister and softer, and, therefore, more impressionable than that of man. No wonder that the old Spanish Benedictine, Benito Jeronimo Feijoo, in his chivalrous *Defensa de la Mujer*, lost all patience with such fantastic theorisers and wrote: "Did I write..... to display my wit, I could easily, by deducting a chain of consequences from received principles, shew that man's understanding, weighed in the balance with female capacity, would be found so light as to kick the beam."[3]

Abandoning the Aristotelian method of envisaging the question under discussion, our modern philosophers have recourse to the recent sciences of biology and psycho-physiology to prove what they, too, assume to be true — viz, woman's incurable mental weakness. Like their predecessors, they are dominated by passion, prejudice, the errors of countless centuries, and, like them, they approach the subject on which they are to pronounce judgement, with minds warped by long ages of imperious instincts, ignorant preconceptions and social bias. They will quote the opinions of Proudhon and Schopenhauer — as if they had the value of mathematical demonstrations — on the mental inferiority of women, and will declare with unblushing assurance that no woman has ever produced a single work of any kind of enduring worth. With the German pessimist, they will blatantly declare, taken as a whole, "women are and remain thoroughgoing Philistines and quite incurable."[4] With the French socialist they will assert, as if it were an axiomatic truth, that "thought in every living being is proportional to force" — that "physical force is not less necessary for thought than for muscular labour."

They have apparently no more doubt respecting the truth of these assumptions than had their predecessors, the Aristotelians, respecting their assumptions of the four elements and their first qualities. Their process of reasoning is somewhat as follows: "Woman is smaller and weaker than man. This is a matter of simple observation, confirmed by the teachings of physiology. Therefore, woman is physically and intellectually inferior to man. Therefore she is incapable of any of those great conceptions and achievements in science or philosophy which have so distinguished the male sex in every age of the world's history. That she is thus weaker and inferior physically and intellectually and forever incapacitated from successfully competing with man in the intellectual arena is a fatality for which, we are gravely told, there is no remedy, and to which women, consequently, must resign themselves as to one of the inexorable laws of nature."

It would be difficult to cite a more preposterous example of ratiocination. If it were true that there is a necessary relation between vigour of body and vigour of mind; that mental power is proportional to physical power; that thought is but a special form of energy and capable of transformation, like heat, light and electricity; that it, like the various physical forces, has its chemical and mechanical equivalents; that psychic work corresponds to a certain amount of chemical or thermic action; that intellectual capacity in man is proportional to muscular strength; it would follow that the great leaders of thought and action through the ages have been Goliaths in stature and Herculeses in strength. But so far is this conclusion from being warranted that it is almost the reverse of the truth. For many, if not the majority, of the great geniuses of the world in every age have been either men of small frame or men of delicate and precarious health.

Among the men of genius who were noted for their diminutive stature were Plato, Aristotle, Alexander the Great, Archimedes, Epicurus, Horace, Albertus Magnus, Montaigne, Lipsius, Spinoza, Erasmus, Lalande, Charles Lamb, Keats, Balzac and Thiers. Many others were remarkable for their spare form. Among these in the prime of life were Aristotle, Demosthenes, Cicero, St Paul, Kepler, Pascal, Boileau, Fenelon, D'Alembert, Napoleon, Lincoln and Leo XIII. Others, like AEsop, Brunelleschi, Leopardi, Magliabecchi, Parini, Scarron, Talleyrand, Pope, Goldsmith, Byron, Sir Walter Scott, to mention only a few of the most eminent, were either hunchbacked, lame, rachitic or clubfooted.

Others, still, were the victims of chronic ill health, or of nervous disorders of the most serious character. Virgil was of a delicate and frail constitution. He essayed the bar, but shrank from it and turned to the "contemplation of diviner things." Nor was Horace, though less completely a recluse and more of a *bon vivant*, a strong man. Both of them, as scholars will remember, sought the couch, while Maecenas went off to the tennis court. Pope's life, says Johnson, was a long disease. Johnson himself, though large and muscular, had queer health and a tormenting constitution. Schiller wrote most of his best work while struggling against a painful malady, and Heine's "mattress grave" is proverbial. France furnishes an excellent example in Pascal.[5]

Some of the most noted leaders of thought in our own era were likewise chronic invalids. Among these were the scholarly theologian, E B Pusey, and J A Symonds, the historian of the Renaissance. There was also Herbert Spencer, who was frequently forced by nervous breakdowns to take long periods of absolute rest. More remarkable still was the case of the famous naturalist, Charles Darwin. "It is," writes his son, "a principal feature of his life that for nearly forty years he never knew one day of the health of ordinary men, and that thus his life was one long struggle against the weariness and the strain of sickness."[6] But, notwithstanding his continued ill health and the spinal anaemia from which he suffered, he was able to conduct those epoch-making researches which put him in the forefront of men of science, and to write those famous books which have completely revolutionised our views of nature and nature's laws.

But a still more remarkable illustration of the fact that there is no necessary relation between muscular and mental power, between physical well-being and intellectual energy, is afforded by the illustrious discoverer of the world of the infinitely little, Louis Pasteur. Stricken by hemiplegia shortly after he had begun those brilliant investigations which have rendered him immortal, he remained affected by partial paralysis until the end of his life. His friends had reason to fear that this attack, even if he should survive it, would weaken or extinguish his spirit of initiative, if it did not make further work entirely impossible. But this was far from the case. For a quarter of a century he continued with unabated activity those marvellous labours which are forever associated with his name. And it was after, not before, his misfortune that he made his most famous discoveries in the domain of microbian life, and placed in the hands of physicians and surgeons those infallible means of combating disease which have made him one of the greatest benefactors of suffering humanity. The complete separation of the intellectual from the motor faculties was never more clearly exhibited than in this case, nor was it ever more completely demonstrated by an experiment, whose validity no one could question, that power of mind does not necessarily depend on strength or health of body. It proved, also, in the most telling manner that it is not muscular but psychic force which avails most, whether to the individual or to society. And it showed, at the same time, the utter absurdity of those theories which would fatally connect intellectual with physical debility in woman, and would forever adjudge the physically weaker sex to be of hopeless inferiority in all things of the mind.

What has been said of men achieving renown, notwithstanding ill health, may likewise be affirmed of women. The case of Elizabeth Barrett Browning is scarcely less remarkable than that of Darwin. In spite of being a chronic invalid the greater part of her life, she attained a position in letters reached by but few of her contemporaries. The same almost may be said of the three Bronte sisters. The deadly seeds of consumption were sown in their systems in early youth, but, although fully aware that life had "passed them by with averted head," they were, through their indomitable wills, able to send forth from their bleak home in the wild Yorkshire moors works of genius that still instruct and delight the world.

From the foregoing it is clear that valetudinarianism, if it prove anything, proves not that it renders intellectual effort impossible, but that it serves as a discipline for the soul. It forces the mind to husband its strength, and thus enables it to accomplish by economy and concentration of effort that which the same mind in a healthy body, with the distractions of society and the allurements of life, would be unable to accomplish. It exemplifies in the most striking manner the truth of what Socrates says in Plato's *Republic* about the beneficent action of the "bridle of Theages," preventing an infirm friend of his from embracing politics and keeping him true to his first love — philosophy.

Failing to show any necessary connection between superior physique and intellectual capacity, between health of body and mental activity, between the amount of food consumed and the degree of intelligence, the class of thinkers whose theories are now under consideration found themselves forced to abandon the argument based on robust health and physical strength and seek elsewhere for support of their views. This, they soon announced, was found in the greater cranial capacity and greater brain weight of the male as compared with that of the female. Following up this fancied clew, anthropologists the world over began measuring skulls and weighing brains in order to determine the supposed ratio of sex-difference.

The results of these investigations were far from corroborating the preconceived notions of those who had fancied a necessary correlation between mental capacity and size of cranium, between the weight of encephalon and degree of intelligence. For it was soon discovered that cranial capacity depended on many causes — many of them unknown — and that people having the largest skulls were often far from being the ones dowered with the greatest intellectual power. It was found, for instance, that climate was a determining factor — that the inhabitants of northern regions have larger heads than those who live farther south. Thus the Lapps, in proportion to their stature, have the largest heads in Europe. After these come in order the Scandinavians, the Germans, the French, the Italians, the Arabs.

It was found also that the least cranial capacity of the ancient Egyptians coincides with the most brilliant period of their civilisation — that of the eighteenth dynasty. Measurements of skulls unearthed at Pompeii showed that the heads of the Romans who lived two thousand years ago were larger than the heads of the Romans of today. Similarly, the skulls of the lake-dwellers of Switzerland were larger than those of the Swiss people of the present time, while the average circumference of the skulls measured in the catacombs of Paris is more than an inch greater than that of the Parisians who have died during the last half century. The circumference of the skulls of a large number of mound-builders, excavated some years ago near Carrollton, Illinois, exceeded that of the average head of white men in New York of our day by nearly three inches. This shows that the culture of the white race during long centuries has not developed its cranial capacity to equal that of the uncultured Indians who flourished in the Mississippi valley untold generations ago.

The skulls of Quaternary men were likewise very voluminous, although they belonged to a race whose mental manifestations were infantile in the extreme. Even the celebrated Engis skull, one of the most ancient in existence, has been described by the late Professor Huxley as well formed and considerably larger than the average of the European skulls of today, not only in the width and height of the forehead, but also in the cubic capacity of the whole. Furthermore, the eminent craniologist, Broca, has proved that the illiterate peasants of Auvergne have a much greater cranial capacity than that of the learned and cultured denizens of Paris. And, as if to show conclusively that there is no necessary connection between intellectual capacity and size of cranium, authentic measurements disclose the fact that some of the most gifted men the world has known had small heads. Among these were Dante and Voltaire. The skull of the latter is one of the smallest which has thus far been observed.

What has been said regarding the relation of cranial volume to intellectual capacity, as revealed by the measurements of the skulls of ancient and modern, savage and civilised peoples may likewise be predicated of the differences in the sizes of the crania of men and women. No argument as to the greater or less intelligence of either sex can be based on mere craniometric determinations. "At the best, cranial capacity is but a rough indication of brain size; and to measure brain size by the external size of the skull furnishes still rougher and more fallacious approximations, since the male skull is more massive than the female."

Even the slight morphological differences between male and female skulls — some anthropologists deny that there are any at all — afford no more ground for conclusions in favour of the superiority of one or the other sex than the relative differences in size. Such trifling differences as do exist exhibit, as Virchow has pointed out, an approximation of men to the savage, simian and senile type, and an approach of women to the infantile type. Havelock Ellis, commenting on this difference, pertinently remarks, "It is open to a man in a Pharisaic mood to thank God that his cranial type is far removed from the infantile. It is equally open to woman in such a mood to be thankful that her cranial type does not approach the senile."[7]

But much stress as has been laid on physical power, health and cranial capacity, as determining factors of intellectual capacity and sexual differences, far greater stress has been laid on conclusions deducible from the relative brain weights of different classes of people as well as of different sexes. It was assumed that by a critical study of the brain, by careful weighings of many brains of both sexes and of many races, it would be easy to secure conclusive evidence that the size and weight of the brain increase with the amount of intelligence of the individual. It was also assumed that function not only makes the organ, but also develops it. Brain became synonymous with mind. A large brain implied vigour of thought; a small brain was evidence of mental inferiority.

Physiology had demonstrated unquestionably that the muscles of the body are enlarged by exercise. It was assumed by those who are wont to measure mind in terms of matter that the brain, being the organ of thought, was also developed by exercise. It was also assumed that the development of the brain was in a direct ratio to its activity. The greater its activity the greater its mass, and the greater the mass the greater the degree of intelligence. In other words, it was assumed that there was an exact and invariable proportion between weight of brain and amount of brain power.

None of the theories which have already been adverted to have been so full of assumptions and prejudices or vitiated by so many fallacies and over-hasty generalisations as this. No subject has possessed a greater fascination for anthropologists, and no subject has been prolific in more diverse and conflicting conclusions. Many men of science who, in other matters, were noted for their care in weighing evidence, before formulating theories, completely lost the scientific spirit when they began to weigh brains and to draw conclusions respecting the relations of brain weight and mental power, and to establish ratios between the character of the convolutions of the organ of thought and the degree of intelligence of its possessor.

Contrary to what is generally believed, a large brain is not always an indication of superior capacity or intelligence. There have been, it is true, a number of men of genius who were the possessors of large brains, but there have also been others whose brains were of but medium weight.

The largest known brains of intellectual workers were those of Cuvier, the noted zoologist, and Turgenev, the distinguished novelist. The brain of the Frenchman weighed 1830 grams, while that of the Russian totalled 2012 grams. Among other large brains — even larger than Cuvier's — were those of a bricklayer, which weighed 1900 grams, and of an ordinary labourer, which reached 1924 grams. The largest brains on record were that of an ignorant labourer named Rustan, which weighed 2222 grams; that of a weak-minded London newsboy, which weighed 2268 grams, and that of a twenty-one-year-old epileptic idiot, which had the unheard of weight of 2850 grams.[8]

The seven largest recorded female brains were three weighing 1580 grams each, one of which belonged to a medical student of marked ability, while the other two belonged to quite undistinguished women. There were two others weighing 1587 each, one of which belonged to an insane woman. Still heavier than these by far were the brains of an insane woman who died of consumption, and of a dwarfed Indian squaw. The brain of the first weighed 1742 grams; while that of the second was no less than 2084 grams.

From the foregoing examples it is evident that a large brain is far from being a certain index of mental capacity or of superior intelligence. It is frequently the very reverse. If, for instance, it fail to receive the necessary supply of blood, it will be inert or disordered and will prove to be a dangerous possession rather than a precious endowment. Epileptics usually have brains that are large relatively to the size of the body. And, while it is probably true that the great thinkers and men of action of the world have, in most instances, had comparatively large brains, it is also true that the brain weights of but few of them exceeded 1500 grams, while those of many fall below 1200 grams.

Thus the brain of Gambetta, "the foremost Frenchman of his time," weighed only 1159 grams, while the weight of the brain of Napoleon I was 1502 grams — barely equal to that of a negro described by the anthropologist Broca, and but little superior to that of a Hottentot mentioned by Dr Jeffries Wyman.[9]

The late Dr Joseph Simms found the average brain weight of sixty persons who were either imbeciles, idiots, criminals or men of ordinary mind to be 1792 grams, while that of sixty famous men was 1454 grams, a difference in favour of men not noted for intellectual greatness of 338 grams. These figures are far from showing that large brains are a necessary concomitant of mental capacity.

In view of these and many similar facts, we are not surprised that the eminent German anatomist and anthropologist, Rudolph Wagner, should declare that "very intelligent men do not differ strikingly in brain weight from less gifted men," and that the noted French physician, Esquirol, should assert that "no size or form of head or brain is incident to idiocy or superior talent."

So far as civilised races are concerned, there can be no doubt that the absolute weight of the male is greater than that of the female brain. According to the investigations of seven of the most notable anthropologists, who have given special attention to the subject under consideration, and who, collectively, have carefully weighed many thousands of brains, the average brain weight of men in Europe is 1381 grams, while that of women is 1237 grams. This shows a difference between the average weight of the brain in man and woman of 144 grams.

But, if it must be conceded that the absolute weight of man's brain is greater than woman's, is it likewise true that the relative weight is greater? This is a question which demands an answer, as it is impossible to come to any just conclusion respecting the intellectual capacity of woman expressed in terms of brain weight, unless we can affirm with certainty that men's brains are relatively, as well as absolutely, larger than those of women.

Speaking of the relative weight of brain in man implies a term of comparison. Several methods of estimating the sexual proportions of brain mass have been suggested, but only two of them have met with any favour. These are determining the ratio of brain weight to body weight or body height.

According to the investigations of anthropologists of acknowledged authority, the average brain weight of woman is to that of man in England and France as 90 is to 100. The average stature of men and women in the same countries is as 93 to 100. This gives man an excess of brain weight over that of woman of something more than an ounce. But this slight difference in weight has been considered sufficient to constitute it "a fundamental sexual distinction." When, however, it is considered that men are not only taller but also larger than women, this apparent advantage of an ounce in favour of the male entirely disappears, and the result is that the relative amount of brain mass in the two sexes is practically equal.

Because of the manifest inaccuracy of the stature criterion, many eminent anthropologists have prepared to estimate sexual differences in brain weight by adopting the method based on the ratio of brain mass to body weight. According to this method, women are found to possess brains which are equal to or even somewhat larger than those of men. If the comparative excess of non-vital tissue in the form of fat in woman be eliminated and estimates be based only on the active organic mass of her body, as compared with the same mass in man, the excess of brain weight in woman over that in man will be still more marked.

A careful study, then, of the brain as a whole, far from proving woman's inferiority to man, rather proves her superiority. The same may be said regarding sexual distinctions based on certain parts of the brain.

Some years ago it was positively asserted that the development of the frontal lobe exhibited a pronounced difference in the two sexes. It was said to be much greater in man than in woman and was regarded as a distinguishing characteristic of the male sex. This was in keeping with the generally accepted assumption that this portion of the brain is the seat of the higher intellectual processes. Further investigation, however, showed that there was practically no sexual difference in the frontal lobe of the brain, or, if there was a difference, it was probably in favour of woman.

It has also become recognised that there is no valid reason for considering the anterior portion of the brain as the seat of the higher mental functions. It is possible, but in the present state of science it can neither be affirmed nor denied. So far as our present knowledge goes, it seems more likely that the whole of the brain, especially the sensori-motor regions of its middle part, have a part in mental operations. At all events, it can certainly be affirmed that Huschke's distinction of man and woman into *homo frontalis et homo parietalis* is utterly devoid of foundation in fact.

Many anthropologists have fancied that a certain index of the degree of intelligence is to be found in the convolutions of the brain. The tortuous foldings of the female brain, it is asserted, are less ample, less pronounced and less beautiful. "Behold," they exclaim, "a most positive evidence of inferiority." These men overlook the fact that certain animals, notably the elephant and divers species of cetaceans, have cerebral convolutions that are more complex than those of man. If, then, brain convolutions were, as claimed, a certain index of the degree of intelligence, the whale or the elephant, and not man — *pace* Shakespeare — would be "the paragon of animals."

But men of science are by no means at one on this alleged sexual difference in brain convolutions. On the contrary, there are many eminent physiologists and anatomists who contend that the superficies of brain convolutions in women is relatively greater than in men. For those who believe — and they are probably the majority at present — that the seat of mental activity is in the gray matter of the brain, this greater brain surface, due to its convolutions, would be a decided compensation for woman's relatively smaller brain volume.[10]

In whatever way, then, we consider the brains of men and women, whether we compare the ratio of brain weight to height of body or to weight of body, or compare the relative amounts of gray matter in the two sexes, the advantage, in spite of her smaller body, is distinctly in favour of woman.

From the preceding considerations it seems clear that there is no ground from the point of view of brain anatomy for considering one sex as superior to the other. They evince, too, that quality as well as quantity of brain tissue must be considered in all our discussions on the relations between the volume of brain and the intelligence of its possessor. Whales and elephants have much larger brains than men, but they nevertheless stand far below him in intelligence.

It must be remembered, also, that the brain is not only an organ of mental function. It is likewise the centre of the entire nervous system, and its volume, therefore, must correspond with the size and number of nerve trunks under its control. In man, as in animals, the brain elements are to a great extent but sensori-motor delegates whose function is the regulation and government of every part of the body. The superior size of the whale's brain, as compared with that of man, can readily be understood when we reflect on the much greater amount of territory which these sensori-motor delegates represent. When this fact is borne in mind it will be found that the whale's brain, relatively to that of man, is extremely small. For while the ratio of man's brain weight to that of his body is as 1 to 36, the ratio of the whale's brain weight to its immense body is but 1 to 3,000.

As an evidence that quality often counts for more than quantity, brain anatomists would do well to reflect on the marvellous intelligence displayed by ants and termites, those mites of animated nature which so excited the admiration of the naturalist Pliny and caused Darwin to declare, "The brain of an ant is one of the most marvellous atoms of matter in the world, perhaps more so than the brain of man."[11]

Moreover, when discussing the relative brain weights of the two sexes, we must not lose sight of the fact that we have, with the solitary exception of the eminent Russian mathematician, Sónya Kovalevsky,[12] no record of the brain weights of any eminently intellectual woman. The brains of scores of men of genius and exceptional mentality have been weighed, but we are utterly ignorant of the weight of brain of such women as Maria Gaetana Agnesi, Madame de Staël, Maria Theresa, Sophie Germain, George Sand, Harriet Martineau, George Eliot, Eleanor Ormerod, Mary Somerville, and others of the same calibre. The only data so far available, regarding the average brain weight of women, are such as have been obtained from the inmates of hospitals, prisons and pauper institutions. And yet we are asked to accept the average based on such data as a fair term of comparison with the average male brain weight as increased by the superior weight of brain of such men as Cuvier and Turgenev. And this is called science![13]

The attempt, then, to prove by weighing and measuring and studying brains that man is the intellectual superior of woman has been an ignominious failure. The old belief that woman is by nature and cerebral organisation less intelligent than man is not borne out by the investigations of those best qualified to pronounce an opinion on the subject. To assert, as so many do, that woman was created man's intellectual inferior is begging the question. Science can adduce no proof of such a gratuitous statement. Broca, the most eminent of French anthropologists, regarded as an absurdity the attempt to establish a necessary relation between the development of intelligence and the volume and weight of the encephalon. With the ripe knowledge of his mature years he was inclined to believe that the apparent difference in intelligence in the two sexes was owing, not to a difference of brain organisation, but rather to a difference of education, physical as well as mental, and that, with equal opportunities for intellectual and physical development, the present sexual differences that we have been considering — differences which are due not to nature but to the long ages of restraint and subjection under which women have lived — would gradually be lessened, and that men and women would eventually approach that equality which characterises them in the state of nature.[14]

Realising the impossibility of arriving, by the study of brain sizes and structure, at any satisfactory conclusion respecting the relative intellectual capacities of men and women, seekers after truth cast about for other methods that were free from the errors and fallacies of those which had proved so unreliable. The attempt to base the alleged mental inferiority of woman upon the facial angle of Camper, the metafacial angle of Serres, the craniofacial angle of Huxley, the sphenoidal angle of Welcker, or the nasobasal angle of Virchow had issued in utter failure, and had proved for the thousandth time that it is easier to formulate theories than to establish their validity. It was evident, notwithstanding the assertions of certain materialistic theorists, that the brain did not secrete thought as the liver secretes bile; it was evident, too, that intelligence could not be estimated in terms of any kind of mechanical units. Psycho-physiologists had no sort of dynamometer for measuring brain power as they would measure muscular energy. By means of the plethysmograph they might determine the amount of blood sent to the brain in a given time, but they had no psychometer of any description which would enable them to estimate the quantity, much less the quality, of psychic force such a blood supply was competent to produce.

Many, of course, still remained adherents of the old view that woman must ever remain the mental inferior of man because she is by nature physically weaker. These persons, however, seemed to lose sight of the fact that women who lead a rational life — who are not the slaves of fashion or the victims of luxury — have little to complain of on the score of physical weakness. This is evidenced by the life and habits of the women of the people, as well as by the tasks performed by women among savage tribes, who in health and strength are little, if at all, inferior to their male companions.

The late Professor Huxley, in referring to this subject, exhibited his usual acumen and sanity in such matters when he indited the following paragraph:

> We have heard a great deal lately about the physical disabilities of women. Some of these alleged impediments, no doubt, are really inherent in their organisation, but nine-tenths of them are artificial — the products of their mode of life. I believe that nothing would tend so effectually to get rid of these creations of idleness, weariness and that 'over-stimulation of the emotions' which in plainer spoken days used to be called wantonness, than a fair share of healthy work, directed toward a definite object, combined with an equally fair share of healthy play, during the years of adolescence; and those who are best acquainted with the acquirements of an average medical practitioner will find it hardest to believe that the attempt to reach that standard is like to prove exhausting to an ordinarily intelligent and well-educated woman.[15]

Substantially the same views are held by Mrs Henry Fawcett and Dr Mary Putnam Jacobi, whose rare experience and knowledge give their opinions on the subject under consideration special weight and value.

After men of science had tried the various theories above enumerated and found them wanting, they finally bethought themselves of investigating the relative intellectual standing of male and female students in coeducational institutions, and inquiring into their comparative capacity for different branches of knowledge, as made known by their professors and by the results of oral and written examinations. Considering the simplicity of this method and the fact that it is the more rational way to reach reliable conclusions, the wonder is that it was not thought of sooner. It excludes the bias of prepossessions and preconceived theories and lends itself to the discussion of results based on incontestable facts.

The first coeducational institution in which the intellectual capacity of women, in competition with men, was fairly tested was, strange to say, in the Royal College of Science for Ireland. This was somewhat more than half a century ago. When the time of examinations came, both the men and women students were handed the same examination papers. At the public distribution of prizes, at the close of the session, "the ladies," in the words of a Dublin paper, "vindicated the genius of their sex by carrying off the highest prizes." In zoology, botany, physics, chemistry and mathematics they proved themselves the peers, and frequently the superiors, of their male competitors.

> The success of the female students disturbed, of course, very much the preconceived notions of some people, who had always taken for granted that the female intellect was inferior to the male; and, not being able to combat the stubborn facts that appeared from time to time in the newspapers, when the results of the examinations were published, they tried to account for them."[16]

These cavillers, however, soon discovered that there was no way of accounting for the disconcerting fact which confronted them, except by confessing that their theory regarding the mental inferiority of women was not substantiated by fact. This unexpected demand for the unconditional surrender of their long-cherished theory of male superiority was a crushing and humiliating blow to their pride of intellect, but there was no remedy for it, nor was it accompanied by any balm of consolation that they, at the time, felt disposed to regard as adequate compensation for their lost prestige — a prestige which their overweening sex had claimed from time immemorial.

Similar experiments under even more trying conditions were subsequently made in the United States and in other parts of the world, and everywhere with the same results. In the universities of Switzerland, France, England, Germany and Russia women, when given a fair opportunity, were able to demonstrate to the satisfaction of all unprejudiced judges that the long-vaunted superiority of the male intellect was a myth; that intelligence, like genius, has no sex.

One of the most interesting and comprehensive investigations ever undertaken regarding this long-debated question was made some years ago by Arthur Kirchhoff, an enterprising German journalist.[17] It consisted in collecting and collaborating the opinions of more than a hundred of the most distinguished professors of the Fatherland, besides the opinions of a number of eminent writers and teachers in girls' high schools. These constitute a volume of nearly four hundred pages, and embody the views on the capacity of woman for science of professors of theology, jurisprudence, anatomy, physiology, surgery, psychology, history, gynaecology, psychiatry, philology, philosophy, art, mathematics, physics, astronomy, chemistry, zooogy, botany, geology, palaeontology and technology. The investigation, indeed, covered every branch of knowledge and evoked the deliberate views of those who were looked upon as the leading representatives of German thought and culture.

This book possesses a special value from the fact that, of all peoples in Europe, the Germans have been the most refractory to the claims of women to be received at the universities on the same footing as men. The German professors, naturally, share the conservatism of their countrymen, and, like them, are wedded to routine when there is question of introducing innovations into their social, political or educational systems. One would anticipate, then, that, when called upon to give their honest opinions respecting the intellectual capacity of women, as compared with that of men, their answer would be decidedly in favour of the sterner sex. "For," they will ask, "have not all the achievements in science which have given the Fatherland such prestige in the eyes of the world been due entirely to men? Have the women of Germany ever undertaken the solution of any great scientific problem, or have they ever made any notable contribution to scientific advancement? They have not."

Yet, notwithstanding all these facts, notwithstanding all traditions and prejudices and social bias, the unexpected has happened, even in conservative, old-fashioned Germany. The German professor may be tenacious of preconceived views; he may be a stickler for ancient customs and usages; nevertheless, when he is called upon to give a question a categorical answer which can be arrived at by observation or experiment, he may generally, in spite of his likes or dislikes, be counted on to give a decision in accord with the principles of legitimate induction. He may have his prejudices — and who has not? — but, when one appeals to him in the name of science and justice, he will rarely be found wanting. Regardless of all personal consideration, he will feel that loyalty to science, of which he is the avowed devotee, requires him to consider a question proposed to him as he would a scientific problem — something to be decided solely by such evidence as may be available.

To the exceeding gratification of the believers in the intellectual equality of the sexes, this proved to be the case in Herr Kirchhoff's investigation. The answers of the German professors, contrary to what most people would have anticipated, were, by a surprising majority, in favour of women. But their answers were in keeping with the changed educational conditions in Germany, as well as in other parts of the civilised world. Had Herr Kirchhoff undertaken his investigation a few decades earlier, the result would undoubtedly have been different, for women were then excluded from the universities and the professors had not had an opportunity of accurately testing their intellectual capacities. But having, during the latter part of the nineteenth century, had them as students in their lecture halls and laboratories, where they were able to study their mental powers and determine the value of their work by strict scientific methods, they were in a better position to express an opinion on the question at issue than would, a few years previously, have been possible.

Accordingly, even the declared enemies of the woman's movement among the German professorate were forced to admit the intellectual equality of the two sexes. For they, too, as well as men of science in other parts of Europe, had been measuring skulls and weighing brains; they, too, had been studying woman's mental calibre in the light of the new psychology; they, too, had been watching her work in the various departments of the university; and, notwithstanding all their observations and experiments, they were unable to detect any difference between men and women in brain organisation or in intellectual capacity. And, as might have been foreseen, results harmonised perfectly with those arrived at by investigators in other parts of the world — namely, that in things of the mind there is perfect sexual equality.

Among the hundred and more professors whose opinions are given in Herr Kirchhoff's book there were, of course, a few who were not prepared to subscribe to the findings of the great majority of their colleagues. But the reasons they assign for dissent were, at least in some instances, little better founded than that of a certain professor of chemistry in the University of Geneva, who, a few years ago, gravely declared that women have no aptitude for science because, forsooth, in chemical manipulations they break more test-tubes than men. Verily, "a Daniel come to judgement."

What probably more deeply impressed the German professors than anything else was the marked talent and taste of many of the women students for the abstract sciences, especially for the higher mathematics. For it had always been asserted that these branches of knowledge were beyond woman's capacity and that she had an instinctive antipathy for abstruse reasoning and for abstractions of all kinds. When, however, they discovered women whose delight was to discuss the theory of elliptic functions or curves defined by differential equations; when they found a mathematical genius like Sónya Kovalevsky speculating on the fourth dimension, and carrying away from the mathematicians of the world the most coveted prize of the French Academy of Sciences, they were forced to confess that another of their illusions was dissipated, and to acknowledge that they had no longer anything on which to base their long and fondly cherished opinion of the mental inequality of the sexes.

As an evidence of the extraordinary change that had been effected among the conservative Germans in the course of a few years respecting their attitude toward the admission of the "Academic Woman" to the universities, and, consequently, toward her intellectual capacity, it will suffice to reproduce a sentence from the elaborately expressed opinion of Dr Julius Bernstein, professor of physiology in the University of Halle. "After reflection on the subject," he declares, "I am convinced that neither God nor religion, neither custom nor law, and still less science, warrants one in maintaining any essential difference in this respect between the male and the female sex."[18]

The controversy of centuries regarding woman's intellectual capacity was now virtually settled beyond all peradventure. Woman had conquered, and her final victory had been won in the heart of the enemy's country, yea, even in what was thought to be the impregnable fortress of her relentless foes. It was achieved where the proud Teuton male had imagined that he was unapproachable and beyond compare — in the laboratories and lecture rooms of his great universities — more irresistible, in his estimation, than the Kaiser's trained legions in battle array.

It finally dawned upon the leaders of thought in the Fatherland, as it had but shortly before dawned upon philosophers and men of science in other lands, that the reputed sexual difference in intelligence was not due to difference in brain size or brain structure, or innate power of intellect, but rather to some other factors which had been neglected, or overlooked, as being unessential or of minor importance. These factors, on further investigation, proved to be education and opportunity.

As far back as 1869 that keen observer and philosopher, John

Stuart Mill, had expressed himself on the subject in the following words:

> Like the French compared with the English, the Irish with the Swiss, the Greeks or Italians compared with the German races, so women compared with men may be found, on the average, to do the same things with some variety in the particular kind of excellence. But that they would do them fully as well, on the whole, if their education and cultivation were adapted to correcting instead of aggravating the infirmities incident to their temperament, I see not the smallest reason to doubt.[19]

It would be difficult to find a better illustration of the sluggishness of the male as compared with the female mind than the tardiness of men of science in arriving at a sane conclusion respecting the subject of this chapter. For five hundred years ago Christine de Pisan arrived at the same conclusion which the learned professors of Germany reached only in the last decade of the nineteenth century. Discussing in *La Cite des Dames* the question at issue she writes as follows: "I say to thee again, and doubt never the contrary, that if it were the custom to put the little maidens to the school, and they were made to learn the sciences as they do to the men-children, that they should learn as perfectly, and they should be as well entered into the subtleties of all the arts and sciences as men be. And peradventure, there should be more of them, for I have teached heretofore that by how much women have the body more soft than the men have, and less able to do divers things, by so much they have the understanding more sharp there as they apply it."

Christine de Pisan's statement is virtually a challenge demanding the same educational opportunities for women as were accorded to men. But it was a challenge that men did not see fit to accept until full five centuries had elapsed, and until it was no longer possible to deny giving satisfaction to the long-aggrieved half of humanity. It was also an appeal to experiment and an appeal, likewise, to the teachings of history in lands where women have enjoyed the same educational advantages as men.

Having reviewed the many disabilities which so long retarded woman's intellectual advancement, and considered some of the objections which were urged against her capacity for scientific pursuits, we are now prepared to consider the appeal of Christine de Pisan and deal with it on its merits. This we shall do by a brief survey of woman's achievements in the various branches of science in which she has been accorded the same intellectual opportunities that were so long the exclusive privilege of her male compeer.

## Notes & References

1) An edition of this work, based on an old manuscript in La Bibliotheque Nationale of Paris, in French, is announced to appear in France at an early date. An interesting account of this precious volume has recently been published by Mlle Mathilde Laigle, PhD, under the title of *Le Livre de Trois Vertus de Christine de Pisan et son Milieu Historique et Litteraire*. It is to be hoped that some enterprising English publisher will soon favour us with a reprint of the quaint old, but none the less valuable, volume, *The Boke of the Cyte of Ladyes*.
2) Quando la genialita compare nella donna e sempre associata a grandi anomalie: e la più grande e la somiglianza coi maschi — la virilità. *L'Uomo di Genio*, sesta edizione, p 261, Torino, 1894.
3) *An Essay on the Learning, Genius and Abilities of the Fair Sex, Proving Them Not Inferior to Man*, p 142, London, 1774.
4) Schopenhauer, *Studies in Pessimism*, p 115, London, 1891.
5) *The Literary Advantages of Weak Health*, in the *Spectator* for October 1894.
6) *The Life and Letters of Charles Darwin*, edited by his son, Francis Darwin, Vol I, p 136, New York, 1888.
7) *Man and Woman*, p 94, London, 1898.
8) Cf *Das Hirngewicht des Menschen*, pp 21 and 137, by Theodor L W von Bischoff, Bonn, 1880, and Dr G van Walsem in *Neurologisches Centralblatt*, pp 578-80, Leipsic, July 1, 1899.
9) *L'Anthropologie*, pp 336-337, by Paul Topinard, Paris, 1876.
10) The importance of gray matter in mental processes has evidently been greatly overestimated, for it has been found to be thicker in the brains of negroes, murderers and ignorant persons than it was

in the encephalon of Daniel Webster. It is also much thicker in the brains of dolphins, porpoises and other cetaceans than it is in the most intellectual of men.

11) *The Descent of Man*, Vol I, p 145, London, 1871.

12) The brain of Sónya Kovalevsky was not weighed until it had been four years in alcohol. Prof Gustaf Retzius then wrote an elaborate account of it and estimated that its weight, at the time of Sónya's death, was 1385 grams. The brain-weight of her illustrious contemporary, Hermann von Helmholtz, was 1440 grams. But when the body-weights of these two eminent mathematicians are borne in mind — Sónya was short and slender — it will be seen that the relative amount of brain tissue was greater in the woman than in the man. Cf *Das Gehirn des Mathematikers Sónja Kovalewski in Biologische Untersuchungen*, von Prof Dr Gustaf Retzius, pp 1-17, Stockholm, 1900.

13) The reader who desires more detailed information respecting the brain-weights of men and women of various races and the relation of brain-weight to intelligence may consult with profit the following works and articles: *Memoires d'Anthropologie de Paul Broca*, 5 volumes, Paris, 1871-1888; *Alte und Neue Gehirn Probleme nebst einer 1078 Falle umfassenden Gehirngewichstatistik aus den Kgl. pathologisch-anatomischen Institut zu München*, von W W Wendt, München, 1909; Gehirngewicht und Intelligenz, by Dr F K Walter, Rostok, 1911; *Gehirngewicht und Intelligenz*, by Dr J Dräseke, Hamburg, in *Archiv für Rassen und Gesellschafts Biologe*, pp 499-522, 1906; *Brain Weights and Intellectual Capacity*, by Joseph Simms, MD, in the *Popular Science Monthly*, December 1898, and *The Growth of the Brain*, by H H Donaldson, London, 1895.

14) "Quand on songe à la difference qui separe de notre temps l'education intellectuelle de l'homme de celle de la femme, on se demande si ce n'est pas cette influence qui retrecit le cervaux et le crane feminins, et si, les deux sexes etant livres a leur spontaneite, leur cervaux ne tendraient pas à se ressembler, aussi qu'il arrive chez les sauvages." *Bulletin de la Societe d'Anthropologie*, p 503, Paris, July 3, 1879.

15) *Times*, London, July 8, 1874. Cf Ch XVII, on "Adolescent Girls and Their Education," in *Adolescence*, Vol II, by G Stanley Hall, New York, 1904.

16) *The Study of Science by Women in the Contemporary Review* for March 1869.

17) *Die Akademische Frau. Gutachten hervorragender Universitäten-professoren, Frauenlehrer und Schriftsteller über die Befähigung*

*der Frau zum wissenschaftlichen Studium and Berufe herausgegeben von Arthur Kirchhoff*, Berlin, 1897.

18) "Ich komme beim Nachdenken hieruber zu der Ueberzeigung, dass kein Gott und keine Religion, kein Herkommen und kein Gesetz, aber ebensowenig die Wissenschaft uns das Recht erteilen, in dieser Beziehung zwischen dem mannerlichen und weiblichen Geschlect einen principiellen Unterschied zu statuiren." *Die Akademische Frau*, p 41.

19) *The Subjection of Women*, p 91, London, 1909.

# Women in Mathematics

"ALL ABSTRACT SPECULATIONS, all knowledge which is dry, however useful it may be, must be abandoned to the laborious and solid mind of man..... For this reason women will never learn geometry."

In these words Immanuel Kant, more than a century ago, gave expression to an opinion that had obtained since the earliest times respecting the incapacity of the female mind for abstract science, and notably for mathematics. Women, it was averred, could readily assimilate what is concrete, but, like children, they have a natural repugnance for everything which is abstract. They are competent to discuss details and to deal with particulars, but become hopelessly lost when they attempt to generalise or deal with universals.

De Lamennais shares Kant's opinion concerning woman's intellectual inferiority and does not hesitate to express himself on the subject in the most unequivocal manner. "I have never," he writes, "met a woman who was competent to follow a course of reasoning the half of a quarter of an hour — *un demi quart d'heure*. She has qualities which are wanting in us, qualities of a particular, inexpressible charm; but, in the matter of reason, logic, the power to connect ideas, to enchain principles of knowledge and perceive their relationships, woman, even the most highly gifted, rarely attains to the height of a man of mediocre capacity."

But it is not only in the past that such views found acceptance. They prevail even today to almost the same extent as during the ages of long ago. How far they have any foundation in fact can best be determined by a brief survey of what woman has achieved in the domain of mathematics.

Athenaeus, a Greek writer who flourished about AD 200, tells us in his *Deipnosophistae* of several Greek women who excelled in mathematics, as well as philosophy, but details are wanting as to their attainments in this branch of knowledge. If, however, we may judge from the number of women — particularly among the hetaerae — who became eminent in the various schools of philosophy, especially during the pre-Christian era, we must conclude that many of them were well versed in geometry and astronomy as well as in the general science of numbers. Menagius declares that he found no fewer than sixty-five women philosophers mentioned in the writings of the ancients;[1] and, judging from what we know of the character of the studies pursued in certain of the philosophical schools, especially those of Plato[2] and Pythagoras, and the enthusiasm which women manifested in every department of knowledge, there can be no doubt that they achieved the same measure of success in mathematics as in philosophy and literature.[3]

The first woman mathematician, regarding whose attainments we have any positive knowledge, is the celebrated Hypatia, a Neo-platonic philosopher, whose unhappy fate at the hands of an Alexandrian mob in the early part of the fifth century has given rise to many legends and romances which have contributed not a little toward obscuring the real facts of her extraordinary career. She was the daughter of Theon, who was distinguished as a mathematician and astronomer and as a professor in the school of Alexandria, which was then probably the greatest seat of learning in the world. Born about the year 375 AD, she at an early age evinced the possession of those talents that were subsequently to render her so illustrious. So great indeed was her genius and so rapid was her progress in this branch of knowledge under the tuition of her father that she soon completely eclipsed her master in his chosen specialty.

There is reason to believe — although the fact is not definitely established — that she studied for a while in Athens in the school of philosophy conducted by Plutarch the Younger and his daughter Asclepigenia. After her return from Athens, Hypatia was invited by the magistrates of Alexandria to teach mathematics and philosophy. Here in brief time her lecture room was filled by eager and enthusiastic students from all parts of the civilised world. She was also gifted with a high order of eloquence and with a voice so marvellous that it was declared to be "divine."

Regarding her much vaunted beauty, nothing certain is known, as antiquity has bequeathed to us no medal or statue by which we could form an estimate of her physical grace. But, be this as it may, it is certain that she commanded the admiration and respect of all for her great learning, and that she bore the mantle of science and philosophy with so great modesty and self-confidence that she won all hearts. A letter addressed to "The Muse," or to "The Philosopher" — *Ôç Öéëïóïöù* — was sure to be delivered to her at once. Small wonder, then, to find a Greek poet inditing to her an epigram containing the following sentiment:

> When I see thee and hear thy word I thee adore; it is the ethereal constellation of the Virgin, which I contemplate, for to the heavens thy whole life is devoted, O august Hypatia, ideal of eloquence and wisdom's immaculate star.[4]

But it was as a mathematician that Hypatia most excelled. She taught not only geometry and astronomy, but also the new science of algebra, which had but a short time before been introduced by Diophantus. And, singular to relate, no further progress was made in the mathematical sciences, as taught by Hypatia, until the time of Newton, Leibnitz and Descartes — more than twelve centuries later.

Hypatia was the author of three works on mathematics, all of which have been lost, or destroyed by the ravages of time. One of these was a commentary on the *Arithmetica* of Diophantus. The original treatise — or rather the part which has come down to us — was found about the middle of the fifteenth century in the Vatican Library, whither it had probably been brought after Constantinople had fallen into the possession of the Turks. This valuable work, as annotated by the great French mathematicians Bachet and Fermat, gives us a good idea of the extent of Hypatia's attainments as a mathematician.

Another of Hypatia's works was a treatise on the *Conic Sections* by Apollonius of Perga — surnamed "The Great Geometer." Next to Archimedes, he was the most distinguished of the Greek geometricians; and the last four books of his conics constitute the chief portions of the higher geometry of the ancients. Moreover, they offer some elegant geometrical solutions of problems which, with all the resources of our modern analytical method, are not without difficulty. The greater part of this precious work has been preserved and has engaged the attention of several of the most illustrious of modern mathematicians — among them Borelli, Viviani, Fermat, Barrow and others. The famous English astronomer, Halley, regarded this production of Apollonius of such importance that he learned Arabic for the express purpose of translating it from the version that had been made into this language.

A woman who could achieve distinction by her commentaries on such works as the *Arithmetica* of Diophantus, of the *Conic Sections* of Apollonius, and occupy an honoured place among such mathematicians as Fermat, Borelli, and Halley, must have had a genius for mathematics, and we can well believe that the glowing tributes paid by her contemporaries to her extraordinary powers of intellect were fully deserved. If, with Pascal, we see in mathematics "the highest exercise of the intelligence," and agree with him in placing geometers in the first rank of intellectual princes — *princes de l'esprit* — we must admit that Hypatia was indeed exceptionally dowered by Him whom Plato calls "The Great Geometer."

There is still a third work of this ill-fated woman that deserves notice — namely, her *Astronomical Canon,* which dealt with the movements of the heavenly bodies. It is the general opinion that this was but a commentary on the tables of Ptolemy, in which event it is still possible that it may be found incorporated in the work of her father, Theon, on the same subject.

In addition to her works on astronomy and mathematics, Hypatia is credited with several inventions of importance, some of which are still in daily use. Among these are an apparatus for distilling water, another for measuring the level of water, and a third an instrument for determining the specific gravity of liquids — what we should now call an areometer. Besides these apparatus, she was likewise the inventor of an astrolabe and a planisphere.

One of her most distinguished pupils was the eminent Neo-platonist philosopher, Synesius, who became the Bishop of Ptolemais in the Pentapolis of Libya. His letters constitute our chief source of information respecting this remarkable woman. Seven of them are addressed to her, and in four others he makes mention of her. In one of them he writes: "We have seen and we have heard her who presides at the sacred mysteries of philosophy." In another he apostrophises her as "My benefactress, my teacher, — *magistra* — my sister, my mother."

In science Hypatia was among the women of antiquity what Sappho was in poetry and what Aspasia was in philosophy and eloquence — the chiefest glory of her sex. In profundity of knowledge and variety of attainments she had few peers among her contemporaries, and she is entitled to a conspicuous place among such luminaries of science as Ptolemy, Euclid, Apollonius, Diophantus and Hipparchus.[5]

It is a matter of regret to the admirers of this favoured daughter of the Muses that she is absent from Raphael's *School of Athens*; but, had her achievements been as well known and appreciated in his day as they are now, we can readily believe that the incomparable artist would have found a place for her in this masterpiece with the matchless form and features of his beloved Fornarina.

After the death of Hypatia the science of mathematics remained stationary for many long centuries. Outside of certain Moors in Spain, the only mathematicians of note in Europe, until the Renaissance, were Gerbert, afterward Pope Silvester II, and Leonardo da Pisa. The first woman to attract special attention for her knowledge of mathematics was Heloise, the noted pupil of Abelard. According to

Franciscus Ambrosius, who edited the works of Abelard and Heloise in 1616, the famous prioress of The Paraclete was a prodigy of learning, for besides having a knowledge of Latin, Greek and Hebrew, which was something extremely rare in her time, she was also well versed in philosophy, theology and mathematics, and inferior in these branches only to Abelard himself, who was probably the most eminent scholar of his age.[6]

Many Italian women, as we have seen in a preceding chapter, were noted for their proficiency in the various branches of mathematics. Some of the most distinguished of them flourished during the seventeenth and eighteenth centuries. Among these were Elena Cornaro Piscopia, celebrated as a linguist as well as a mathematician; Maria Angela Ardinghelli, translator of the *Vegetable Statics* of Stephen Hales; Cristina Roccati, who taught physics for twenty-seven years in the Scientific Institute of Rovigo, and Clelia Borromeo, fondly called by her countrymen *gloria Genuensium* — the glory of the Genoese. In addition to a special talent for languages, she possessed so great a capacity for mathematics and mechanics that no problem in these sciences seemed to be beyond her comprehension.[114] Then there was also Diamante Medaglia, a mathematician of note, who wrote a special dissertation on the importance of mathematics in the curriculum of studies for women, *Alle matematiche, alle matematiche prestino l'opera loro le donne, onde non cadano in crassi paralogismi* — "To mathematics, to mathematics," she cries, "let women devote attention for mental discipline."[7]

The most illustrious, by far, of the women mathematicians of Italy was Maria Gaetana Agnesi, who was born in Milan in 1718 and died there at the age of eighty-one. At an early age she exhibited rare intelligence and soon distinguished herself by her extraordinary talent for languages. At the age of five she spoke French with ease and correctness, while only six years later she was able to translate Greek into Latin at sight and to speak the former as fluently as her own Italian. At the early age of nine she startled the learned men and women of her native city by discoursing for an hour in Latin on the rights of women to the study of science. This discourse — *Oratio* — was not, as usually stated, her own composition, but a translation from the Italian of a discourse written by her teacher of Latin. That a child of nine years should speak in the language of Cicero for a full hour before a learned assembly and without once losing the thread of her discourse was, indeed, a wonderful performance, and we are not surprised to learn that she was regarded by her countrymen as an infant prodigy.[8]

In addition to Italian, French, Latin and Greek, she was acquainted with German, Spanish and Hebrew. For this reason she was, like Elena Cornaro Piscopia, the famous "Venetian Minerva," called Oracolo Settilingue — Oracle of Seven Languages.[9]

But it was in the higher mathematics that Maria Gaetana was to win her chief title to fame in the world of learning. So successful had she been in her prosecution of this branch of science that she was, at the early age of twenty, able to enter upon her monumental work — *Le Instituzioni Analitiche* — a treatise in two large quarto volumes on the differential and integral calculus. To this difficult task she devoted ten years of arduous and uninterrupted labour. And if we may credit her biographer, she consecrated the nights as well as the days to her herculean undertaking. For frequently, after working in vain on a difficult problem during the day, she was known to bound from her bed during the night while sound asleep and, like a somnambulist, make her way through a long suite of rooms to her study, where she wrote out the solution of the problem and then returned to her bed. The following morning, on returning to her desk, she found, to her great surprise, that while asleep she had fully solved the problem which had been the subject of her meditations during the day and of her dreams during the night. Could the psychiatrist who so loves to deal with obscure mental phenomena find a more interesting case to engage his attention or one more worthy of the most careful investigation?

Finally Maria Gaetana's *opus majus* was completed and given to the public. It would be impossible to describe the sensation it produced in the learned world. Everybody talked about it; everybody admired the profound learning of the author, and acclaimed her: "Il portento del sesso, unico al Mondo" — the portent of her sex, unique in the world. By a single effort of her genius she had completely demolished that fabric of false reasoning which had so long been appealed to as proof positive of woman's intellectual inferiority, especially in the domain of abstract science. Maria Gaetana's victory was complete, and her victory was likewise a victory for her sex. She had demonstrated once for all, and beyond a quirk or quibble, that women could attain to the highest eminence in mathematics as well as in literature, that supreme excellence in any department of knowledge was not a question of sex but a question of education and opportunity, and that in things of the mind there was essentially no difference between the male and the female intellect.

The world saw in Agnesi a worthy accession to that noble band of gifted women who count among their number a Sappho, a Corinna, an Aspasia, a Hypatia, a Paula, a Hroswitha, a Dacier, an Isabella Rosales who, in the sixteenth century, successfully defended the most difficult theological theses in the presence of Paul III and the entire college of cardinals. And so delighted were the women — especially those in Italy — with the signal triumph of their eminent sister that they defied the traducers of their sex — *muliebris sapientiae infensissimis hostibus* — to continue any longer their unreasonable campaign against the rights of women which were based on the intellectual equality of the two sexes.

So highly did the French Academy of Science value Agnesi's achievement that she would at once have been made a member of this learned body had it not been against the constitutions to admit a woman to membership. M. Motigny, one of the committee appointed by the Academy to report on the work, in his letter to the author, among other things, writes: "Permit me, Mademoiselle, to unite my personal homage to the plaudits of the entire Academy. I have the pleasure of making known to my country an extremely useful work which has long been desired, and which has hitherto" — both in France and in England — "existed only in outline. I do not know any work of this kind which is clearer, more methodic or more comprehensive than your *Analytical Institutions.* There is none in any language which can guide more surely, lead more quickly, and conduct further those who wish to advance in the mathematical sciences. I admire particularly the art with which you bring under uniform methods the divers conclusions scattered among the works of geometers and reached by methods entirely different."

As an indication of the exceptional merit of Agnesi's work, even long after its publication in 1748, it suffices to state that the second volume of the *Instituzioni Analitiche* was translated into French in 1775 by Antelmy and annotated by the Abbe Bossuet, a member of the French Academy and a collabourator of D'Alembert on the mathematical part of the famous *Encyclopaedie.*

A still greater proof of the estimation in which Agnesi's work was held by men of science is the fact that it was translated in its entirety into English by the Rev John Colton, Lucasian Professor of Mathematics in the University of Cambridge, and published in 1801, fifty-two years after it had appeared in Italian. His impression of the methods followed by the Milanese *savante* was so favourable that, in the words of a contemporary writer, it "gave rise to his very spirited resolution of learning a new language at an advanced period of life, that he might make himself perfect master of them."[10]

Gratifying, however, as were the tributes of admiration and appreciation which came to Agnesi from all quarters, from learned societies, from eminent mathematicians, from sovereigns — the Empress Maria Theresa sent her a splendid diamond ring and a precious crystal casket bejewelled with diamonds — that which touched her most deeply was, undoubtedly, the recognition which she received from the great Maecenas of his age, Pope Benedict XIV. As Cardinal Lambertini and Archbishop of Bologna, he had taken a conspicuous part in the honours showered on Laura Bassi when she received her doctorate, and was specially delighted when she was made professor of physics in his favoured university. Being himself familiar with the higher mathematics, he recognised at once the exceptional merit of Maria Gaetana's work and showed his appreciation of it not only by letters and presents, but also by having her, *motu proprio*, appointed by the Bolognese senate as professor of higher mathematics in the University of Bologna.

In advising her of this appointment, he writes her that he had in view the honour of the University in which he had always taken a special interest, and that the appointment carried with it no obligation of thanks on her part but rather on his — *che porta seco ch'ella non deve ringraziar Noi, ma che Noi dobbiamo ringraziar lei*. The interest that this wise and broad-minded pontiff exhibited in the advancement of learned women and the rewards he was ever ready to accord to their achievements in science and literature — especially in the cases of Laura Bassi and Maria Gaetana Agnesi — is in keeping with the policy pursued by his predecessors, and accounts in great measure for that large number of learned women in Italy who, since the opening of the first universities, have been the glory of their sex and country.

But ardent as was the desire of the Supreme Pontiff to have Agnesi occupy the chair of mathematics, and numerous as were the appeals of her friends and the members of the university faculty to have her accept the appointment that carried with it such signal honour, she could never be induced to leave her beloved Milan. For, after completing her masterpiece, she resolved to retire from the world and devote the rest of her life to the care of the poor, the sick and the helpless in her native city. She did not, however, as is so frequently asserted, enter the convent and become a nun.[11] Dur ing many years after her retirement from the world, she lived in her own home, a part of which she had converted into a hospital. During the last fifteen years of her life she had charge of the Pio Albergo Trivulzio — a large institution founded by Prince Trivulzio for the aged poor who were without home or assistance.

She had devoted ten years of the flower of her life to the writing of her *Instituzioni Analitiche* — prepared primarily for the benefit of one of her brothers who had a taste for mathematics — and, after it was finished, she entered upon that long career of heroic charity which was terminated only at her death at the advanced age of eighty-one.

One loves to speculate regarding Maria Gaetana's possible achievements if she had continued during the rest of her life that science in which, during a few short years, she had won such distinction. She had made her own the discoveries of Newton, Leibnitz, Roberval, Fermat, Descartes, Riccati, Euler, the brothers Bernouilli, and had mastered the entire science of mathematics then known. Her pinions were trimmed for essaying loftier flights than any hitherto attempted, and her intellect was prepared, as one of her scientific friends expressed it, "for fixing the limits of the infinite." But while the world of science was still sounding her praises and predicting for her still greater triumphs in the field of analysis, it learned with surprise and sorrow that she had bid adieu to those studies in which she had achieved such extraordinary success, and had consecrated her life to the service of the poor and the afflicted. She disappeared completely from those literary and scientific reunions where she had so long been the most conspicuous figure, and was thenceforth known only as the ministering angel of the suffering and the abandoned. For half a century hers was a life of the most heroic charity and self-abnegation. Very readily, therefore, we can understand why a recent representative of the scientific world should desire to see her name placed on the calendar of saints.[12]

Had Agnesi devoted her entire life to science instead of abandoning it just when she was prepared to do her best work, she might today be ranked among such supreme mathematicians as Lagrange, Monge, Laplace and the Bernouillis, all of whom were her contemporaries. Even as it was, she has been placed beside Cardan, Leibnitz and Euler for her remarkable powers of analysis of infinitesimals, while the best proof of the literary value of her *Instituzioni Analitiche* is the fact that it has been selected by the famous society Della Crusca as a *testo di lingua*—a work considered as a classic of its kind and used in the preparation of the great authoritative dictionary of the Italian language.

But by consecrating herself to charity she probably accomplished far more for humanity and for the well-being of her sex than if she had elected to continue her work in the higher mathematics. There had been many learned women in Italy before her time and many since; many who were distinguished as Hellenists, as Latinists, as polyglots, as mathematicians — women like the Roccati, the Borghini, the Brassi, the Ardinghelli, the Barbapiccola, the Caminer Turra, the Tambroni; but Maria Gaetana Agnesi surpasses them all, not only in knowledge, but as a potent influence for the diffusion of culture and the spirit of brotherhood, for the expansion of benevolence and charity, and, above all, for the elevation of woman. She was also, as her latest and best biographer beautifully expresses it, "an inspired *condottiera* who, in the field of civility, anticipated the conquests of these latter days." She was, indeed, as her epitaph informs us, *pietate, doctrina, beneficentia insignis,* and as such she will live in the memory of our race as long as men shall admire genius and love virtue.

In the year following the publication of Agnesi's *Instituzioni Analitiche* was recorded the premature and tragic death of the distinguished French mathematician, the Marquise Émilie du Chatelet. She has been described as a "thinker and scientist, precieuse and pedant, but not the less a coquette — in short, a woman of contradictions."[13] To most readers she is better known by reason of her liaison with Voltaire, of whom she is regarded as a mere satellite, than for her work in science. But she was far more than a satellite that shone by the light received from the sage of Ferney. For there can be no doubt that she was a highly gifted woman who, besides having a thorough knowledge of several languages, including Latin, possessed a special talent for mathematics. It was said of her that "she read Virgil, Pope and algebra as others read novels," and that she was able "to multiply nine figures by nine others in her head." No less an authority than the illustrious Ampere declared her to be "a genius in geometry."

Among her teachers in mathematics were Clairaut, Koenig, Maupertuis, Pere Jaquier and Jean Bernouilli, the immediate predecessors of such distinguished mathematicians as Monge, Lagrange, d'Alembert and Laplace. At her Chateau of Cirey, where she and Voltaire spent many years together, she was visited by learned men from various parts of Europe. Among these was the Italian scholar, Francisco Algarotti, who was the author of a work entitled *Newtonism for Women*. And as Mme. du Chatelet was an ardent admirer of Newton, the author of the *Principia* soon became a strong bond of union between her and the brilliant Italian. She called the savants who frequented her chateau at Cirey the *Émiliens* and purposed writing memoirs to be entitled *Emiliana* — a design, however, which she was never able to execute.

The first work of importance from the pen of the Marquise was entitled *Institutions de Physique*. In it she gave an exposition of the philosophy of Leibnitz and dissertations on space, time and force. In the discussion of the last topic she seems to have anticipated some of the later conclusions of science respecting the nature of energy.

Her most noted achievement, however, was her translation of Newton's *Principia*, the first translation into French of this epoch-making work. To translate this masterpiece from its original Latin, it was necessary that the Marquise, in order to make it intelligible to others, should have a thorough understanding of it herself. To the translation she added a commentary, which shows that Mme du Chatelet had a mathematical mind of undoubted power. She laboured assiduously on this great undertaking for many years and completed it only shortly before her death; but it was not published until ten years after her demise.

In his *Élogie Historique* on the Marquise's translation of the *Principia*, Voltaire, in his usual flamboyant style, declares "Two wonders have been performed: one that Newton was able to write this work, the other that a woman could translate and explain it." In an effort to express in a single sentence all his admiration for his talented friend he does not hesitate to state: "Never was woman so learned as she, and never did anyone less deserve that people should say of her, 'She is a learned woman.'" Again he refers to her with characteristic Frenchiness as "a woman who has translated and explained Newton, in one word a very great man — *en un mot un tres grand homme*."[14]

But, although the extent of her attainments and her ability as a mathematician were unquestionable, she fell far short of her great contemporary, Gaetana Agnesi, both in the depth and breadth of her scholarship and in her power of infinitesimal analysis. As to her moral character, she was infinitely inferior to the saintly savante of Milan. She was by inclination and profession an Epicurean and an avowed sensualist. In her little treatise, *Reflexions sur le Bonheur*— Reflections on Happiness — she unblushingly asserts "that we have nothing to do in this world except procure for ourselves agreeable sensations." Considering her profligate life, bordering at times on utter *abandon*, we are not surprised that one of her countrymen has characterised her as "*Femme sans foi, sans moeurs, sans pudeur*" — a woman without faith, without morals, without shame.[15]

Anna Barbara Reinhardt of Winterthur in Switzerland was another woman of exceptional mathematical talent. She is remarkable for having extended and improved the solution of a difficult problem that specially engaged the attention of Maupertuis. According to so competent an authority as Jean Bernouilli, she was the superior, as a mathematician, of the Marquise de Chatelet.

Of a more original and profound mathematical mind was Sophie Germain, a countrywoman of the Marquise du Chatelet. Hers was the glory of being one of the founders of mathematical physics. A pupil of Lagrange and a co-worker with Biot, Legendre, Poisson and Lagrange, she has justly been called by De Prony "the Hypatia of the nineteenth century."

Her success, however, was not achieved without overcoming many and great difficulties. In the first place, she had to overcome the opposition of her family, who were decidedly averse to her studying mathematics. "Of what use," they asked, "was geometry to a girl?" But in trying to extinguish her ardour for mathematics they but augmented it. Alone and unaided she read every work on mathematics she could find. The study of this science had such a fascination for her that it became a passion. It occupied her mind day and night. Finally her parents, becoming alarmed about her health and resolved to force her to take the necessary repose, left her bedroom without fire or light, and even removed from it her clothing after she had gone to bed. She feigned to be resigned; but when all were asleep, she arose and, wrapping herself in quilts and blankets, she devoted herself to her favourite studies, even when the cold was so intense that the ink was frozen in her ink-horn. Not infrequently she was found in the morning chilled through, having been so engrossed in her studies that she was not aware of her condition. Before such a determined will, so extraordinary for one of her age, the family of the young Sophie had the wisdom to permit her to dispose of her time and genius according to her own pleasure. And they did well. Like the great geometer of Syracuse, Archimedes, who had ever been her inspiration in the study of mathematics, she would have died rather than abandon a problem which, for the time being, engaged her attention.

She first attracted the attention of savants by her mathematical theory of Chladni's figures. By the order of Napoleon, the Academy of Science had offered a prize for the one who would "Give the mathematical theory of the vibration of elastic surfaces and compare it with the results of experiment." Lagrange declared the problem insoluble without a new system of analysis, which was yet to be invented. The consequence was that no one attempted its solution except one who, until then, was almost unknown in the mathematical world; and this one was Sophie Germain.

Great was the surprise of the savants of Europe when they learned that the winner of the *grand prix* of the Academy was a woman. She became at once the recipient of congratulations from the most noted mathematicians of the world. This eventually brought her into scientific relations with such eminent men as Delambre, Fourier, Cauchy, Ampere, Navier, Gauss[16] and others already mentioned.

It was in 1816, after eight years of work on the problem, that her last memoir on vibrating surfaces was crowned in a public seance of the *Institut de France*. After this event Mlle. Germain was treated as an equal by the great mathematicians of France. She shared their labours and was invited to attend the sessions of the *Institut*, which was the highest honour that this famous body had ever conferred on a woman.

The noted mathematician, M Navier, was so impressed with the extraordinary powers of analysis evinced by one of Mlle Germain's memoirs on vibrating surfaces that he did not hesitate to declare that "it is a work which few men are able to read and which only one woman was able to write."

Biot, in the *Journal de Savants*, March 1817, writes that Mlle Germain is probably the one of her sex who has most deeply penetrated the science of mathematics, not excepting Mme du Chatelet, *for here there was no Clairaut.*[17]

Like Maria Gaetana Agnesi, Mlle. Germain was endowed with a profoundly philosophical mind as well as with a remarkable talent for mathematics. This is attested by her interesting work entitled *Considerations Generales sur l'État des Sciences et des Lettres aux Differentes Époques de Leur Culture*. All things considered, she was probably the most profoundly intellectual woman that France has yet produced. And yet, strange as it may seem, when the state official came to make out the death certificate of this eminent associate and co-worker of the most illustrious members of the French Academy of Sciences he designated her as a *rentiere* — *annuitant* — not as a *mathematicienne*. Nor is this all. When the Eiffel tower was erected, in which the engineers were obliged to give special attention to the elasticity of the materials used, there were inscribed on this lofty structure the names of seventy-two savants. But one will not find in this list the name of that daughter of genius, whose researches contributed so much toward establishing the theory of the elasticity of metals — Sophie Germain. Was she excluded from this list for the same reason that Agnesi was ineligible to membership in the French Academy — because she was a woman? It would seem so. If such, indeed, was the case, more is the shame for those who were responsible for such ingratitude toward one who had deserved so well of science, and who by her achievements had won an enviable place in the hall of fame.[18]

Four years after the birth of Sophie Germain was born in Jedburgh, Scotland, one whom an English writer has declared was "the most remarkable scientific woman our country has produced."

She was the daughter of a naval officer, Sir William Fairfax; but is best known as Mary Somerville. Her life has been well described as an "unobtrusive record of what can be done by the steady culture of good natural powers and the pursuit of a high standard of excellence in order to win for a woman a distinguished place in the sphere naturally reserved for men, without parting with any of those characteristics of mind, or character, or demeanour which have ever been taken to form the grace and the glory of womanhood."[19]

The surroundings of her youth were not conducive to scientific pursuits. On the contrary, they were entirely unfavourable to her manifest inclinations in that direction. Having scarcely any of the advantages of a school education, she was obliged to depend almost entirely on her own unaided efforts for the knowledge she actually acquired. She, like Sophie Germain, was essentially a self-made woman; and her success was achieved only after long labour and suffering and in spite of the persistent opposition of family and friends.

When she was about fifteen years old, the future Mrs Somerville received her first introduction to mathematics; and then, strange to say, it was through a fashion magazine. At the end of a page of this magazine, "I read," writes Mrs Somerville, "what appeared to me to be simply an arithmetical question; but in turning the page I was surprised to see strange-looking lines mixed with letters, chiefly X's and Y's, and asked 'What is that?'" She was told it was a kind of arithmetic, called algebra.

Her interest was at once aroused; and she resolved forthwith to seek information regarding the curious lines and letters which had so excited her curiosity. "Unfortunately," she tells us, "none of our acquaintances or relatives knew anything of science or natural history; nor, had they done so, should I have had courage to ask of them a question, for I should have been laughed at."

Finally she was able to secure a copy of a work on algebra and a Euclid. Although without a teacher she immediately applied herself to master the contents of these two works, but she had to do so by stealth in bed after she had retired for the night. When her father learned of what was going on, he said to the girl's mother, "Peg, we must put a stop to this, or we shall have Mary in a straightjacket one of these days." The mother, who had no more sympathy with her daughter's scientific pursuits than had the father, and, fully convinced, like the great majority of her sex, that woman's duties should be confined to the affairs of the household, strove to divert her daughter's mind from her "unladylike" pursuits. But her efforts were ineffectual. The young woman, in spite of all obstacles and opposition, contrived to continue her cherished studies; and, through her uncle, the Rev Dr Somerville, afterward her father-in-law, she was able to become proficient in both Latin and Greek. When she was thirty-three years of age she became the happy possessor of a small library of mathematical works. "I had now," she writes, "the means, and pursued my studies with increased assiduity; concealment was no longer necessary, nor was it attempted. I was considered eccentric and foolish, and my conduct was highly disapproved of by many, especially by some members of my own family."[20]

In March 1827, Mrs Somerville received a letter from Lord Brougham, who had heard of her remarkable acquirements, begging her to prepare for English readers a popular exposition of Laplace's great work — *Mecanique Celeste*. She was overwhelmed with astonishment at this request, for her modesty made her diffident of her powers; and she felt that her self-acquired knowledge of science was so far inferior to that of university men that it would be sheer presumption for her to undertake the task proposed to her. She was, however, finally persuaded to make the attempt, with the proviso that her manuscript should be consigned to the flames unless it fulfilled the expectations of those who urged its production.

In less than a year her work, to which she gave the name of *The Mechanism of the Heavens*, was ready for the press. But it was far more than a translation and epitome, as originally intended by its projector, Lord Brougham; for, in addition to the views of Laplace, it contained the independent opinions of the translator respecting the propositions of the illustrious French savant. No sooner was the work published than Mrs. Somerville found herself famous. She had, as Sir John Herschel expressed it, "written for posterity," and her book placed her at once among the leading scientific writers and thinkers of the day. She was elected an honorary member of the Royal Astronomical Society at the same time as Caroline Herschel, they being the first two women thus honoured. Her bust, by Chantry, was placed in the great hall of the Royal Society, and she was made a member of many other scientific societies in Europe and America. In recognition of her services to science she was granted by the government a pension of £200 a year — a sum which was shortly afterward increased to £300. In addition to all this, Mrs. Somerville had the satisfaction of learning that her work was so highly esteemed by Dr Whewell, the great master of Trinity, that it was, chiefly on his recommendation, introduced as a textbook in the University of Cambridge and prescribed as "an essential work to those students who aspire to the highest places in the examinations." What Mme du Chatelet had done for Newton, Mrs Somerville did for Laplace.

Among other books from the pen of this highly gifted woman is her *Connection of the Physical Sciences* and a work entitled *Physical Geography*, which, together with the *Mechanism of the Heavens*, was the object of the "profound admiration" of Humboldt. Then there is a number of very abstruse monographs on mathematical subjects, one of which is a treatise of two hundred and forty-six pages *On Curves and Surfaces of Higher Orders*, which, she tells us, she "wrote *con amore* to fill up her morning hours while spending the winter in Southern Italy."

Her last work was a treatise *On Molecular and Microscopic Science* embodying the most recondite investigations on the subject. This book, begun after she had passed her eightieth birthday, occupied her for many years and was not ready for publication until she was close upon her ninetieth year. Her last occupations, continued until the day of her death at the advanced age of ninety-two, were the reading of a book on *Quaternions* and the review and completion of a volume *On the Theory of Differences*.

Like her illustrious friend, the great Humboldt, Mary Somerville was possessed of extraordinary physical vigour, and, like him, she retained her mental powers unimpaired until the last. And like her great rival in mathematics, Maria Gaetana Agnesi, she was always "beautifully womanly." Her scientific and literary occupations did not cause her to neglect the duties of her household or to disregard "the graceful and artistic accomplishments of an elegant woman of the world." Her daughter Martha writes of her: "It would be almost incredible were I to describe how much my mother contrived to do in the course of the day. When my sister and I were small children, although busily engaged in writing for the press, she used to teach us for three hours in the morning, besides managing her house carefully, reading the newspapers — for she was always a keen and, I must add, a liberal politician — and the most important new books on all subjects, grave and gay. In addition to this, she freely visited and received her friends..... Gay and cheerful company was a pleasant relaxation after a hard day's work."[21]

The life of Mary Somerville, like that of Gaetana Agnesi, proves that the pursuit of science is not, as so often asserted, incompatible with domestic and social duties. It also disposes of the fallacy, so generally entertained, that intellectual labour is detrimental to the health of women and antagonistic to longevity. The truth is that it is yet to be demonstrated that intellectual work, even of the severest kind, is, *per se*, more deleterious to women than to those of the stronger sex.

Scarcely less remarkable as a mathematician was Mrs Somerville's distinguished contemporary, Janet Taylor, who was known as the "Mrs Somerville of the Marine World." She was the author of numerous works on navigation and nautical astronomy which in their day were highly prised by seafaring men. In recognition of her valuable services to the marine world she was placed on the civil list of the British government.

As an eminent mathematician as well as a "representative of the highest intellectual accomplishments to which women have attained," Sónya Kovalevsky will ever occupy an honoured place among the votaries of science. In many respects this richly endowed daughter of Holy Russia was *par excellence* the woman of genius of the latter half of the nineteenth century.

She was born in Moscow in 1850, but although her career was brief it was one of meteoric splendour. At an early age she exhibited an unusual talent for mathematics and an unquenchable thirst for knowledge. Not being able to obtain in her own country the educational advantages she desired, she resolved at the age of eighteen to go to Germany with a view of pursuing her studies there under more favourable auspices.

She first matriculated in the University of Heidelberg, where she spent two years in studying mathematics under the most eminent professors of that famous old institution. Thence she went to Berlin. She could not enter the University there, as its doors were closed to female students; but she was fortunate enough to prevail on the illustrious Professor Weierstrass, regarded by many as the father of mathematical analysis, to give her private lessons. He soon discovered to his astonishment that this child-woman had "the gift of intuitive genius to a degree he had seldom found among even his older and more developed students." Under this eminent mathematician Sónya spent about three years, at the end of which period she was able to present to the University of Gottingen three theses which she had written under the direction of her professor. The merit of her work and the testimonials which she was able to present from Weierstrass, Kirchhoff and others were of such supreme excellence that she was exempted from an oral examination and was enabled, by a very special privilege, to receive her doctorate without appearing in person.

Not long after receiving her doctor's degree — one of the first to be granted to a woman by a German university — she was offered the chair of higher mathematics in the University of Stockholm. She was the first woman in Europe, outside of Italy, to be thus honoured. But her appointment had to be made in the face of great opposition. No other university, it was urged by the conservatives, had yet offered a professor's chair to a woman. Strindberg, one of the leaders of modern Swedish literature, wrote an article in which he proved, "as decidedly as that two and two make four, what a monstrosity is a woman who is a professor of mathematics, and how unnecessary, injurious and out of place she is."[22]

The fame that came to Sónya through her achievements in the German and Swedish universities was immensely enhanced when, on Christmas eve, 1888, "at a solemn session of the French Academy of Sciences, she received in person the *Prix Bordin* — the greatest scientific honour which any woman had ever gained; one of the greatest honours, indeed, to which any one can aspire."

She became at once the heroine of the hour and was thenceforth "a European celebrity with a place in history." She was fêted by men of science whithersoever she went and hailed by the women of the world as the glory of her sex and as the most brilliant type of intellectual womanhood.

Mme Kovalevsky's printed mathematical works embrace only a few memoirs including those which she presented for her doctorate and for the *Prix Bordin*. But brief as they are, all of these memoirs are regarded by mathematicians as being of special value. This is particularly true of the memoirs, which secured for her the *Prix Bordin*; for it contains the solution of a problem that long had baffled the genius of the greatest mathematicians.

The prize had been opened to the competition of the mathematicians of the world, and the astonishment of the committee of the French Academy was beyond expression when it was found that the successful contestant was a woman.[23]

Everyone admired her varied and profound knowledge, but, above all, her amazing powers of analysis. A German mathematician, Kronecker, did not hesitate to declare that "the history of mathematics will speak of her as one of the rarest investigators."[24]

Shortly before her premature death, she had planned a great work on mathematics. All who are interested in the intellectual capacities and achievements of woman must regret that she was unable to complete what would undoubtedly have been the noblest monument of woman's scientific genius. She was then in the prime of life and perfectly equipped for the work she had in mind. Considering the extraordinary receptive and productive power of this richly dowered woman, there can be little doubt, had she lived a few years longer, that she would have produced a work that would have caused her to be ranked among the greatest mathematicians of the nineteenth century.

It is pleasant to record that this woman of masculine mind, masculine energy and masculine genius, far from being mannish or unwomanly, was, on the contrary, a woman of a truly feminine heart; and that, although a giantess in intellectual attainments, she was in grace and charm and delicacy of sentiment one of the noblest types of beautiful womanhood. She could with the greatest ease turn from a lecture on *Abel's Functions* or a research on Saturn's rings to the writing of verse in French or of a novel in Russian or to collaborating with her friend, the Duchess of Cajanello, on a drama in Swedish, or to making a lace collar for her little daughter, Fouzi, to whom she was most tenderly attached.[25]

Little more than a quarter of a century has elapsed since Strindberg, expressing the sentiment of the great majority of the men of his time, declared that a woman professor of mathematics is a monstrosity. But during this short period what a change has been effected in the attitude of the world toward women who devote themselves to the study and the teaching of science! Women mathematicians are found today in all civilised countries, and no sane person now considers it any more "unwomanly" or more "monstrous" for them to study or teach mathematics than for them to teach music or needlework. Yet more. They are now frequent contributors to mathematical magazines and to the official bulletins of learned societies, and not infrequently they are on the editorial staffs of publications devoted exclusively to mathematics. They are also found as computers in some of the largest astronomical observatories, where the speed and accuracy of their work have evoked the most favourable comment.

Of women in America, who have distinguished themselves by their work in the higher mathematics, it suffices to mention the name of Miss Charlotte Angas Scott, recently deceased, who was for years professor of mathematics in the College of Bryn Mawr. Her writings on various problems of the higher mathematics show that she faithfully followed in the footsteps of her illustrious predecessors, —Hypatia, Agnesi, du Chatelet, Germain, Somerville and Kovalevsky.

## Notes & References

1) "Ipse mulieres Philosophas in libris Veterum sexaginta quinque reperi," *Historia Mulierum Philosopharum*, p 3, Amstelodami, 16z92.

2) Plato had inscribed above the entrance of *his school, Ïõäåéò áãåùìåôñçôïò åéóéôç. Let no one enter here who is not a geometer.*

3) Menagius in referring to this matter, *op cit*, p 37, writes as follows: "Meritrices Graecas plerasque humanioribus literis et mathematicis disciplinis operam dedisse notat Athenaeus."

4) The sentiment of the Greek epigram is well expressed in the following Latin verses:
Quando intueor te, adoro, et sermones,
Virginis domum sideream intuens.
E coelis enim tua sunt opera,
Hypatia casta, sermonum venustas,
Impollutum astrum sapientis doctrinae.

5) Among modern works on Hypatia may be mentioned *Hypatia, die Philosophin von Alexandria*, by St Wolt, Vienna, 1879; *Hypatia von Alexandria*, by W A Meyer, Heidelberg, 1886; *Ipazia Alessandrina*, by D Guido Bigoni, Venise, 1887, and *De Hypatia*, by B Ligier, Dijon, 1879.

6) Ambrosius in his preface to the works of Abelard and Heloise refers to the latter as "Clarum sui sexus sidus et ornamentum," and declares "necnon mathesin, philosophiam et theologiam a viro suo edocta, illo solo minor fuit."

7) Mazzuchelli says of her in his *Museo*, "Sembra non avervi nella Natura cosa la piu intralciata ed oscura nelle storie, ne finalemente la piu astrusa nelle matematiche e nelle mecchaniche, che a lei conta non sia e palese, e che sfugga la capacita del suo spirito." *Dizionario Biografico*, Vol I, p 122, by Ambrogio Levati, Milano, 1821.

8) *Delle Donne Illustri Italiane del XIII al XIX Secolo*, p 268, Roma.

9) The full title of this celebrated discourse is *Oratio qua ostenditur Artium liberalium studia a Faemineo sexu neutiquam abhorere, habita a Maria de Agnesis Rhetoricae Operam Dante, Anno Aetatis suae nono nondum exacto, die 18, Augusti, 1727*. It is found at the end of a work entitled *Discorsi Academici di varj autori Viventi*

*intorno agli Stuj delle Donne in Padova*, 1729. This subject, it may be remarked, frequently engaged the attention of Maria Gaetana as she advanced in years, for we find it among the questions discussed in her *Propositiones Philosophicae*, pp 2 and 3, Mediolani, 1738.

10) M Charles de Brosses, in his *Lettres Familieres ecrites de l'Italie en 1739 et 1740*, speaks of Agnesi in terms that recall the marvellous stories which are related of Admirable Crichton and Pico della Mirandola. "She appeared to me," he tells us, "something more stupendous — *una cosa piu stupenda* — than the Duomo of Milan." Having been invited to a *conversazione* for the purpose of meeting this wonderful woman, the learned Frenchman found her to be a "young lady of about eighteen or twenty." She was surrounded by "about thirty people..... many of them from different parts of Europe." The discussion turned on various questions of mathematics and natural philosophy.

"She spoke," writes de Brosses, "wonderfully well on these subjects, though she could not have been prepared beforehand any more than we were. She is much attached to the philosophy of Newton; and, it is marvellous to see a person of her age so conversant with such abstruse subjects. Yet, however much I was surprised at the extent and depth of her knowledge, I was still more amazed to hear her speak Latin..... with such purity, ease and accuracy, that I do not recollect any book in modern Latin written in so classical a style as that in which she pronounced these discourses.... The conversation afterwards became general, everyone speaking in the language of his own cxountry, and she answering in the same language; for, her knowledge of languages is prodigious."

11) At the conclusion of an elaborate review of Colton's translation of Agnesi's *Instituzioni Analitiche* in the *Edinburgh Review* for January 1804, the writer expresses himself as follows: "We cannot take leave of a work that does so much honour to female genius, without earnestly recommending the perusal of it to those who believe that great talents are bestowed by nature exclusively on man, and who allege that women, even in their highest attainments, are to be compared only to *grown children*, and have, in no instance, given proofs of original and inventive powers, of a capacity for patient research, or for profound investigation. Let those who hold these opinions endeavour to follow the author of the *Analytical Institutions* through the long series of demonstrations, which she has contrived with so much skill and explained with such elegance and perspicuity. If they are able to do so, and to compare her work with others of the same kind, they will probably retract their former

opinions, and acknowledge that, in one instance at least, intellectual powers of the highest order have been lodged in the brain of a woman.

"At si gelidus obstiterit circum praecordia sanguis; and if they are unable to attend this illustrious female in her scientific excursions, of course, they will not see the reasons for admiring her genius that others do; but they may at least learn to think modestly of their own."

12) It is surprising how many legends have obtained respecting the life of Agnesi after the publication of her *Instituzioni Analitiche*. Thus, the writer of the article in the *Edinburgh Review*, above quoted, declares that "she retired to a convent of *blue nuns*" — a statement that has frequently been repeated in many of our most noted encyclopaedias.

In a *Prospetto Biografico delle Donne Italiane*, written by G C Facchini and published in Venice in 1824, it is stated that Maria Gaetana was selected by the Pope to occupy "the chair of mathematics which had been left vacant by the death of her father," while Cavazza in his work "*Le Scuole dell,*" *Antico Studio Bolognese*, pp 289-90, published in Milan in 1896, assures us that Gaetana Agnesi taught analytical geometry in the University of Bologna for full forty-eight years. The facts are that neither the father nor the daughter ever taught even a single hour either in this or in any other university. Cf *Maria Gaetana Agnesi*, p 273 *et seq*, by Luisa Anzoletti, Milano, 1900. This is far the best life of Milan's illustrious daughter that has yet appeared. The reader may also consult with profit the *Elogio Storico* di Maria Gaetana Agnesi, by Antonio Frisi, Milano, 1799, and *Gli Scrittori d'Italia*, of G Mazzuchelli, Vol I, Par I, p 198 *et seq*, Brescia, 1795.

13) M Rebiere, in *his Les Femmes dans la Science*, p 13, Paris, 1897, writes, "Ne pourrait-on aller plus loin et canoniser notre Agnesi? J'estime, moi profane, que ce serait une sainte qui en vaudrait bien d'autres."

14) *An Eighteenth Century Marquise: A Study of Émilie du Chatelet*, p 5, by F Hamel, New York, 1911.

15) Preface to Mme du Chatelet's translation of the *Principia* of Newton, Paris, 1740.

16) Voltaire's last tribute, "The Divine Émilie," or, as Frederick II was wont to call her, "Venus-Newton," concluded with the following verses:

L'Univers a perdu la sublime Émilie;
Elle aimait les plaisirs, les arts, la verite;

Les dieux, en lui donnant leur ame et genie,
N'avaient garde pour eux que l'immortalite.

The universe has lost the sublime Émilie; she loved pleasure, the arts, truth; the gods, in giving her their soul and genius, retained for themselves only immortality.

For further information of this extraordinary woman, see *Lettres de la Mme du Chatelet, Reunies pour la premiere fois*, par Eugene Asse, Paris, 1882.

17) At the beginning of her correspondence with Gauss, Legendre and Lagrange Mlle Germain concealed her sex under a pseudonym, "in order," as she declared, "to escape the ridicule attached to a woman devoted to science" — *craignant le ridicule attache au titre de femme savante*. She, too, suffered from the widespread effects of Moliere's *Les Femmes Savantes*, as had many a gifted woman before her time and as have many others of a much later date.

18) This celebrated mathematician, as is well-known, was a collaborator with Mme du Chatelet in her translation of Newton's *Principia*.

19) For further information respecting this remarkable woman the reader is referred to *Oeuvres Philosophiques de Sophie Germain Suivies de Pensees et de Lettres Inedites et Precedees d'une Étude sur sa Vie et ses Oeuvres*, by H Stupy, Paris, 1896. One may also consult Todhunter's *History of the Theory of Elasticity and of the Strength of Materials*, Vol I, pp 147-60, Cambridge, 1886, in which is given a careful resume of Mlle Germain's mathematical memoirs on elastic surfaces.

20) *Saturday Review*, January 10, 1874.

21) *Personal Recollections, From Early Life to Old Age, of Mary Somerville*, p 80, Boston, 1874.

x22) *Personal Recollections*, ut sup, p 5.

23) *Sónya Kovalevsky, Her Recollections of Childhood, With a Biography*, by Anna Carlotta Leffler, p 219, New York, 1895.

24) "The prise was doubled to five thousand francs, on account of the 'quite extraordinary service rendered to mathematical physics by this work,' which the Academy of Sciences pronounced 'a remarkable work.' The competing dissertations were signed with mottoes, not with names, and the jury of the Academy made the award in utter ignorance that the winner was a woman. Her dissertation was printed, by order of the Academy, in the *Memoires des Savants Etrangers*. In the following year Mme Kovalevsky received a prise of fifteen hundred kroner from the Stockholm Academy for two works connected with the foregoing."

25) Men of science will realise the capacity of this gifted Russian woman

as a mathematician when they learn that she gave in the University of Stockholm courses of lectures on such subjects as the following: Theory of derived partial equations; theory of potential functions; applications of the theory of elliptic functions; theory of Abelian functions, according to Weierstrass; curves defined by differential equations, according to Poincare; application of analysis to the theory of whole numbers. How many men are there who give more advanced mathematical courses than these?

# Women in Astronomy

URANIA, THE MUSE of astronomy, was a woman; and, although most of her devotees have been men, the number of the gentler sex who have achieved success in the cultivation of the science of the stars has been much larger than is usually supposed.

There is reason to believe that woman's interest in astronomy dates back to early Egyptian and Babylonian times when the star-gazers in the fertile valley of the Nile and on the broad plains of Chaldea were so active, and when they made so many important discoveries respecting the laws and movements of the heavenly bodies. According to Plutarch, Aganice, the daughter of Sesostris, King of Egypt, tried to predict future events by the aid of celestial globes and by the study of the constellations. Her observations, however, were in the interests of astrology rather than of astronomy, as we now understand the science.

The first woman whose name has come down to us, who deserved to be regarded as an astronomer, was most probably Aglaonice, the daughter of Hegetoris of Thessaly. By means of the lunar cycle known as the Saros, a period discovered by the Chaldean astronomers and embracing a little more than eighteen years, during which the eclipses of the moon and sun recur in nearly the same order as during the preceding period, this Greek woman was able to predict eclipses. The people among whom she lived regarded her as a sorceress; but she flouted them all, and declared that she was able to make the sun and moon disappear at will.

The first woman, however, to attain eminence as an astronomer was undoubtedly Hypatia, that universal genius of the ancient world, who seemed equally at home in literature, philosophy and mathematics, and who may justly be regarded as one of the most highly gifted women that has ever lived. In Alexandria, where she was born and lived, this accomplished daughter of Theon taught not only philosophy, but also algebra, geometry and astronomy. One of her pupils, Synesius, who became Bishop of Ptolemais, informs us that she was the inventor of two important astronomical instruments: an astrolabe and a planisphere. In addition to two mathematical works, a *Treatise on the Conics of Apollonius* and a *Commentary on the Arithmetic of Diophantus,* which was in reality a treatise on algebra, she was the author of an *Astronomical Canon,* which contained tables regarding the movements of the heavenly bodies. It is generally supposed that this was an original work; but there are some who think it was but a commentary on the tables of Ptolemy. In this latter case Hypatia's work may still exist in connection with that of her father, Theon, on the same subject.[1]

If the works of Hypatia had not been destroyed by the ravages of time, they would undoubtedly prove that she fully merited all the encomiums bestowed on her by antiquity for her genius; and they would also prove, we may well believe, that she deserved to be ranked not only with the eminent mathematicians upon whose works she commented, but also with such masters of astronomic science as Ptolemy, Eratosthenes and Aristarchus.

After the tragic death of Hypatia many centuries elapsed before any other woman attracted attention for her work in astronomy. Indeed, so neglected was the study of the heavens between the time of Hypatia and the Arab prince and astronomer, Albategni, who flourished during the latter part of the ninth century and the early part of the tenth, that only eight observations, it is asserted, were recorded during this long period. The works and observations of Albategni, it may be remarked, have a particular interest from the fact that they form a connecting link between those of the Alexandrine astronomers and those of modern Europe.

Antoine Hamilton, in his *Gaufrey*—a parody on *The Thousand*

*and One Nights* — tells of a Saracen princess, *Fleur d'Épine*, who, before she was fifteen years of age, was able not only to speak Latin and Romance, but who was also "better acquainted than any woman in the world with the movements of the stars and the moon."

> Et du cours des etoiles et de la lune luisant
> Savoit moult plus que fame de chest siecle vivant.

If any woman between the time of Hypatia and Galileo deserved such high praise for her astronomical knowledge it was certainly Saint Hildegard, the famous Benedictine abbess of Bingen on the Rhine. She has well been called "the marvel of the twelfth century," not only on account of her sanctity, but also on account of her extraordinary attainments in every branch of knowledge then cultivated.

When treating of the sun, Hildegard tells us that it is in the centre of thc firmament and holds in place the stars that gravitate around it, as the earth attracts the creatures which inhabit it. This view of a twelfth century nun is indeed remarkable. For, in her time, the earth was by everyone considered as the centre of the firmament, while universal gravitation — the sublime discovery of Newton — had not as yet entered into the scientific theories of that epoch.

Hildegard likewise anticipates subsequent discoveries regarding the alternation of the seasons. "If," she writes, "it is cold in the winter time on the part of the earth which we inhabit, the other part must be warm, in order that the temperature of the earth may always be in equilibrium." That she should have arrived at this conclusion before navigators had visited the southern hemisphere is truly astonishing.[2]

"The stars," she continues, "have neither the same brightness nor the same size. They are kept in their course by a superior body." Here again is her idea of universal gravitation.

These stars, she further declares, are not immovable, but they traverse the firmament in its entirety. And to make clearer her conception of the motion of the stars, she compares this motion to that of the blood in the veins. To hear one of this early period speaking of blood coursing through the veins and thus traversing the whole body of man seems to presage, in a remarkable manner, the beautiful discoveries of Cesalpino and Harvey regarding the circulation of the blood.

The most celebrated astronomer of the early Renaissance was John Müller, of Konigsburg, better known as Regiomontanus. In his observatory in Nuremberg he was ably assisted by his wife who exhibited a special interest in astronomy. At the end of the sixteenth century, Sophia Brahe, the youngest sister of Tycho Brahe, following in the footsteps of her illustrious brother, attained great celebrity as an astronomer.

More distinguished for her astronomical work than either of these two women was Maria Cunitz, a Silesian, who, from her tenderest years, displayed extraordinary zeal for study and who eventually became mistress of seven languages, among which were Latin, Greek and Hebrew. She also cultivated poetry, music and painting; but her favourite studies were mathematics and astronomy. At the solicitation of her husband, she undertook the preparation of an abridgment of the *Rudolphine Tables*. Her work, under the name of *Urania Propitia*, was published after her death by her husband, and gained for the talented authoress the name of "The second Hypatia."[3]

Shortly after the completion of *Urania Propitia*, a French woman, Jeanne Dumee, distinguished herself by writing a work on the theory of Copernicus entitled *Entretiens sur l'Opinion de Copernic Touchant la Mobilite de la Terre*. So far as known, this work was never published, but the original manuscript is still preserved in the National Library of Paris. The authoress deems it necessary it apologise for writing on a subject that is usually considered foreign to her sex and to explain why she was ambitious to discuss questions to which the women of her time never gave any thought. It was that she might "prove to them that they are not incapable of study, if they wish to make the effort, because between the brain of a woman and that of a man there is no difference."[4]

How often before had not women endeavoured to prove the

equality of brain power of the two sexes, and how often since have they bent their efforts in this direction! And yet the majority of men still remain sceptical about such equality.

Among the contemporaries of Jeanne Dumee were two other women who gained more than ordinary distinction by their attainments in astronomy. These were Mme de la Sabliere, in France, and Maria Margaret Kirch, of Germany.

Mme de la Sabliere evinced from an early age a special aptitude for science, especially for physics and astronomy. She studied mathematics under the eminent mathematician, Roberval, and at the age of thirty was famous. Her home became the resort of learned and eminent men, including some of the most noted characters of the age. Among these was Sobieski, King of Poland. But it is as the friend and protectress of La Fontaine and as the object of Boileau's satire that she is best known.

For a woman to devote herself to the study of science so soon after the appearance of Moliere's *Les Femmes Savantes* argued more than ordinary courage. But for her to become distinguished for her scientific acquirements was almost tantamount to defying public opinion. The great majority of men had come to regard learned women in the same light as those who were so mercilessly derided in the *Precieuses Ridicules*; and they had, accordingly, no hesitation in treating them as unbearable pedants. No one could have made less parade of her learning than Mme de la Sabliere, or striven more successfully to conceal her admirable gifts. But this was not sufficient. She was known to have devoted special study to science, particularly to astronomy, and this was sufficient to make her the target of the satirists of her time.

By an act that wounded the self-love of Boileau this Venus Urania, as she has been called, soon found herself the victim of the satirist's well-directed shafts. The poet does not name her, but refers to her as

> Cette savante
> Qu'estime Roberval et que Sauveur frequente..... .....

this learned woman whom Roberval esteems and whom

Sauveur frequents. And with the view of pricking the object of his spleen in her most sensitive part, he tells, in his *Satire contre les Femmes,* how she, with astrolabe in hand, spends her nights in making observations of the planet Jupiter and how this occupation has had the effect of weakening her sight and ruining her complexion.[5]

Mme de la Sabliere does not, however, seem to have been greatly perturbed by the ungracious effusions of the satirist, for she continued her cultivation of astronomy as before the poet's ill-natured outburst. She probably found ample compensation in the writings of La Fontaine, who addressed her as his muse and proclaimed her as one in whom were combined manly beauty and feminine grace — *beaute d'homme avec grace de femme.*

Maria Kirch, born at Panitch, near Leipsic, in 1670, was the wife of a Berlin astronomer, Gottfried Kirch. After her marriage she, like her three sisters-in-law, became her husband's pupil in astronomy. In 1702, as his assistant in observations and calculations, she was fortunate enough to discover a comet. She was the friend of Leibnitz, and was by him presented to the court of Prussia. It is a matter of regret to those of her own sex that this comet was not, as it should have been, named after its discoverer.

The death of Herr Kirch, which took place in 1710, caused no interruption in Frau Kirch's astronomical occupations. Among the evidences of her activity is a work which she wrote in 1713 on the conjunction of Jupiter and Saturn in the year following. In our day the conjunction of planets is for the laity a mere matter of curiosity, while for professional astronomers it is quite devoid of particular interest. But it was not so in the time of Maria Kirch, for then astronomy was so intimately associated with astrology that mankind attributed to such special positions of the planets a certain occult and capricious influence on the destiny of the earth and its inhabitants. As theoretical astronomy progressed, such erroneous notions were abandoned, because it was then recognised that the conjunction of the superior planets was not something fortuitous, but something that was reproduced at fixed periods by the known movements of these bodies. Writers on the subject made it a point to warn the public that they had nothing in common with astrologers. Among these was Christopher Thurm, who published a work on the conjunction of Jupiter and Saturn in 1681. Similarly, the book of Maria Kirch contains only astronomical calculations and nothing more — a fact that redounds to the honour of the author and to the age in which she lived.

The daughters of Maria Kirch, even long after their mother's death, continued to occupy themselves with astronomy. They calculated for the Berlin Academy of Sciences its *Almanac* and *Ephemeris*, which were among the sources of revenue of this learned body.

During the same period a number of French and Italian astronomers had female collaborators in their own families. Celsus, the celebrated professor of Upsala, and a pupil of the son of Gottfried Kirch, had been accorded a most cordial reception, while passing through Paris on his way to Bologna, by De L'Isle who had a sister who was devoted to astronomy. On his arrival in Italy he found that his new master, the director of the observatory at Bologna, had two sisters, Teresa and Maddalena, both of great learning, who, like their brother, were engaged in the study of the heavens and collaborated with him in the preparation of the *Ephemeris* of Bologna. This caused Celsus, in a letter to Kirch, to declare "I begin to believe that it is the destiny of all the astronomers whom I have had the honour of becoming acquainted with during my journey to have learned sisters. I have also a sister, although not a very learned one. To preserve the harmony, we must make an astronomer of her."[6]

The Polish astronomer, Hevilius, who had an observatory at Dantzig, is noted for having made the most accurate observations that had been known before the adaptation of the telescope to astronomical instruments. He is also noted for his *Prodromus Astronomiae*, a catalogue of 1,888 stars; for his *Selenographia*, containing accurate descriptions and drawings of the moon in her different phases and librations, and for his *Machina Coelestis*, which contained the results of forty years of observations and labour. Much of his success and eminence, however, was due to his intelligent and devoted wife, Elizabeth, who, during twenty-seven years, was a zealous collaborator and should share the credit usually given to her husband. It was she who, after his death, edited and published their joint work, the *Prodromus Astronomiae*.

Among the women most distinguished in the eighteenth century for astronomical pursuits was the Marquise du Chatelet, who was likewise famous for her knowledge of mathematics. It was she who accomplished the difficult task of translating Newton's *Principia* into French. "This translation," writes Voltaire, "which the most learned men of France should have made and which the others should study, was undertaken by a woman and completed to the astonishment and glory of her country."[7]

France was at this time devoted to the doctrines of Descartes and to his theory of elementary vortices; and Voltaire, who had been deeply impressed by the admirable simplicity of Newton's theory of universal attraction as a means of explaining the seemingly complex motions of the heavenly bodies, resolved to make his countrymen acquainted with the teachings of the great English geometer and, at the same time, dethrone Descartes in the French Academy. It was, indeed, a huge undertaking; but, thanks to the ability which Mme du Chatelet displayed in translating and elucidating Newton's immortal masterpiece, he lived to see his dream realised.

How proud Mme du Chatelet's countrywomen must have been of her! How they must have rejoiced in her success and acclaimed her as the intellectual glory of her sex! How they must have pointed to her work as a triumphant refutation of the age-old belief in woman's incapacity for mathematics and all abstract science! How they must have been elated to find one of their number successfully executing a task which would have taxed the powers of the most eminent mathematicians of France! How they must have associated her truly notable performance with similar achievements of Hypatia and Maria Gaetana Agnesi and discerned in it concrete evidence of the falsity of all those imputations of mental inferiority which had been fostered by "man's huge egotism and woman's carefully coddled superstition." How they must have been encouraged by her achievement and spurred on to emulate her by similar contributions to the advancement of science!

That is what we think now; but the light and frivolous women who constituted the leaders of society in Mme du Chatelet's day, and who were devoured by envy and jealousy of one who was so much their superior in intellect were not so minded. Far from sympathising with her work, they proved to be her most virulent critics and most pronounced enemies. Neither Moliere nor Boileau could have heaped more ridicule on the pedantic women of their time than was meted out to the translator of the *Principia* by certain noble dames of provincial chateaux or by distinguished habituees of prominent Parisian salons.

Thus the petulant *ennuyee*, Mme de Staël, in a letter to her friend, Mme du Deffand, writing of Mme du Chatelet, who was then her guest at Sceaux, tells us that "she is now passing in review her principles. This is a task she performs every year, else they might, perhaps, make their escape and run to such a distance that she would never be able to recover any of them. I verily believe that they are in durance vile while in her possession, as they were certainly not born with her. She does well to keep a strict watch over them."[8]

And, in her turn, Mme du Deffand, who was wont to pose as the intimate friend of Mme du Chatelet, did not hesitate to write and circulate a pen portrait of this friend — and that after the unhappy woman was in her grave — which for bitter reviling and brutal vilification has probably never been equalled. A witty Frenchman observed of this portrait that it reminded him of an observation once made by a medical acquaintance of his concerning one of his patients: " 'My friend fell ill; I attended him. He died; I dissected him.' "[9]

Among other women astronomers of the eighteenth century who deserve mention are Mme du Pierry, the Duchesse Louise of Saxe-Gotha, and Mme Hortense Lepaute.

According to Lalande, Mme du Pierry was the first woman

professor of astronomy in Paris. He dedicated to her his *Astronomie des Dames,* and incorporated in his own works many of her memoirs on astronomical subjects. She devoted much time to calculating eclipses with a view to accurately determining the motion of the moon, and was, besides, the author of numerous astronomical tables which exhibit patient research and unquestioned skill.

The Duchesse Louise had a great reputation as a rapid and accurate computer, and was celebrated for the number and variety of her computations. Her modesty, however, prevented her from publishing anything or even having her work quoted.

Considering, however, the amount and character of her work, the most eminent woman astronomer that France has yet produced was, without doubt, Mme Hortense Lepaute, the wife of the royal clockmaker of France. She first distinguished herself by her investigations on the oscillations of pendulums of different lengths, an account of which is to be found in her husband's valuable work, *Traite d'Horlogerie,* published in 1755.

In 1759 Lalande, who was then the Director of the Paris Observatory, engaged Mme Lepaute and the celebrated mathematician, Clairaut, to determine the amount of the attraction of Jupiter and Saturn on Halley's comet, whose return was expected in that year. So difficult was this problem, and so numerous were the complications involved, that Lalande frankly confesses that he would not have dared to undertake its solution without Mme Lepaute's assistance. For it necessitated calculating for every degree, and for one hundred and fifty years the distances and forces of each of the planets with reference to the comet. "It would be difficult," declares Lalande, "to realise the courage which this enterprise required, if one did not know that for more than six months we calculated from morning until night, sometimes even at meals, and that at the end of this enforced labour I was stricken by a malady which affected me during the rest of my life." Clairaut was so impressed by Mme Lepaute's energy and skill during this time that he declared "her ardour was surprising," and he did not hesitate to call her *La savante calculatrice* — the learned computer.[10]

The eclipse of 1762 also engaged Mme Lepaute's attention, as did also the annular eclipse of 1764. The latter was a curious phenomenon for France, as it had never before been observed. Mme Lepaute calculated it for the whole of Europe and published a chart showing its path for every quarter of an hour. She also published another chart for Paris, in which were exhibited the different phases of the eclipse.

On the occasion of the different eclipses which she had calculated, Mme Lepaute recognised the advantage of having a table of parallactic angles. She accordingly prepared a very extended table of this kind which was published by the French government. Besides this table, she was the author of numerous memoirs on astronomical subjects. Among them was one embracing calculations based on all the observations which had been made on the transit of Venus in 1761.

"In 1759," again writes Lalande, "I was given charge of the *Connaissance des Temps,* a work which the Academy of Sciences published every year for the use of astronomers and navigators, the calculations for which gave occupation to several persons. I had the good fortune to find in Mme Lepaute a co-worker without whom I should not have been able to undertake the labour required. She continued in this occupation until 1774, when another Academician assumed this laborious task. But she thereupon began work on the *Ephemeris,* of which the seventh volume in quarto, which appeared in 1774, goes to 1784, and of which the eighth, published in 1783, extends to the year 1792. In this latter volume she made, unaided, all the computations for the sun, the moon and all the planets.

"This long series of calculations finally enfeebled her eyesight, which had been excellent, and she was in the last years of her life obliged to discontinue them."[11]

In view of her extraordinary and long-continued work in her chosen specialty, M Lalande was quite warranted in stating that "Mme Lepaute is the only woman in France who has acquired veritable knowledge in astronomy; and she is now replaced only by Mme du Pierry, who has published divers astronomical calculations, and who has deserved to have dedicated to her *L'Astronomie des Dames,* which appeared in 1786."

It is gratifying to know that the beautiful Japan Rose — originally called *Pautia,* but changed to *Hortensia* by Jussieu — was named after this distinguished woman. It is also gratifying to be assured that her engrossing work in astronomy in no wise caused her to neglect her home duties or to lose that sweetness of character and delicacy of refinement for which she was noted before she entered upon the absorbing and taxing career of astronomical computer.

The wife of Lalande's nephew, Mme Lefrançais de Lalande, proved herself in many respects a worthy successor of Mme Lepaute. "My niece," writes her uncle, Jerome Lalande, "aids her husband in his observations and draws conclusions from them by calculation. She has reduced the observations of ten thousand stars, and prepared a work of three hundred pages of horary tables — an immense work for her age and sex. They are incorporated in my *Abrege de Navigation.*

"She is one of the rare women who have written scientific books. She has published tables for finding the time at sea by the altitude of the sun and stars. These tables were printed in 1791 by the order of the National Assembly..... In 1799 she published a catalogue of ten thousand stars, reduced and calculated."

This distinguished observer and computer had a daughter in whom her grand-uncle was particularly interested. "This daughter of astronomy," he tells us, "was born the twentieth of January 1790, the day on which we at Paris saw for the first time the comet which Miss Caroline Herschel had just discovered. The child was accordingly named Caroline; her godfather was Delambre."

The discoverer of the comet referred to was, in many ways, a most remarkable woman. She was the sister of Sir William Herschel, the illustrious pioneer of modern physical astronomy and the virtual founder of sidereal science, as we know it today. She was also the aunt of Sir John Herschel, who was the only rival of his uncle, Sir William, as an explorer of the heavens.

But she was far more than a mere relative of these immortal leaders in astronomic science. She herself was an astronomer of distinction, and is known, in the annals of astronomy, as the discoverer of no fewer than eight comets. Great, however, as was her skill as an observer and computer, it was as her brother's assistant that she is entitled to the most distinction. Her affection for him was as unbounded as her devotion to his life work was abiding and productive of great results. For fifty years, after joining him in England — they both had been born and bred in Hanover — she was ever at his side, to assist him in his labours and to cheer him by her words of counsel and encouragement. She helped him to grind and polish the mirrors that were used in his epoch-making reflectors. This was a most arduous task; for, at that time, there was no machinery sufficiently exact for grinding specula, and, as a consequence, the work had all to be done by hand. So interested was the great astronomer in his work, when polishing his larger specula, that he forgot all about the passage of time, and on these occasions his sister was constantly obliged, as she herself informs us, "to feed him by putting the victuals by bits into his mouth by way of keeping him alive." When finishing his seven-foot reflector he was on one occasion found so intent on his work that "he had not taken his hands from it for sixteen hours together."

In our day, when all kinds of astronomical apparatus are made by machinery, it is difficult for us to realise what stupendous labour was required to produce those giant telescopes with which the Herschels made their great discoveries and by which they, at the same time, revolutionised the science of the stars. For they had not only to design and make the specula, but also the mountings of the mirrors as well. And, in order to obtain the money required for material and workmen, they were obliged to make telescopes for sale. This meant an immense loss of precious time that would otherwise have been devoted to the study of the heavens.

After long years of struggle, during which the devoted brother and sister overcame countless difficulties of every kind, their condition was somewhat ameliorated by financial aid from the government and by William's appointment to the position of astronomer royal with a salary of £200 a year. When Sir William Watson heard that this limited sum had been granted by George III to the discoverer of Georgium Sidus — the planet now known as Uranus — he exclaimed, "Never bought monarch honour so cheap."

Shortly afterwards Caroline was appointed as assistant to her brother at a salary of £50 a year. This we should now consider but a nominal sum, but she managed to live on it. When she received the first quarterly payment of twelve pounds she wrote in her memoirs, "It was the first money I ever in all my lifetime thought myself to be at liberty to spend to my liking." Her appointment as assistant to her brother is notable from the fact that she was the first woman in England, if not in the world, to hold such a position in the government service.

Miss Herschel held this official appointment until Sir William's death in 1822. When not acting as her brother's assistant or secretary, she devoted her time to what she quaintly called "minding the heavens." It was during this period that she made her most important discoveries. As assistant, however, to so indefatigable an observer as Sir William Herschel, she had but little time for sweeping the heavens, for, when at home, Sir William "was invariably accustomed to carry on his observations until daybreak, circumstances permitting, without regard to seasons; it was the business of his assistant to note the clocks and to write down the observations from his dictations as they were made. Subsequently she assisted in the laborious numerical calculations and reductions, so that it was only during his absence from home or when any other interruption of his regular course of observation occurred that she was able to devotc herself to the Newtonian sweeper, which she used to such good purpose. Besides the eight comets by her discovered, she detected several remarkable nebulae and clusters of stars, previously unnoticed, especially the superb nebulae known as No 1, Class V, in Sir William Herschel's catalogue. Long practice taught her to make light of her work. 'An observer at your twenty-foot when sweeping,' she wrote many years after, 'wants nothing but a being who *can* and *will* execute his commands with the quickness of lightning; for you will have seen that in many sweeps six or twice six objects have been secured and described in one minute of time.' "[12]

It was her quick, intelligent action, combined with a patience, enthusiasm and powers of endurance that were most extraordinary, that made Caroline Herschel so valuable as an assistant to her brother, and enabled him to achieve the unique position which is his among the world's greatest astronomers. Had she been able to devote all her time to "minding the heavens," it is certain that she would have made many more discoveries than are now credited to her; but her service to astronomy would have been less than it was as the auxiliary of her illustrious brother. No two ever did better "teamwork;" no two were ever more devoted to each other or exhibited greater enthusiasm in the task to which they so heroically devoted their lives.[13]

In addition to her arduous and engrossing duties as secretary and assistant to her brother, Caroline found time to prepare a number of works for the press. Among these were a *Catalogue of Eight Hundred and Sixty Stars Observed by Flamsteed but not Included in the British Catalogue* and *A General Index of Reference to Every Observation of Every Star in the Above-mentioned British Catalogue*. She had the honour of having these two works published by the Royal Society. Another, and a more valuable work, was *The Reduction and Arrangement in the Form of Catalogue, in Zones, of All the Star-Clusters and Nebulae Observed by Sir W Herschel in His Sweeps*. It was for this catalogue that a gold medal was voted to her by the Royal Astronomical Society in 1828 — a production that was characterised as "a work of immense labour" and "an extraordinary monument to the unextinguished ardour of a lady of seventy-five in the cause of abstract science." To her nephew, Sir John Herschel, it proved invaluable, as it supplied the needful data "when he undertook the review of the nebulae of the northern hemisphere." It was also a fitting prelude to Sir John's *Cape Observations,* a copy of which great work she received from her nephew nearly twenty years subsequently, after he had completed his famous observations of the southern heavens in his observatory at the Cape of Good Hope.

"By a most striking and happy coincidence," writes Mrs John Herschel, "she, whose unflagging toil had so greatly contributed to its successful prosecution in the hands of her beloved brother, lived to witness its triumphant termination through the no less persistent industry and strenuous labour of his son; and her last days were crowned by the possession of the work which brought to its glorious conclusion Sir William Herschel's vast undertaking — *The Survey of the Heavens.*"

That Miss Herschel's labours in the cause of astronomy were appreciated by her contemporaries is evidenced by the honours of which she was the recipient. The first of these honours came in the form of a gold medal, unanimously awarded by the Royal Astronomical Society for her reduction of twenty-five hundred nebulae "discovered by her illustrious brother, which may be considered as the completion of a series of exertions probably unparalleled either in magnitude or importance in the annals of astronomical labour."

It was on this occasion, when referring to the immensity of the task which Sir William Herschel had undertaken, that the vice-president of the society paid a deserving tribute to the great astronomer's devoted sister, in which is found the following statement:

> Miss Herschel it was who by right acted as his amanuensis; she it was whose pen conveyed to paper his observations as they issued from his lips; she it was who noted the right ascensions and polar distances of the objects observed; she it was who, having passed the night near the instrument, took the rough manuscripts to her cottage at the dawn of day and produced a fair copy of the night's work on the following morning; she it was who planned

the labour of each succeeding night; she it was who reduced every observation, made every calculation; she it was who arranged everything in systematic order; and she it was who helped him to obtain his imperishable name.[14]

Besides this gold medal from the Royal Astronomical Society, Miss Herschel also received two others, one from the King of Denmark and the other from the King of Prussia. The latter was accompanied by a most eulogistic letter from Alexander von Humboldt, who informed her that the medal was awarded her "in recognition of the valuable services rendered by her as the fellow worker of her immortal brother, Sir William Herschel, by discoveries, observations and laborious calculations."

In 1835, when she was eighty-five years of age, Miss Herschel had the signal honour of being elected, along with Mrs Somerville, an honorary member of the Royal Astronomical Society. As they were the first two women in England to receive such recognition for their contributions to science, it seems desirable to reproduce here an extract from the report of the council of the society regarding the bestowal of an honour which marked so distinct a change in England of the attitude that should be taken toward women who excelled in intellectual achievements. The extract reads as follows:

> Your council has no small pleasure in recommending that the names of two ladies distinguished in different walks of astronomy be placed on the list of honorary members. On the propriety of such a step, in an astronomical point of view, there can be but one voice; and your council is of the opinion that the time is gone by when either feeling or prejudice, by whichever name it may be proper to call it, should be allowed to interfere with the payment of a well-earned tribute of respect. Your council has hitherto felt that, whatever might be its own sentiment on the subject, or however able and willing it might be to defend such a measure, it had no right to place the name of a lady in a position the propriety of which might be contested, though

> upon what it might consider narrow grounds and false principles. But your council has no fear that such a difference could now take place between any men whose opinion could avail to guide the society at large; and, abandoning compliment on the one hand and false delicacy on the other, submits that, while the tests of astronomical merit should in no case be applied to the works of a woman less severely than to those of a man, the sex of the former should no longer be an obstacle to her receiving any acknowledgment which might be held due to the latter. And your council, therefore, recommends this meeting to add to the list of honorary members the names of Miss Caroline Herschel and Mrs Somerville, of whose astronomical knowledge, and of the utility of the ends to which it has been applied, it is not necessary to recount the proofs.[15]

Three years after this splendid recognition of Miss Herschel's astronomical labours she was elected an honorary member of the Royal Irish Academy.

But exceptional as were the honours conferred on her by sovereigns and learned societies, none of them afforded her the extreme satisfaction that she experienced on the receipt of a copy, shortly before her death, of her nephew's epochal *Cape Observations*; for, as has well been said, "nothing in the power of man to bestow could have given such pleasure on her death-bed as this last crowning completion of her brother's work." We are told that a copy, just from the press, of his immortal work, *De Orbium Celestium Revolutionibus*, in which he had established the heliocentric theory of the planetary system, was placed in the hands of Copernicus on the day of his death, just a few hours before he expired. He seemed conscious of what it was; but, after touching it and contemplating it for a moment, he lapsed into a state of insensibility which soon terminated in death. With Miss Herschel the case was different. Although in her ninety-seventh year, she still retained possession of all her faculties and was fully able to appreciate the volume which told of the crowning of her brother's life work — a volume which must have given her additional satisfaction when she recalled her fifty years of loyal service at her brother's side as his associate and ministering angel in the greatest work ever undertaken by a single man in the history of astronomy.

Caroline Herschel died at the advanced age of ninety-seven years and ten months, retaining to the last her interest in astronomy which had occupied her mind for more than three-quarters of a century.

Her epitaph, composed by herself, is engraved on a heavy stone slab which covers her grave and contains the following words: "The eyes of her who is glorified were here below turned to the starry heavens. Her own discoveries of comets and her participation in the immortal labours of her brother, William Herschel, bear witness of this to future ages."

Space precludes any extended reference to Miss Herschel's distinguished associate in the Royal Astronomical Society, Mrs Somerville, whose masterly translation and exposition of Laplace's *Mecanique Celeste* secured for her so enviable a place among the mathematicians of her time, and placed all English students of mathematical astronomy under such deep obligations. It is true that she ever manifested a lively interest in celestial phenomena; but it is rather as a mathematician than as an astronomer that she will be remembered by the devotees of science.

The first American woman to win distinction in astronomy was Miss Maria Mitchell. Born in the island of Nantucket in 1818, she, at an early age, displayed remarkable talent for astronomy and mathematics. Her first instructor was her father, who, besides being a school teacher, had from his youth been an enthusiastic student of astronomy, and that, too, at a time when very little attention was given to its study in this country, and when the observatory of Harvard College consisted of only a little projection to an old mansion in Cambridge, in which there was a small telescope.

At the age of thirteen little Maria counted seconds by the chronometer for her father while he observed the annular eclipse of the sun in 1831; and from that time on she was his assiduous co-worker in the study of the heavens. After teaching school for some years, she became the librarian of the Nantucket Atheneum, a position which she held for nearly twenty years. Here she continued the study of her favourite science, and read all the books on astronomy which she could obtain. It was during this period that she read Bowditch's translation of Laplace's *Mecanique Celeste* and Gauss's *Theoria Motus Corporum Caelestium* in the original.

On the evening of October 1, 1847, she was the discoverer of a comet that attracted great attention because it secured for her a medal offered by the King of Denmark in 1831 for the first one who should discover a telescopic comet. The same comet was observed by Father de Vico in Rome two days subsequently, by Dawes in England on October seventh, and by Madame Rümker, wife of the director of the observatory of Hamburg, on the eleventh of the same month. As there was no Atlantic cable in those days, it was not known who was the fortunate winner of the prize until nearly a year afterward, when word was received from Denmark announcing that the priority of Miss Mitchell's discovery had been recognised and that she would be the recipient of the prize, which, for a while, it was thought would go to De Vico or Madame Rümker.[16]

In 1849 Miss Mitchell was appointed a compiler for the *Nautical Almanac*, a position she held for nineteen years. During the same period she was employed by the United States Coast Survey.

When Vassar College was opened in 1865 for the higher education of women, Miss Mitchell was called to fill the chair of astronomy and to be the first director of the observatory. In this position she soon succeeded in giving astronomy a prominence that it never had had before in any other college for women, and in but few for men.

Miss Mitchell was a member of several learned societies and the author of a number of papers containing the results of her observations on Jupiter and Saturn and their satellites. But she is notable chiefly for being the first woman astronomer in the United States and for training up a number of young women who have followed in her footsteps as enthusiastic astronomers. She held her position at Vassar until 1889, when she died, a few months before her seventy-first birthday.

Since the pioneer days of Miss Caroline Herschel, the number of women throughout the world who have achieved distinction in astronomy has rapidly augmented. One of the most noted of these was Caterina Scarpellini, niece of Feliciano Scarpellini, professor of astronomy in Rome, restorer of the Academy of the Lyncei, and founder of the Capitoline Observatory. Born in 1808, she manifested at an early age a decided taste for astronomy, which was carefully developed by her uncle. She it was who organised the Meteorologico Ozonometric station in Rome and edited its monthly bulletin. She exhibited a special interest in shooting stars and prepared the first catalogue of these meteors observed in Italy. In 1854 she discovered a comet. She has also left valuable studies on the probable influence of the moon on earthquakes — studies which brought her distinction from several of the learned societies of Europe. In 1872 the Italian government decreed her a gold medal for her statistical labours in science. Since her death her countrymen have recognised the value of her contributions to science by erecting a statue to her memory.

Another woman who has won enduring fame in the annals of astronomy is Miss Dorothea Klumpke, of San Francisco. While yet quite young, she and her sisters were taken to Europe to be educated. There she soon became proficient in a number of languages, and then devoted herself to the study of mathematics and astronomy. After securing her baccalaureate and licentiate in Paris, she applied for admission as a student to the Paris observatory. "The directors of the observatory consulted the statutes. No woman had hitherto proposed herself as a colleague, but there was no rule opposing it. They themselves approved, and gave her a telescope to make her own observations. After a time she completed the work begun by Mme Kovalevsky on the rings of Saturn, which she made the subject of her thesis, and, when she had become Doctor of Science, she was given a decoration by the Institute and made an *Officier de l'Academie.*"

After Miss Klumpke had brilliantly defended her thesis in the Sorbonne, M. Darboux, the president of the jury, complimented the young American doctor on her splendid work and concluded a notable address in her honour in the following laudatory words:

> The great names of Galileo, of Huyghens, of Cassini, of Laplace, without speaking of those of my illustrious colleagues and friends, are attached to the history of every serious step forward made in this attractive and difficult theory of Saturn's rings. Your work constitutes another valuable contribution to the same subject and places you in an honourable rank beside those women who have consecrated themselves to the study of mathematics. In the last century Maria Agnesi gave us a treatise on the differential and integral calculus. Since then Sophie Germain, as remarkable for her literary and philosophical talent as for her faculty for mathematics, won the esteem of the great geometricians who honoured our country at the commencement of this century. It is but a few years since the Academy awarded one of its most beautiful prizes which will place the name of Mme Kovalevsky beside those of Euler and Lagrange in the history of discoveries relative to the theory of the movement of a solid body about a fixed point..... And you, mademoiselle, your thesis is the first which a woman has presented and successfully defended before our faculty for the degree of doctor in mathematics. You worthily open the way, and the faculty unanimously makes haste to declare you worthy of obtaining the degree of doctor.

Besides her thesis just referred to, Miss Klumpke is the author of numerous communications to scientific journals and learned societies regarding her researches on the spectra of stars and meteorites and other allied subjects. For many years she was at the head of the bureau in the Paris Observatory for measuring the photographic plates that are to be used in the large catalogue of stars and map of the heavens which are to constitute the crowning achievements of the International Astronomical Congress. She was the first woman to be elected a member of the Astronomical Society of France, and the character of her work as an observer as well as a computer has given her an enviable position among the astronomers of the world.[17]

In America another woman has won renown among astronomers by successfully executing the same kind of work as was entrusted to Miss Dorothea Klumpke in Paris. For many years Mrs W Fleming, with her large corps of women assistants, had charge of the immense collection of astronomical photographs in the Observatory of Harvard University. To her and her staff were assigned the reductions and measurements of the photographic and photometric work done in Cambridge and Arequipa, Peru. She was singularly successful in her studies of photographic plates and made many discoveries which astronomers regard of the greatest importance. By such studies she and her assistants detected many new nebulae, double and variable stars, besides spectra of different types and of rare interest. In addition to this they examined and classified tens of thousands of photographs of stellar spectra, a labour which involved countless details of reduction and measurements of exceeding delicacy and skill.

A complete list of the women who, during the past half century, have devoted themselves to the study of astronomy and who have contributed to its advancement by their observations and writings would be a very long one. Among those, however, whose labours have attracted special notice, mention must be made of the Misses Antonia C Maury, Florence Cushman, Louisa D Wells, Mabel C Stephens, Eva F Leland, Anna Winlock, Annie J Cannon and Henrietta S Leavitt, all of whom are on the staff of the Harvard Observatory.

Then, too, there are many women who occupy important positions as professors or assistant professors in our colleges and universities. Chief among these in the United States are Sarah F Whiting, of Wellesley; Mary W Whitney, of Vassar; Mary E Boyd, of Smith; Susan Cunningham, of Swarthmore, and Annie S. Young, of

Mt Holyoke. Nor must we forget such able computers as Mrs Margaretta Palmer, of Yale, and Miss Hanna Mace, the clever assistant of the late Simon Newcomb in the Naval Observatory in Washington.

In the Old World among the women who, during the last few decades, have materially contributed to the progress of astronomy, either as observers and computers or as writers, are Miss Alice Everett, who has done splendid work in the observatories of Greenwich and Potsdam, Misses M A Orr, Mary Ashley, Alice Brown, Mary Proctor — daughter of the late astronomer, R A Proctor — Agnes M. and Ellen M Clerke, and Lady Huggins, of England; Mmes. Jansen, Faye, and Flammarion, in France; the Countess Bobinski, in Russia; and Miss Pogson, in the Observatory of Madras, India.

In conclusion, it is but just to observe that women's work in astronomy has by no means been confined to their contributions as observers, writers and computers. Reference must also be made to the financial aid which they have given to various observatories and learned societies for the furtherance of astronomical research both in the New and the Old World. It must suffice here to recall the endowment at Harvard University of the Henry Draper Memorial, by Mrs Henry Draper, in order that the work of photographing stellar spectra, which occupied her husband's later years, might be continued under the most favourable auspices, and the munificent sum of fifty thousand dollars given by Miss C Bruce, of New York, for the construction of a large telescope especially designed for photographing faint stars and nebulae. The photographs taken with this instrument will be used in the preparation of the great chart of the heavens which is to be the joint production of the chief observatories of the world.

## Notes & References

1) To a friend, who expressed surprise at her fluttering to and fro between mathematics and literature, she made a reply which deserves a place here, as it gives a better idea than anything else of the wonderful versatility of this gifted daughter of Russia. "I

understand," she writes, "your surprise at my being able to busy myself simultaneously with literature and mathematics. Many who have never had an opportunity of knowing any more about mathematics confound it with arithmetic, and consider it an arid science. In reality, however, it is a science which requires a great amount of imagination, and one of the leading mathematicians of our century states the case quite correctly when he says that it is impossible to be a mathematician without being a poet in soul. Only, of course, in order to comprehend the accuracy of this definition, one must renounce the ancient prejudice that a poet must invent something which does not exist, that imagination and invention are identical. It seems to me that the poet has only to perceive that which others do not perceive, to look deeper than others look. And the mathematician must do the same thing. As for myself, all my life I have been unable to decide for which I had the greater inclination, mathematics or literature. As soon as my brain grows wearied of purely abstract speculations it immediately begins to incline to observations on life, to narrative, and *vice versa*, everything in life begins to appear insignificant and uninteresting, and only the eternal, immutable laws of science attract me. It is very possible that I should have accomplished more in either of these lines, if I had devoted myself exclusively to it; nevertheless, I cannot give up either of them completely."

From Ellen Key's *Biography of the Duchess of Cajanello*, quoted in Anna Leffler's biography of Sónya Kovalevsky, ut sup, pp 317-18.

2) Cf the preceding chapter, p 140. See also *Histoire de l'Astronomie Ancienne*, Vol I, p 317, by M Delambre, Paris, 1817.

3) "Calor etiam solis in hieme maior est sub terra quam super terram, quod si tunc frigus tantum esset sub terra quam super terram, vel si in Aestate calor tantus esset sub terra quantus est super terram, de immoderatione ista terra tota scinderetur." *Hildegardis Causae et Curae*, p 7, Lipsiae, 1903.

4) *Commentaire de Theon d'Alexandrie*, p X, translated by the Abbe Halma, Paris, 1882.

5) "Enfin de leur faire connoistre qu'elles ne sont pas incapable de l'estude, si elles s'en vouloient donner la peine puisqu'entre le cerveau d'une femme et celui d'un homme il n'y a aucune difference." Cf *Journal de Savans*, Vol III, p 304, Amsterdam, 1687.

6) D'ou vient qu'elle a l'oeil trouble et le teint si terni?
C'est que sur le calcul, dit-on, de Cassini,
Un astrolabe à la main, elle a, dans la gouttiere,
A suivre Jupiter passe la nuit entiere.

7) "Celebre inter observatores hujus Aevi nomen adeptus est

Godfredus Kirchius, astronomus nuper regius in Societate Scienciarum Berlinensi; mense Julio A, 1710 mortuus. Ejus vidua, Maria Magdalena Winckelmannia, non minore in observando et calculo astronomico dexteritate pollet, ac in utroque laboure maritum, cum viveret, fideliter juvit..... quod laudi ducitur foeminae ea animo comprehendisse, quae sine ingenii vi studiique assiduitate non comprehenduntur," *Acta Eruditorum*, pp 78, 79, Lipsiae, 1712.

8) *Preface Historique* to *Principes Mathematiques de la Philosophie Naturelle* par feue Madame la Marquise du Chastellet, Vol I, p V, Paris, 1759.

9) *The Unpublished Correspondence of Madame du Deffand*, Vol I, pp 202-03, London, 1810.

10) Mme du Deffand's venomous letter, somewhat abridged, reads as follows:

Imagine a tall, hard and withered woman, narrow-chested, with large limbs, enormous feet, a very small head, a thin face, a pointed nose, two small sea-green eyes, her colour dark, her complexion florid, her mouth flat, her teeth set far apart and very much decayed; there is the figure of the beautiful Émilie, a figure with which she is so well pleased that she spares nothing for the sake of setting it off. Her manner of dressing her hair, her adornments, her top-knots, her jewellery, all are in profusion; but, as she wishes to be lovely in spite of nature, and as she wishes to appear magnificent in spite of fortune, she is obliged, in order to obtain superfluities, to go without necessaries such as under-garments and other trifles.

She was born with sufficient intellect, and the desire to appear as though she had a great deal made her prefer to study the most abstract sciences rather than more general and pleasant branches of knowledge. She thought she would gain a greater reputation by this peculiarity and a more decided superiority over other women. She did not limit herself to this ambition. She wished to be a princess as well, and she became so, not by the grace of God nor by that of the King, but by her own act. This absurdity went on like the others. One became accustomed to regard her as a princess of the theatre, and one almost forgot that she was a woman of rank.

Madame worked so hard to appear what she was not that no one knew what she really was. Even her faults were perhaps not natural. They may have had something to do with her pretensions, her want of respect with regard to the state of princess, her dullness in that of *savante*, and her stupidity in that of a *jolie femme*.

However much of a celebrity Mme du Chatelet may be, she would not be satisfied if she were not celebrated, and that is what she desired in becoming the friend of M. de Voltaire. To him she owes the *eclat* of her life, and it is to him that she will owe immortality. See *Lettres de la Marquise du Deffand à Horace Walpole*, Vol, pp 200-201, Paris, 1824.

As a contrast to this atrocious caricature, it is but due to the memory of Mme du Chatelet to give her portrait by Voltaire, to whom she was ever the beautiful, the charming Urania, the

Vaste et puissante genie,
Minerve de la France, immortelle Émilie."

It is contained in the following verses:

"L'esprit sublime et la delicatesse,
L'oubli charmante de sa propre beaute
L'amitie tendre et l'amour emporte
Sont les attraits de ma belle maitresse.

If the whole truth were known, it would, doubtless, be found somewhere between the above extreme and contradictory views, and the cause of the caustic statements of Mesdames de Staël and du Deffand would probably be found to be quite accurately expressed in the first part of Voltaire's *Epistle on Calumny*, which was written about the beginning of his particular relationship with "the divine Émilie." The first lines of this epistle, as translated by Smollett, are:

Since beautiful, 'twill be your fate,
Emelia, to incur much hate;
Almost one-half of human race
Will even curse you to your face;
Possesst of genius, noblest fire,
With fear you will each breast inspire;
As you too easily confide,
You'll often be betrayed, belied;
You ne'er of virtue made parade,
To hypocrites no court you've paid,
Therefore, of Calumny beware,
Foe to the virtuous and the fair.

11) In his work on *Comets*, Clairaut at first gave Mme Lepaute full credit for her work which had been of such inestimable service to himself; but, in order to gratify a woman who, having pretensions without knowledge, was very jealous of the superior attainments of Mme Lepaute, he had the weakness subsequently to suppress his generous tribute to merit. Commenting on this strange conduct of his assistant, Lalande expresses himself as follows: "We know that it is not rare to see ordinary women depreciate those who have knowledge, tax them with pedantry and contest their merit in order to avenge themselves upon them for their superiority. The latter are

so few in number that the others have almost succeeded in making them conceal their acquirements."

12) *Bibliographie Astronomique*, pp 676-87, par Jerome de la Lande, Paris, 1803.

13) *Memoirs and Correspondence of Caroline Herschel*, p 144, by Mrs John Herschel, London, 1879.

14) So sensitive was Miss Herschel in her old age regarding the reputation of her brother, William, who had always been her idol and the one in whom she had concentrated all her affection, that she came to look askance at every person and thing that seemed calculated to dull the glory of his achievements. Thus her niece, in writing to Sir John Herschel, after her death, declares: "She looked upon progress in science as so much detraction from her brother's fame; and, even your investigations would have become a source of estrangement had she been with you." In a letter to Sir John Herschel, written four years before her death, she exhibits, in an amusing fashion, her jealous spirit anent the great telescope of Lord Rosse. "They talk of nothing here at the clubs," she writes, "but of the great mirror and the great man who made it. I have but one answer for all — *Der Kerl ist ein Narr* — the fellow is a fool." Even "Every word said in her own praise seemed to be so much taken away from the honour due to her brother. She had lived so many years in companionship with a truly great man, and in the presence of the unfathomable depths of the starry heavens, that praise of herself seemed childish exaggeration." And notwithstanding the honour and recognition which she received from learned men and learned societies for her truly remarkable astronomical labours, her dominant idea was always the same — "I am nothing. I have done nothing. All I am, all I know, I owe to my brother. I am only a tool which he shaped to his use — a well-trained puppy-dog would have done as much." *Op cit*, pp ix, 335 and 346.

15) *Op cit*, p 224.

16) *Memoirs and Correspondence of Caroline Herschel*, ut sup, pp 226-27.

17) *Maria Mitchell, Life, Letters and Journals*, compiled by Phebe Mitchell Kendall, p 267 *et seq*, Boston, 1896.

# Women in Physics

PHYSICS, BEING ONE of the inductive sciences, received little attention until modern times. True, the Greeks were familiar with some of the fundamental facts of the mechanics of solids and fluids, and had some notions respecting the various physical forces; but their knowledge of what until recently was known as natural philosophy was extremely limited. Aristotle, Pythagoras and Archimedes were among the most successful investigators of their time respecting the laws and properties of matter, and contributed materially to the advancement of knowledge regarding the phenomena of the material universe; but the sum total of their information of what we now know as physics could be embodied in a few pages.

In view of the foregoing facts, we should not expect to find women engaged in the study, much less in the teaching, of physical science during ancient times. And yet, if we are to credit Boccaccio, who bases his statements on those of early Greek writers, there was at least one woman that won distinction by her knowledge of natural philosophy as early as the days of Socrates. In his work, *De Laudibus Mulierum,* which treats of the achievements of some of the illustrious representatives of the gentler sex, the genial author of the *Decameron* gives special praise to one Arete of Cyrene for the breadth and variety of her attainments. She was the daughter of Aristippus, the founder of the Cyrenaic school of philosophy, and is represented as being a veritable prodigy of learning. For among her many claims to distinction she is said to have publicly taught natural and moral philosophy in the schools and academies of Attica for thirty-five years, to have written forty books, and to have counted among her pupils one hundred and ten philosophers. She was so highly esteemed by her countrymen that they inscribed on her tomb an epitaph which declared that she was the splendour of Greece and possessed the beauty of Helen, the virtue of Thirma, the pen of Aristippus, the soul of Socrates, and the tongue of Homer.[1]

This is high praise, indeed, but, when we recollect that Arete lived during the golden age of Greek learning and culture, that she had exceptional opportunities of acquiring knowledge in every department of intellectual effort; when we recall the large number of women who, in their time, distinguished themselves by their learning and accomplishment, and reflect on the advantages they enjoyed as pupils of the ablest teachers of the Lyceum, the Portico, and the Academy; when we remember further that they lived in an atmosphere of intelligence such as has since been unknown; when we call to mind the signal success that rewarded the pursuit of knowledge by the scores of women mentioned by Athenaeus and other Greek writers; when we peruse the fragmentary notices of their achievements as recorded in the pages of more recent investigators regarding the educational facilities of a certain class of women living in Athens and the eminence which they attained in science, philosophy and literature, we can realise that the character and amount of Arete's work as an author and as a teacher have not been overestimated.

Living in an age of prodigious mental activity, when women, as well as men, were actuated by an abiding love of knowledge for its own sake, there is nothing surprising in finding a woman like Arete commanding the admiration of her countrymen by her learning and eloquence. For was not the learned and eloquent Aspasia her contemporary? And did not Theano, the wife of Pythagoras, take charge of her husband's school after his death; and does not antiquity credit her with being not only a successful teacher of philosophy, but also a writer of books of recognised value? Such being the case, what is there incredible in the statements made by ancient writers regarding the literary activity of Arete, and about her eminence as a teacher of science and philosophy? She was but one of many of the Greek women of her age that won renown by their gifts of intellect and by their contributions to the educational work of their time and country.

Better known than Arete, but probably not superior to her as a teacher or writer, was the illustrious Hypatia of Alexandria. She, too, like her distinguished predecessor in Athens, was an instructor in natural philosophy, as well as other branches of science. Of her we know more than we do of the daughter of Aristippus, but even our knowledge of the acquisitions and achievements of Hypatia is, unfortunately, extremely meagre. We do, however, know from the historian, Socrates, and from Synesius, bishop of Ptolemais, who was her pupil, that she was one of the most richly dowered women of all time. Born and educated in Alexandria when its schools and scholars were the most celebrated in the world, she was even at an early age regarded as a marvel of learning. For, not satisfied with excelling her father, Theon, in mathematics, of which he was a distinguished professor, she, as Suidas informs us, devoted herself to the study of philosophy with such success that she was soon regarded as the ablest living exponent of the doctrines of Plato and Aristotle. "Her knowledge," writes the historian, Socrates, "was so great that she far surpassed all the philosophers of her time. And succeeding Plotinus, in the Platonic school which he had founded in the city of Alexandria, she taught all the branches of philosophy with such signal success that students flocked to her in crowds from all parts."[2] Her home, as well as her lecture room, was the resort of the most noted scholars of the day, and was, with the exception of the Library and the Museum, the most frequented intellectual centre of the great city of learning and culture. Small wonder, then, that her contemporaries lauded her as an oracle and as the most brilliant luminary in Alexandria's splendid galaxy of thinkers and scholars — *sapientis artis sidus integerrimum.*

Among the many inventions attributed to Hypatia, besides the planisphere and astrolabe which she designed for the use of astronomers, are several employed in the study of natural philosophy. Probably the most useful of these is an areometer mentioned by her pupil Synesius. He calls it a hydroscope and describes it as having the form and size of a flute, and graduated in such wise that it can be used for determining the density of liquids. That Hypatia was thoroughly familiar with the science of natural philosophy, as then known, there can be no doubt. That she also contributed materially to its advancement, as well as to that of astronomy, in which she always exhibited a special interest, there is every reason to believe.[3]

After the death of Hypatia, the study of natural philosophy was almost entirely neglected for more than a thousand years. The first woman in modern times to attract attention by her discussion of physical problems was the famous Marquise du Chatelet, although she was better known as a mathematician and as the translator into the French of Newton's *Principia*. In her chateau at Cirey she had a well-equipped physical cabinet in which she took special delight. But in her time, as in that of Hypatia, natural philosophy was far from being the broad experimental science which it has become through the marvellous discoveries made in heat, light, electricity and magnetism during the last hundred years, as well as through those countless brilliant investigations which have led up to our present doctrine of the correlation and conservation of the various physical forces. There was then no occasion for those delicate instruments of precision which are now found in every physical laboratory by means of which the man of science is able to investigate phenomena and determine laws that were quite unknown until a few years ago.

In the time of Mme du Chatelet, as during the century following, natural philosophy consisted rather in the mechanical and mathematical than in the physical study of nature. This is illustrated by the title of the great work on the translation of which she spent the best years of her life — Newton's immortal *Philosophiae Naturalis Principia Mathematica*.

The Marquise's first scientific work was an investigation regarding the nature of fire. The French Academy of Sciences had offered a prize for the best memoir on the subject. Among the contestants for the coveted honour were the chatelaine of Cirey and the celebrated Swiss mathematician, Leonard Euler. The Marquise was unsuccessful in the contest, but her paper was of such value that the eminent physicist and astronomer, Arago, was able to characterise it as an "elegant piece of work, embracing all the facts relating to the subject then known to science and containing among the experiments suggested one which proved so fecund in the hands of Herschel." In this remarkable *Memoire sur le Feu,* which is printed in the *Collections* of the Academy, the Marquise anticipates the results of subsequent researches of others by maintaining that both heat and light have the same cause, or, as we should now say, are both modes of motion.

The second book written by this remarkable woman is entitled *Institutions de Physique,* and was dedicated to her son, for whose benefit it was primarily written. It deals specially with the philosophy of Leibnitz and discusses such questions as force, time and space. Her views respecting the nature of the force called *vis viva,* which was much discussed in her time, are of particular interest, as they are not only opposed to those which were held by Descartes and Newton, but also because they are in essential accord with those now accepted in the world of science.

All things considered, the Marquise du Chatelet deservedly takes high rank in the history of mathematical physics. In this department of science she has had few, if any, superiors among her own sex. And, when we recollect that she laboured while the foundations of dynamics were still being laid, we shall more readily appreciate the difficulties she had to contend with and the distinct service which her researches and writings rendered to the cause of natural philosophy among her contemporaries.

The first woman to occupy a chair of physics in a university was the famous daughter of Italy, Laura Maria Catarina Bassi. She was born in Bologna in 1711 — but five years after the birth of Madame du Chatelet — and from her most tender years she exhibited an exceptional facility for the acquisition of knowledge.

After she had, through the assistance of excellent masters, become proficient in French and Latin, she took up the study of logic, metaphysics and natural philosophy. In all these branches of learning her progress was so rapid that it far exceeded the fondest expectations of her parents and teachers. Thanks to a wonderful memory and a highly developed reasoning faculty, she was able, while still a young maiden, to prove herself the possessor of knowledge that is ordinarily obtained only in the maturity of age and after long years of systematic study.

When she had attained the twenty-first year of her age she was induced by her family and friends — much against her own inclination, however — to take part in a public disputation on philosophy. Her entering the lists against some of the most distinguished scholars of the time was made the occasion for an unusual demonstration in her honour. The hall of the university in which such intellectual jousts were generally held was too small for the multitude that was eager to witness the young girl's formal appearance among the scholars and the notables of the old university city. It was, accordingly, arranged that the disputation should be held in the great hall of the public Palace of the Senators.

Among the vast assemblage present at the disputation were Cardinal Grimaldi, the papal legate; Cardinal Archbishop Lambertini, afterwards Pope Benedict XIV; the gonfalonier, senators, literati from far and near, leading members of the nobility and representatives of all the religious orders.

When the argumentation began the young girl found herself pitted against five of the most distinguished scholars of Bologna. But she was fully equal to the occasion and passed the ordeal to which she was subjected in a manner that excited the admiration and won the plaudits of all present. Cardinal Lambertini was so impressed with the brilliant defence which she had made against the five trained dialecticians and the evidence she gave of varied and profound learning that he paid her a special visit the next day in her own home to renew his congratulations on her signal triumph and to encourage her to continue the prosecution of her studies.

In less than a month after this interesting event Laura Bassi, in response to the expressed desire of the whole of Bologna, presented herself as a candidate for the doctorate in philosophy. This was the occasion for a still more brilliant and imposing ceremony. It was held in the spacious Hall of Hercules in the Communal Palace, which was magnificently decorated for the splendid function. In addition to the distinguished personages who had been spectators of the fair student's triumph a few weeks before, there was present in the vast audience the noted French ecclesiastic, Cardinal Polignac, who was on his way from Rome to France.

The heroine of the hour, dressed in a black gown, was ushered into the great hall, preceded by two college beadles and accompanied by two of the most prominent ladies of the Bolognese nobility. She was given a seat between the chancellor and the prior of the university, who, in turn, were flanked by the professors and officials of the institution.

After the usual preliminaries of the function were over the prior of the university, Doctor Bazzani, rose and pronounced an eloquent discourse in Latin to which Laura made a suitable response in the same language. She was then crowned with a laurel wreath exquisitely wrought in silver, and had thrown round her the *vajo*, or university gown, both symbols of the doctorate. After this the young doctor proceeded to where the three cardinals were seated, and in delicately chosen words, also in Latin, expressed to them her thanks for the honour of their presence. All then withdrew to the apartments of the gonfalonier, where refreshments were served in sumptuous style, after which the young *Laureata*, accompanied by a numerous cortege and applauded by the entire city, was escorted to her home.

So profound was the impression made on the university senate by the deep erudition of Laura Bassi that it was eager to secure her services in its teaching body. But, before she could be offered a chair in the institution, long-established custom required that she should pass a public examination on the subject matter which she was to teach. Five examiners were chosen by lot, and all of them proved to be men whose names, says Fantuzzi, "will always be held by our university in glorious remembrance." They had all to promise under oath that the candidate for the chair should have no knowledge before the examination of the questions which were to be asked, and that the test of the aspirant's qualifications to fill the position sought should be absolutely free from any suspicion of favouritism or partiality.

Notwithstanding the difficulties she had to confront, Laura acquitted herself with even greater credit than on former occasions of a similar character. There was no question in the mind of any one present at the examination of the candidate's ability to fill the chair of physics, and it was, accordingly, offered to her by acclamation.

The first public lecture of the gifted young *dottoressa* was made the occasion of a demonstration such as the old walls of the university had rarely witnessed. Her lecture room was thronged by the elite of the city, as well as by a large class of enthusiastic students. All were charmed by her eloquence and amazed at the complete mastery she evinced of the subject she had selected for discussion. From that day forth her reputation as a scholar and a teacher was established, and her lectures were attended by appreciative students from all parts of Europe. She was especially popular with the students from Greece, Germany and Poland, and her popularity, far from waning, waxed greater with the passing years.

At the time of Laura's entering upon her professional career the senate of Bologna had a medal coined in her honour, on the obverse of which was her name and effigy, while on the reverse there was an image of Minerva, with the inscription, *Soli cui fas vidisse Minervam.*

Far from interrupting her studies, which had hitherto been the joy of her life, Laura's university work gave new zest to the literary and scientific pursuits which had always such a fascination for her. Among the subjects that specially engaged her attention were studies so diverse as Greek and the higher mathematics. She was particularly interested in the great physico-mathematical work of Newton, and did not rest until she had thoroughly mastered the contents of his epoch-making *Principia.*

A few years after she had become a member of the university faculty Laura was a European celebrity, and no one eminent by learning or birth passed through Bologna without availing himself of the opportunity of making the acquaintance of so extraordinary a woman. Men of science and letters vied with princes and emperors in doing honour to one who was looked upon by many as being, like Arete of old, endowed with a soul and a genius far above that of ordinary mortals, and as being the possessor of a talent that indicated something superhuman.

Laura Bassi was in constant correspondence with the most celebrated scholars of Europe, and more especially with those who had attained eminence in her special line of work. Among the letters received from her illustrious correspondents were two from Voltaire. They were written shortly after the author had been refused admittance into the French academy. He then bethought himself of securing membership in the Academy of Sciences of Bologna. This, he reasoned, would be a splendid tribute to the versatility of his genius and would, at the same time, be a biting satire on the demigods of French literature who had dared to exclude him from their society.

That he might not meet the same refusal on the part of the Academy of Bologna as he had experienced in Paris, Voltaire determined not to rely entirely on the good will of the male members of the Bolognese academy. He accordingly resolved to enlist the services of Laura Bassi, who was one of the leading members of this distinguished body, and trust to her influence in his behalf on the hearts of her colleagues.

The first letter, written in Italian, is so characteristic of the writer that it will bear reproduction.

"Most Illustrious Lady," he writes from Paris, the 23d of November, 1744, "I have been wishing to journey to Bologna in order to be able one day to tell my countrymen I have seen Signora Bassi; but, being deprived of this honour, let it at least be permitted me to place at your feet this philosophic homage and to salute the honour of her age and of women. There is not a Bassi in London, and I should be more happy to be a member of the Academy of Bologna than of that of the English, although it has produced a Newton. If your protection should obtain for me this title, of which I am so ambitious, the gratitude of my heart will be equal to my admiration for yourself. I beg you to excuse the style of a foreigner who presumes to write you in Italian, but who is as great an admirer of yours as if he were born in Bologna."

The second letter of Voltaire is in response to one received from Laura Bassi announcing that he had been elected to membership in the Bologna Academy. The first sentence of it suffices to indicate its tenor. "Nothing," he writes, "was ever more grateful to me than to receive from your hand the first advice that I had the honour, by means of your favour, of being united by this new link to one who had already bound me to her car by all the chains of esteem and admiration."[4]

Like so many of her gifted sisters of sunny Italy, Laura was in every way "a perfect woman nobly planned." Of a deeply religious nature, she was as pious as she was intelligent, and was throughout her life the devoted friend of the poor and the afflicted. The mother of twelve children, she never permitted her scientific and literary work to conflict with her domestic duties or to detract in the least from the singular affection which so closely united her to her husband and children. She was as much at home with the needle and the spindle as she was with her books and the apparatus of her laboratory. And she was equally admirable whether superintending her household, looking after her children, entertaining the great and the learned of the world, or in holding the rapt attention of her students in the lecture room. She was, indeed, a living proof that higher education is not incompatible with woman's natural avocations; and that cerebral development does not lead to race suicide and all the other dire results attributed to it by a certain class of our modern sociologists and anti-feminists.

Considering her manifold duties as a professor in the university and the mother of a large family, it was scarcely to be expected that Laura Bassi would have much time for writing for the press. She was, however, able to devote some of her leisure moments to the cultivation of the Muses, of whom, Fantuzzi informs us, she was a favourite. Her verses, as well as her contributions to the science of physics, are scattered through various publications, but they suffice to show that the accounts of her transmitted to us by her contemporaries were not exaggerated.[5]

A learned French traveller who visited Laura in Bologna describes her as having a face that was sweet, serious and modest. Her eyes were dark and sparkling, and she was blessed with a powerful memory, a solid judgement, and a ready imagination. "She conversed fluently with me in Latin for an hour with grace and precision. She is very proficient in metaphysics; but she prefers modern physics, particularly that of Newton."

How many of our college women of today could readily carry on a conversation in Latin, if this were the sole medium of communication, or discuss the philosophy of Plato and Aristotle in the tongue of Cicero, or give public lectures on the physico-mathematical discoveries of Descartes and Newton in what was the universal language of the learned world, even less than a century ago?

It must not, however, be inferred from the foregoing statements regarding the great intellectual capacity of Laura Bassi or the enthusiastic demonstrations that were so frequently made in her honour that she was unique in this respect among her countrywomen. Special attention has been called to her as a type of the large number of her sex who, by their learning and culture, graced the courts and honoured the universities of her country for full ten centuries. Scarcely had death removed Laura Bassi from a career in which for twenty-eight years she had won the plaudits of the whole of Europe, when the University of Bologna welcomed to its learned halls two other women who, in their respective lines of research, were fully as eminent as their departed countrywoman. These were Maria dalle Donne, for whom Napoleon established a chair of obstetrics, and Clotilda Tambroni, the famous professor of Greek, of whom a noted Hellenist declared, "Only three persons in Europe are able to write Greek as well as she does, and not more than fifteen are able to understand her."

Burckhardt, in his thoughtful work on the culture of the Italian Renaissance, has a paragraph which expresses, in a few words, what was always the attitude of the Italian father toward the education of his daughter.

"The education of the woman of the upper class was absolutely the same as that of the man. The Italian of the Renaissance did not for a moment hesitate to give his son and daughter the same literary and philosophical training. He considered the knowledge of the works of antiquity life's greatest good, and he could not, therefore, deny to woman participation in such knowledge. Hence the perfection attained by the daughters of noble families in writing and speaking Latin."[6]

This attitude of the members of the nobility toward the education of their daughters was essentially the same as that of the universities of Italy toward women who had a thirst for knowledge. For from the dawn of learning in Salerno to the present there never was a time when women were not as cordially welcomed to the universities as students and professors as were the men; and never a time when the merit of intellectual work was not determined without regard to sex.

In Bologna, where were passed the sixty-seven years of her mortal life, the name of Laura Bassi, like that of her illustrious colleague, Luigi Galvani, is one to conjure with, and a name that is still pronounced with respect and reverence. Her last resting place is in the Church of Corpus Domini, the same sacred shrine in which were deposited all that was mortal of the renowned discoverer of galvanic electricity.[7]

Two years after Signora Bassi was gathered to her fathers there was born near Edinburgh to a Scotch admiral, Sir William George Fairfax, an infant daughter who was destined to shed as much lustre on her sex in the British Isles as the incomparable Laura Bassi had diffused on womankind in Italy during her brilliant career in "Bologna, the learned." She is known in the annals of science as Mary Somerville, and was in every way a worthy successor of her famous sister in Italy, both as a woman and as a votary of science.

Although her chief title to fame is her notable work in

mathematical astronomy, especially her translation of Laplace's *Mechanique Celeste*, she is likewise to be accorded a prominent place among scientific investigators for her contributions to physics and cognate branches of knowledge. Chief among these are her works on the *Connection of the Physical Sciences* and *Physical Geography*. As to the last production, no less an authority than Alexander von Humboldt pronounced it an exact and admirable treatise, and wrote of it as "that excellent work which has charmed and instructed me since its first appearance."

In a letter from the illustrious German savant to the gifted authoress of the two last-named volumes occurs the following paragraph: "To the great superiority you possess and which has so nobly illustrated your name on the high regions of mathematical analysis, you add, Madam, a variety of information in all parts of physics and descriptive natural history. After the *Mechanism of the Heavens*, the philosophical *Connection of the Physical Sciences* has been the object of my profound admiration.... The author of the vast *Cosmos* should more than any one else salute the *Physical Geography* of Mary Somerville..... I know of no work on physical geography in any language that can compare with yours."

Among the other works by Mrs Somerville, treating of physical subjects or of subjects intimately related to physics are *The Form and Rotation of the Earth, The Tides of the Ocean and Atmosphere,* and an abstruse investigation *On Molecular and Microscopic Science*. The last volume was published in 1869, when its author was near her ninetieth year, and bore as its motto St Augustine's sublime words: *Deus magnus in magnis, maximus in minimis* — God is great in great things, greatest in the least.

After Mrs Somerville's death, in 1872, at the advanced age of ninety-two, the number of women who devoted themselves to the study and teaching of physics was greatly augmented. The brilliant success of Laura Bassi and Mary Somerville had not been without results, and their notable achievements as authors and teachers had the effect of stimulating women everywhere to emulate their example, and encouraging them to devote more attention to a branch of science which, until then, had been regarded by the general public as beyond the sphere and capacity of what was assumed to be the intellectually weaker sex.

One of the most eminent scientific women of the present day in England is Mrs Ayrton, the wife of the late Professor W E Ayrton, the well-known electrician. Her chosen field of research, like that of her husband, has been electricity, in which she has achieved marked distinction. Her investigations on the electric arc and on the sand ripples of the seashore won for her the first medal ever awarded to a woman by the Royal Society. When, however, in 1902, she was formally nominated for fellowship in this same society, she failed of election because the council of the society discovered that "it had no legal power to elect a married woman to this distinction."

How different it was in the case of Laura Bassi, who was an active member of all the leading scientific and literary societies of Italy, where from time immemorial women have been as cordially welcomed to membership in its learned societies as to the chairs of its great universities.

The list of the women who in Europe and America are now engaged in physical research and in teaching physics in schools and colleges is a long one, and the work accomplished by them is, in many cases, of a high order of merit. It is only, indeed, during the present generation that such work has been made generally accessible to them; and, considering the success which has already attended their efforts in this branch of science, we have every reason to believe that the future will bring forth many others of their sex who will take rank with such intellectual luminaries as Hypatia, Mme du Chatelet, Laura Bassi and Mary Somerville.

## Notes & References

1) Miss Klumpke, the reader may be interested in knowing, belongs to a singularly gifted family. Her sister, Augusta, is a distinguished physician and an authority on nervous diseases. Hers is the glory to be the first woman permitted, after an exceptionally severe examination, to serve as *interne* in the Paris hospitals. Julia, her youngest sister, who achieved distinction as a violinist with Ysaye,

was one of the first to pass the examination required of women entering the Paris *Lycees*, while Anna, the eldest, has won fame as an artist, and as the friend, heiress and executrix of France's famous daughter, Rosa Bonheur.

2) "Publice philosophiam naturalem et moralem in scholis Academiisque Atticis docuit haec foemina annis XXXV, libros composuit XL, discipulos habuit philosophos CX, obiit anno Aetatis LXXVII, cui tale Athenienses statuere epitaphium:

Nobilis hic Arete dormit, lux Helladis, ore
Tyndaris at tibi par, Icarioti, fide.
Patris Aristippi calamumque animamque dederunt,
Socratis huic linguam Maeonidaeque Dii.

— Boccaccio, *De Laudibus Mulierum*, Lib II.

Cf Wolf's *Mulierum Graecarum quae Oratione Prosa Usae Sunt Fragmenta et Elogia*, pp 283 *et seq*, London, 1739.

3) "Mulier quaedam fuit Alexandriae, nomine Hypatia, Theonis filia. Haec ad tantam eruditionem pervenerat ut omnes sui temporis philosophos longo intervallo superaret, et in Platonicam scholam a Plotino deductam succederet, cunctasque philosophiae disciplinas auditoribus exponeret. Quocirca omnes philosophiae studiosi ad illam undique confluebant." *Socrates, Historiae Ecclesiasticae*, Lib VII, Cap 15.

4) For extracts from the ancient authors regarding Hypatia, as well as for the extant letters to her from her friend and pupil, Synesius, the reader is referred to Wolf's erudite *Mulierum Graecarum quae Oratione Prosa Usae sunt Fragmenta et Elogia*, pp 72-91, ut sup.

5) Ernesto Masi, *Studi e Ritratti*, p 166 *et seq*, Bologna, 1881.

6) Two of her Latin dissertations on certain physical problems were published in the *Commentaries of the Bologna Institute*. One of them is entitled *De Problemate quodam Mechanico*; the other *De Problemate quodam Hydrometrico*. Many of her lectures on physics still exist in manuscript, and it is to be hoped that at least the titles of them may be given in a biography of the learned author which has been long desired and long promised.

7) *Die Cultur der Renaissance in Italien*, Vol I, p 363, 1869.

# Women in Chemistry

THE FIRST WOMAN deserving special mention in the history of chemistry is the wife of the immortal Lavoisier, the most famous of the founders of modern chemical science. While yet in her teens, this remarkable woman gave evidence of exceptional intelligence and will power. She was thoroughly devoted to her husband, and had the greatest admiration for his genius. Her highest ambition was to prove herself worthy of him and to render herself competent to assist him in those investigations that have given him such imperishable renown. With this end in view, she learned Latin and English, and she thus became an accomplished translator from these languages of any chemical works which might aid her spouse in his epoch-making researches. It was she who translated for him the chemical memoirs of Cavendish, Henry, Kirwan, Priestly and other noted English scientific investigators.

Arthur Young, well known in his day as a traveller and author, who in 1787 made the acquaintance of Madame Lavoisier, describes her as a woman full of animation, good sense and knowledge. In referring to a breakfast she had given him, he declares that "unquestionably the best part of the repast was her conversation on Kirwan's *Essay on Phlogiston,* which she was then translating, and on other subjects which a woman of sense, working in the laboratory of her husband, knows so well how to make interesting."

She was an ardent co-worker with her husband in his laboratory and materially aided him in his labours. Under his direction she wrote the results of the experiments that were made, as is evidenced by the records of his work. As a pupil of the illustrious painter,

David, she was naturally skilful in drawing. Besides this, she was a good engraver, and it is to her that are due the illustrations in Lavoisier's great *Traite de Chimie*, which contributed so much toward revolutionising the science of chemistry. It was, indeed, the first work that deserved to be regarded as a textbook of modern chemistry. Among her drawings are two of special interest. They represent her as seated at a table in the laboratory, taking notes, while her husband and his assistant, Seguin, are making an experiment on the phenomena of respiration.[1]

All Mme Lavoisier's writings testify to her great admiration of the genius of her husband. Intimately associated with him in his work, she combated for the triumph of his ideas and sought to make converts to them. One of her most notable converts was the Swiss chemist, de Saussure. "You have, Madame," he writes her, "triumphed over my doubts, at least in the matter of phlogiston, which is the principal object of the interesting work of which you have done me the honour of sending me a copy."

After Lavoisier's tragic death on the guillotine, it was his devoted wife who edited his *Memoirs on Chemistry*, of which Lavoisier had himself projected the publication. The two volumes constituting this work were not for sale, but were gratuitously distributed by the bereaved widow among the most eminent scientific men of the epoch. Cuvier, in acknowledging the receipt of these precious memoirs, declares: "All the friends of science are under obligations to you for your sorrowful determination to publish this collection of papers and to publish them as they were written — a melancholy monument of your loss and theirs — a loss which humanity will feel for centuries."

To realise the importance of the work in which Mme Lavoisier participated, it suffices to recall the fact that her husband, as one of the creators of modern chemistry, was the first to demonstrate the existence of the law of the conservation of matter, which declares that in all chemical changes nothing is lost and nothing is created. The co-discoverer with Scheele and Priestly of oxygen, he was the first one to exhibit the role of this important element in the phenomena of combustion and respiration and the first, also, to lay the foundations of a chemical nomenclature. We are not, then, surprised to learn that Mme Lavoisier's salon, even long after her lamented husband's death, was frequented by the most eminent savants of the time. For here were gathered such scientific luminaries as Cuvier, Laplace, Arago, Lagrange, Prony, Berthollet, Delambre, Biot, Humboldt, and others scarcely less brilliant.

After the conclusion of Mme Lavoisier's work in the laboratory of her husband, little was accomplished by women in chemistry for more than half a century. The reason was simple. Chemistry was not a part of the curriculum of studies for girls either in Europe or America. Even "during the sixties," writes a teacher of one of the prominent female seminaries of the United States, "the study of chemistry was mostly confined to the textbook, supplemented once a year by a course of lectures from an itinerant expert, who with his tanks of various gases produced highly spectacular effects."

When one recollects that the first institution in America — Vassar — for the higher education of women was not opened until 1865, one will understand that there were previously to this date few opportunities for women to study either chemistry or any of the other sciences.

The first scientific institution to open its doors to women was the Massachusetts Institute of Technology. This was on May 11, 1876, when the governing board of the institute decided that "hereafter special students in chemistry shall be admitted without regard to sex." In less than a year after this event every department of this institution was open to women, and any one who could pass the requisite examination was admitted as a student.

Five years, however, before women were formally admitted to the courses of chemistry an energetic young graduate from Vassar, eager to devote her life to the pursuit of science, had, as an exceptional favour, been allowed to enter the Institute as a special student in chemistry. As she was the first woman in the United States to enter a strictly professional scientific school, her entrance marks the beginning of a new epoch in the history of female education. The name of this ardent votary of science was Miss Ellen Swallow, better known to the world as Mrs Ellen H Richards.

Mrs Richards had not devoted herself long to the study of her favourite science before she resolved to apply the knowledge thus gained to the problems of daily life. She saw, among other things, the necessity of a complete reform in domestic economy, and resolutely set to work to have her views adopted and put in practice. She was, in consequence, one of the first leaders of the crusade in behalf of pure food, and her lectures and books on this all-important subject contributed greatly toward the diffusion of exact knowledge respecting the dangers lurking in unwholesome food.

She was likewise one of the first to apply the science of chemistry to an exhaustive study of the science of nutrition — to the study of food and the proper preparation of food materials. In this she was eminently successful, and was able to achieve for home economics what the illustrious Liebig had many years before accomplished for agricultural chemistry — put it on a firm and lasting basis. To her the kitchen was the centre and source of political economy.

The facts of science, indeed, were to Mrs Richards more than mere uncorrelated facts. They are potential agencies of service, and their chief value consists in their enabling us to control our environment in such wise as to secure the maximum of physical well being. Hence her constant insistence on personal cleanliness, on the cleanliness of food, of the house we live in, and, above all, of the kitchen. Hence, also, her preaching, in season and out of season, on the necessity of pure air, pure water and abundance of vitalising sunshine.

We cannot, then, wonder that sanitary chemistry eventually became the life work of Mrs Richards, and that, when the course of sanitary engineering was inaugurated in the Institute of Technology — the first course of its kind in the world — she became an important agent in its development and contributed immensely to its popularity and prestige.

She held the position of instructor of sanitary chemistry in the institute for twenty-seven years. During this time she trained a large number of young men in her chosen specialty, and these, after graduating, engaged in similar work in various parts of the New and the Old World.

The branch of sanitary chemistry to which Mrs Richards devoted most attention was air, water and sewage analysis. In this she was a recognised expert, and her advice and services were sought in all parts of the country. During the last three years of her life she acted, according to her own testimony, as general sanitary adviser to no fewer than two score corporations and schools. In addition to this she was also during this brief period consulted on the subject of foods by nearly two hundred educational and other institutions.

What, however, constituted the greatest contribution of Mrs Richards to the public health was the part she took in the great sanitary survey of the waters of the State of Massachusetts. During this long and laborious investigation she analysed more than forty thousand samples of water. These analyses exhibited the condition of the water from all parts of the state during all seasons of the year and were of the greatest value in solving a number of important problems in state sanitation.

But notwithstanding the drafts made on her time and energy by her classwork in the laboratory and her occupation as sanitary engineer for scores of public and private institutions, she still found leisure to engage in many important movements which had in view the public health and the betterment of sanitary conditions in city and country. It is safe to say that no one ever put her knowledge of chemical science to more practical use or made it more perfectly subserve the public weal than did Mrs Richards. To spread among the masses a knowledge of the principles of sanitation, to make them realise how indispensable to health are pure food, pure water, pure air and life-giving sunshine was her great mission in life, and in this she displayed an energy and a tireless zeal which were an inspiration to all with whom she came into contact.

This indefatigable woman, it is proper to record here, might have distinguished herself as a discoverer in chemical science had she elected to devote her life to original research rather than to utilising the knowledge already available for the welfare of her fellows. Thus, after a careful analysis of the rare mineral samarskite, she found an insoluble residue which led her to believe might contain unknown elements. This view she repeatedly expressed to her co-workers in the laboratory. But she was unwilling to take from what she regarded more important work the time necessary for making investigations which might have given her undying fame as a discoverer. For not long afterward this insoluble residue, in the hands of two French chemists, yielded the exceedingly rare elements, samarium and gadolinium.

Another chemist of a less altruistic nature than Mrs Richards would not have resisted the temptation to achieve distinction in the domain of original research. But where there was so much suffering to be relieved and so much ignorance to be removed regarding the most fundamental principles of sanitation, this philanthropic woman preferred to put to practical use what she called "the considerable body of useful knowledge now lying on our shelves."

Her duty, as she conceived it, is well indicated in the following paragraph, taken from a thoughtful discussion by her of the subject of home economics a short time before her death in 1911. "The sanitary research worker in laboratory and field," she declares, "has gone nearly to the limit of his value. He will soon be smothered in his own work, if no one takes it. Meanwhile children die by the thousands; contagious diseases take toll of hundreds; back alleys remain foul and the streets are unswept; school-houses are unwashed and danger lurks in the drinking cups and about the towels. Dust is stirred up each morning with the feather duster to greet the warm, moist noses and throats of the children. To the watchful expert it seems like the old cities dancing and making merry on the eve of a volcanic outbreak."[2]

From the day in 1873 when Mrs Richards received from the Institute of Technology the degree of Bachelor of Science — a degree which made her not only the first woman graduate of this institution, the sudden and premature extinction of the husband's life gave such a shock to the entire scientific world.

After her marriage the gifted young Polish woman had reached the goal of her ambition. She was able to devote herself exclusively to what was henceforth to constitute her life work in one of the best laboratories of Paris, that of the École de Physique et de Chimie, and that, too, in collaboration with her husband, from whom she was never separated during the entire period of their married life for even a single day.

It was about this time that Mme Curie had her interest aroused by the brilliant discoveries of Rontgen and Becquerel regarding radiant matter. After a long series of carefully conducted experiments on the compounds of uranium and thorium, she, with the intuition of genius, opened up to the world of science an entirely new field of research. But she soon realised that the labour involved in the investigations which she had planned was entirely beyond the capacity of any one person. It was then that she succeeded in enlisting her husband's interest in the undertaking which was to lead to such marvellous results.

Confining their work to a careful analytical study of the residue of the famous Bohemian pitchblend — an extremely complex mineral, largely composed of oxide of uranium — they soon found themselves confronted by most extraordinary radioactive phenomena. Continuing their researches, their labour was rewarded by the discovery of a new element which Mme Curie, in her enthusiasm, named in honour of the land of her birth, polonium.

As their investigations progressed, they became correspondingly difficult. They were dealing with substances which exist in pitchblend residue only in infinitesimal quantities — not more than three troy grams to the ton. The difficulties they had to contend with were enough to discourage the stoutest heart. Few believed in their theories, while the majority of those who had some intimation of the character of their work were persuaded that they were pursuing a phantom. But the indefatigable pair toiled on day and night and continued their experiments through long years of poverty and deferred hopes.

Considering the herculean task in which they were engaged for so many years, we scarcely know which to admire most, their clearness of vision, which made them divine success; their profound knowledge, which guided them in the choice of reagents; or the indomitable perseverance which characterised them in their laborious task and in the countless sacrifices which they were obliged to make before their efforts were crowned with success.

During this long search into the inner heart of nature, Pierre Curie was often so discouraged and depressed that, had he not been sustained by his more sanguine wife, he would time and again have given up his investigations in despair. But Marie Curie never faltered. She never lost faith in their theories or confidence in the outcome of their great undertaking. Before her deft hands and fertile brain difficulties vanished as if under the magic wand of Prospero.

At length, after countless experiments of the most delicate character, after bringing to bear on the solution of the problem before them the most refined methods of chemical analysis, they were rewarded by one of the most extraordinary discoveries recorded in the annals of science. With the announcement of the discovery of radium, the Curies sprang into worldwide fame, and the name of the wonderful woman who had been the prime mover in the supreme achievement was on every lip. Pierre Curie himself declared that more than half of the epochal discovery belonged to his wife. It was she who began the work. It was she who, after her marriage, enlisted in it the cooperation of her husband. It was she whose invincible patience and persistence — typical of the noblest representatives of her race — supported him during periods of doubt and despondency and fanned his flagging spirits to new endeavour. It can indeed be truthfully asserted that had it not been for her penetrating intelligence, her tenacity of purpose and her keenness of vision, which were never at fault, the great victory which crowned their efforts would never have been achieved.[3]

Compare their work with that which was accomplished by their illustrious predecessors, Antoine Laurent Lavoisier, and his words, "No woman need apply."

When one reads of the sad experiences of Mme Curie and Mrs Ayrton with the learned societies of Paris and London, one instinctively asks, "When will the day come when women, in every part of the civilised world, shall enjoy all the rights and privileges in every field of intellectual effort which have so long been theirs in the favoured land of Dante and Beatrice — the motherland of learned societies and universities?" For not until the advent of the day when such exclusive organisations as the Royal Society and the French Academy of Sciences, such ultra-conservative universities as Oxford and Cambridge shall admit women on the same footing as men, will these institutions be more than half serving the best interests of humanity.[6]

Women, it is true, are now eligible to many literary and scientific associations from which they were formerly debarred, and are, in most countries, admitted to colleges and universities whose portals were closed to them until only a few years ago; but until they shall be welcomed to all universities and all societies whose objects are the advancement of knowledge, until they shall participate in the advantages and prestige accruing from connection with these organisations, they will have reason to feel that they are not yet in the full possession of the intellectual advantages for which they have so long yearned — that they have been but partially liberated from that educational disqualification in which they have been held during so many long centuries of deferred hopes and fruitless struggles.

## Notes & References

1) As no satisfactory biography of Laura Bassi has yet been written, most of our knowledge respecting her is limited to that found in Fantuzzi's *Notizie degli Scrittori Bolognesi*, Vol I, pp 384-91, and Mazzuchelli's *Gli Scrittori d'Italia*, Vol II, Part I, pp 527-29, Brescia, 1758.

2) *Lavoisier 1743-1794, d'apres sa Correspondence, Ses Manuscrits, Ses Papiers de Famille et d'Autres Documents Inedits*, p 42 *et seq*, by E Grimaux, Paris, 1896.

3) *The Life of Ellen H Richards*, p 273 *et seq*, by Caroline L Hunt, Boston, 1912.

4) Mme Curie, in an article which she wrote shortly after her discovery of radium, shows that she possesses a genius for inductive science of the highest type. "It was at the close of the year 1897," she writes, "that I began to study the compounds of uranium, the properties of which had greatly attracted my interest. Here was a substance emitting spontaneously and continually radiations similar to Röntgen rays, whereas ordinarily, Röntgen rays can be produced only in a vacuum tube with the expenditure of electrical energy. By what process can uranium furnish the same rays without expenditure of energy and without undergoing apparent modification? Is uranium the only body whose compounds emit similar rays? Such were the questions I asked myself; and it was while seeking to answer them that I entered into the researches which have led to the discovery of radium." Radium and Radioactivity, in *The Century Magazine*, for January, 1904.

4) *Notice sur Pierre Curie*, p 20 *et seq*, by M D Gernez, Paris, 1907, and *Le Radium, Son Origine et ses Transformations*, by M L Houllerigue, in *La Revue de Paris*, May 1, 1911.

5) The day following Pierre Curie's refusal of the decoration offered by the Government, the elder of his two daughters, little Irene, climbed upon her father's knee and put a red geranium in the lapel of his coat. "Now, papa," she gravely remarked, "you are decorated with the Legion of Honour." "In this case," the fond father replied, "I make no objection."

# Women in the Natural Sciences

It is reasonable to suppose that women, who are such lovers of nature, have always had a greater or less interest in the natural sciences, especially in botany and zoology; but the fact remains that the first one of their sex to write at any length on the various kingdoms of nature was that extraordinary nun of the Middle Ages, St Hildegard, the learned abbess of the Benedictine convent of St Rupert, at Bingen on the Rhine. Of an exceptionally versatile and inquiring mind, her range of study and acquirement was truly encyclopaedic. In this respect she was the worthy forerunner of Albert the Great, the famous *Doctor Universalis* of Scholasticism.

Although St Hildegard has much to say about nature in several of her works, the one of chiefest interest to us as an exposition of the natural history of her time is her treatise entitled *Liber Subtilitatum Diversarum Naturarum Creaturarum*. It is usually known by its more abbreviated name, *Physica*, and, considering the circumstances under which it was written, is, in many ways, a most remarkable production. It consists of nine books treating of minerals, plants, fishes, birds, insects and quadrupeds. The book on plants is composed of no fewer than two hundred and thirty chapters, while that on birds contains seventy-two chapters.

In reading Hildegard's descriptions of animated nature we are often reminded of Pliny's great work on natural history; but, so far as known, there is no positive evidence that the learned religieuse had any acquaintance whatever with the writings of the old Roman naturalist. Had she had, the general tenor of her work would have been quite different from what it actually is.

The mystery, then, is, what were the sources of *Physica*? Some have fancied that Hildegard in preparing this made use of the writings not only of Pliny and Virgil, but also of those of Macer, Constantinus Africanus, Walafrid Strabo, Isodore of Seville, and other writers who were in great vogue during the Middle Ages. The general consensus of opinion, however, of those who have carefully studied this interesting problem is that the gentle nun was not acquainted with any of the authors named, except, possibly, Isodore of Seville, whose works were all held in high esteem, especially during the period of Hildegard's greatest literary activity.

Hildegard's *Physica* has a special value for philologists, as well as for students of natural history, for it contains the German names of plants still used by the people of the Fatherland seven hundred years after they were penned by the painstaking abbess of St Rupert's.[1]

Referring to the Saint's work entitled *De Natura Hominis, Elementorum, Diversarumque Creaturarum* — a treatise on the nature of man, the elements and divers created things — no less an authority than Dr Charles Daremberg declares that it will always hold an important place in the history of medical art and of inanimate and animate nature — *insignis semper locus debetitur in artis medicae rerumque naturalium historia*.[2]

He even goes further and affirms that Hildegard was familiar with numerous facts of science regarding which other medieval writers were entirely ignorant. More than this. She was acquainted with many of nature's secrets which were unknown to men of science until recent times, and which, on being disclosed by modern researches, have been proclaimed to the world as new discoveries.[3]

One reason why St Hildegard's writings on botany, zoology and mineralogy are not better known is that few students care to make the effort to master her voluminous works. They require long and assiduous study and a knowledge of her peculiarities of style and expression which is acquired only after patient and persistent labour. But the labour is not in vain, as is evidenced by the numerous monographs which have appeared in recent years, especially in Germany, on the scientific works of this marvellous nun of the twelfth century. All things considered, the Abbess of Bingen may be said to hold the same position in the natural sciences of her time as was held in the physical and mathematical sciences seven hundred years earlier by the illustrious Hypatia of Alexandria.

After the death of St Hildegard, full six centuries elapsed before any one of her sex again achieved distinction in the domain of natural science. And then, strange to relate, the first woman who won fame by her knowledge of science and by her contributions to it, did so in the field where a woman would, one would think, be least disposed to exercise her talent and least likely to find congenial work. It was in the then comparatively new science of human anatomy — a science which had been inaugurated in the famous medical schools of Salerno and which was subsequently so highly developed in the great University of Bologna.

The name of this remarkable woman was Anna Morandi Manzolini. She was born in 1716 in Bologna, where, after a brilliant career in her favourite branch of science, she died at the age of fifty-eight. She held the chair of anatomy in the University of Bologna for many years, and is noted for a number of important discoveries made as the result of her dissections of cadavers.

But she won a still greater title to fame by the marvellous skill which she exhibited in making anatomical models out of indurated wax. They were so carefully fashioned that some of them could scarcely be distinguished from the parts of the body from which they were modelled. As aids in the study of anatomy they were most highly valued and eagerly sought for on all sides. The collection which she made for her own use was, after her death, acquired by the Medical Institute of Bologna and prised as one of its most precious possessions.

Three years after her demise, Luigi Galvani, professor of anatomy in the same university in which Anna had achieved such fame, made use of these wax models for a course of lectures on the organs and structure of the human body.

These famous models, first perfected by Anna Manzolini, were the archetypes of the exquisite wax models of Vassourie as well as of the unrivalled *papier-mache* creations of Dr Auzoux and of all similar productions now so extensively used in our schools and colleges.

Even during the lifetime of the gifted modeller there were demands for specimens of her work from all parts of Italy. From many cities in Europe, even from London and St Petersburg, she received the most flattering offers for her services. So eager was Milan to have her accept a position which had been offered her that the city authorities sent her a blank contract and begged her to name her own conditions. But she could never be induced to leave the home of her childhood and the city which had witnessed and applauded her triumphs of maturer years.

Men of learning and eminence, on passing through Bologna, invariably made it a point to call on the learned *professora* in order to make her acquaintance and to see her wonderful anatomical collection, which was celebrated throughout Europe as *Supellex Manzoliniana*. Among these visitors was Joseph II of Austria. So greatly was His Majesty impressed by Anna's rare intellectual attainments and by her marvellous skill in reproducing the various parts of the "human form divine" that he could not take leave of her without showing his appreciation of them by loading her with gifts worthy of a sovereign.[4]

A contemporary of Anna Manzolini, who also distinguished herself in the preparation of anatomical models, was the French woman, Mlle Biheron. Her facsimiles of parts of the human body were, according to Mme de Genlis, so true to nature that they could not be distinguished from the originals. This led the facetious Chevalier Ringle, after examining a specimen of her handiwork, to declare, "Verily, it is so perfect that it lacks only the odour of the natural object."

While yet prince royal, Gustavus of Sweden visited the French Academy of Sciences in Paris. Here he was entertained by a number of experiments in anatomy. The demonstrator was Mlle Biheron, who is said to have had a veritable passion for both anatomy and surgery. So impressed was Gustavus with the extraordinary skill and knowledge of this gifted daughter of France that he offered her the position of demonstrator of anatomy in the royal University of Sweden.

Other branches of science, apparently quite as alien as anatomy to women's taste and talent, are mineralogy and metallurgy. Yet as early as the first half of the seventeenth century, the Baroness de Beausoleil had achieved a great reputation by her investigations into the mineral treasures of France. Indeed, she may, strange as it may appear, be regarded as the first mining engineer of her native land. She details the qualifications of a mining engineer and tells us he must, among other things, be well versed in chemistry, mineralogy, geometry, mechanics and hydraulics. As for herself, she assures us that she devoted thirty years of unremitting study to these divers branches.

To Mme de Beausoleil is also attributed the glory of awakening her countrymen's interest in the mineral resources of France, and of showing them how their proper exploitation would inure not only to the credit of the nation abroad but also to its prosperity at home.

She was the author of two works which prove that she was a woman of rare attainments combined with exceptional breadth of view and political acumen. She was deeply concerned in the development of the mineral resources of her country and foresaw how greatly they could be made to contribute to the augmentation of the nation's finances.

Her work entitled *La Restitution de Pluton* is a report on the mines and ore deposits of France, and is a document as precious as it is curious. It was addressed to Cardinal Richelieu, and shows how the French monarch could, if the subterranean treasures of the country were properly developed, become the greatest ruler in Christendom and his subjects the happiest of all peoples.

Another report by this energetic and enthusiastic woman is in the same strain. In it she proves how the King of France, by utilising the underground riches of his country, could make himself and his people independent of all other nations.[5]

In these two productions Mme de Beausoleil treats of the science of mining, the different kinds of mines, the assaying of ores and the divers methods of smelting them, as well as of the general principles of metallurgy, as then understood. But, unlike the majority of her contemporaries, this enlightened woman had no patience with those who believed that the earth's hidden treasures could not be discovered without recourse to magic or to the aid of demons. She was unsparing in her ridicule of those who had faith in the existence of gnomes and kobolds, or thought that ore deposits could be located only by divining-rods or similar foolish contrivances which were relics of an ignorant and superstitious age.

The same century that witnessed the exploring activity of the Baroness de Beausoleil saw the beginnings of the notable achievements of a daughter of Germany, well known in the annals of science as Maria Sibylla Merian. Born in Frankfort in 1647, she died in Amsterdam in 1717, after a somewhat checkered career, most of which was devoted to the pursuit of natural history. So fond was she of flowers and insects that it is said they told her all their secrets.

After having familiarised herself with the fauna and flora of her native land, she proceeded to investigate the collections of the principal European cabinets of natural history. This only fired her ambition to see more of the world and study Nature where she is seen in her greatest splendour and luxuriance.

She accordingly resolved to undertake a journey to the equatorial regions of South America. Such a voyage can now be made with comparative ease, but in her days it was fraught with discomforts and dangers of all kinds, and one that no woman thought to venture on unless obliged to do so by stern necessity.

But she was set on investigating animals and plants in their own habitats in the glorious and exuberant flora of the tropics and, accompanied by her two daughters, Helena and Dorothea, she embarked for Surinam. Here, assisted by her daughters, who, like their mother, were both skilful artists, the intrepid naturalist spent two years in studying the wonders of plant and animal life that everywhere greeted her delighted vision. All the time not occupied in research work was devoted to sketching and painting those superb insects that are so abundant in tropical fields and forests.[6]

Returning to Holland with her precious scientific treasures, she began the preparation of a work that will long endure as a monument to her knowledge and industry. It was a magnificent volume in folio on the insects of Surinam. It appeared simultaneously in Dutch and Latin, and was subsequently translated into French.

In illustrating this sumptuous work, Frau Merian was greatly assisted by her younger daughter, Dorothea. The etchings and hand-coloured reproductions of the gorgeous butterflies and flowers of Surinam commanded universal admiration, and marked a new epoch in book-making. Even today this noble volume is eagerly sought by both book-lovers and men of science, for it is not only a work of rare conception and beauty but also one of exceptional accuracy in illustration and statement of fact.[7]

Besides etchings of multiform insects, lizards and batrachians indigenous to Dutch Guiana, there were in this unique volume carefully executed illustrations of plants and trees peculiar to tropical America, such as vanilla, cacao, and the species of manihot which constitutes the staff of life of so large a portion of the population in the basins of the Amazon and the Orinoco.

A new and enlarged edition of this work was published after Frau Merian's death by her daughter Dorothea. The same gifted daughter showed her interest in her parent's work and her devotion to her memory by bringing out a beautifully illustrated edition of her mother's earliest work which treated of the wonderful life-history of silkworms.[8]

The century following that which had celebrated the scientific triumphs of Maria Merian found in Josephine Kablick, born in 1787 in Hohenelbe, Bohemia, a woman who was destined to prove a worthy successor, as a nature-student, of the noted daughter of Frankfort-on-the-Main.

From her tenderest years she exhibited a passionate love for every form of plant life. In addition to this, she had, while yet young, the good fortune of studying under the best botanists of her time.

Soon she became an enthusiastic collector and was in a short time the happy possessor of a herbarium which contained many new species of plants which she had discovered during her frequent botanical excursions. From making collections for her private herbarium, she was gradually led to make collections for the schools and colleges of her native country, as well as for the museums and learned societies of various parts of Europe. Many public institutions owed to her cordial cooperation some of the choicest treasures in their herbaria, and not a few botanical writers of her day found in her an intelligent and sympathetic collaborator.

But Frau Kablick's interest in nature was not confined to plants. She was an assiduous student of palaeontology as well as of botany, and the many fossil animals and plants named in her honour testify to her success in the pursuit of her favourite branches of science.

There was nothing of the conventional blue-stocking about this ardent votary of nature. Strong and healthy, neither wind nor rain interfered with her fieldwork in botany or palaeontology. It was her greatest pleasure to roam through dark forests and scale high mountains in search of new species of plants and fossils. And the success which rewarded her efforts was such that the old and trained naturalists among her male friends had reason to envy her good fortune as an explorer.

But Frau Kablick never permitted her frequent excursions, or her devotion to science, to cause her to neglect the duties of her household. Fortunately, her husband was also an ardent student of nature, and while his wife was devoting her attention to botany and palaeontology, he was making investigations in zoology and mineralogy. They spent fifty happy years together in the pursuit of science and their joint efforts contributed not a little toward the advancement of the branches of science to which they had devoted their lives with such well-directed effort and enthusiasm.

As the fruitful life of Josephine Kablick who had shed such lustre on her sex in Bohemia was drawing to a close, a young woman in Germany, Amalie Dietrich by name, was preparing herself to fill the void which would be occasioned by her predecessor's death. Her first love, as a young girl, was plant life, and this was subsequently accentuated by her husband, who was not only a botanist himself but also one who belonged to a distinguished family of botanists.

A keen observer and an indefatigable collector, Frau Dietrich soon became known throughout Europe as a botanist of marked ability and daring. She was wont, unaccompanied, to climb the highest peaks of the Salzburg Alps, and spend entire weeks there seeking new species of Alpine flora. During the day she explored the deep ravines and clambered along the brambly ledges of beetling precipices, and during the night she sought shelter and repose in the humble hut of some hospitable herdsman.

Valuable, however, as was Amalie Dietrich's work in the Austrian Alps, it was but a preparation for that which some years later she was to enter upon in far-off Australia. Here she devoted twelve of the best years of her life to the cultivation of botany in the virgin soil of Queensland. Here, too, she surprised everyone by her venturesome spirit no less than by her irrepressible zeal in making collections. Heedless of danger, she plunged quite alone into the wilderness and spent days and weeks at a time with the wild aborigines.

But she secured what she went in quest of — a large and valuable collection of plants, containing many new and interesting species. Besides these, she was able to bring back with her to Europe a large mass of zoological specimens as well as countless domestic utensils and implements of warfare and husbandry employed by the savages among whom she so frequently journeyed and with whose manners and customs she eventually became so familiar.

Modest and trustworthy, Frau Dietrich had a host of friends in the scientific world, and the number of plants which bear her name are not only a tribute to her worth, but a striking evidence of the extent of her activity in the pursuit of the science which became the absorbing passion of her life.[9]

Of Russian women who have become specially noted for their contributions to natural science, a very prominent place must be assigned to Sophia Pereyaslawzewa. After receiving the doctorate of science in the University of Zurich, she became director of the biological station at Sebastopol, a position she held with great eclat during twelve years. Here she made numerous important researches on manifold forms of marine life and prepared many works for the press in German and French, as well as in her native Russian. Her *Monographie de Turbellaries de la Mer Noire*, a large and beautifully illustrated volume published at Odessa in 1892, placed her at once among biologists of the first rank. Indeed, so meritorious was this production of the talented daughter of Holy Russia that the Congress of Naturalists in 1893 did not hesitate to recognise its exceptional value by conferring on the fair authoress a special prize.

This gifted biologist has since rendered distinct service in the cause of science by her explorations of the Gulf of Naples and the coasts of France. Her activity is prodigious, and the long list of books and monographs which she has published on the lower forms of marine life in the Black and Mediterranean seas shows that she has a capacity for work that is truly extraordinary.

Here is, probably, the place to make mention of a woman of encyclopaedic mind, Clemence Augustine Royer, who was born in 1830 in Nantes, France. She wrote on such a variety of subjects that it is difficult to classify her. She was in no sense of the word a specialist, and she seems by temperament to have been averse to confining herself to any one branch of knowledge.

Her first work to attract particular attention was one on a topic connected with political economy. A prize had been offered for the discussion of this subject, and the little French woman acquitted herself so well that she had the honour of sharing the prize with the noted Proudhon. She has also written many works on philosophy and physics. Among these are two which attracted considerable notice at the time of their publication. In one of them she attacks the positivism of Comte; in the other she assails Laplace's hypothesis regarding the origin of the material universe.

But the work which made her famous, particularly in France, was her translation into French in 1862 of Darwin's *Origin of Species.* It is safe to say that this version created as much of a sensation in France as the original had caused in Great Britain and America. Her preface to the work of the English naturalist, in which she indicates the results which flow from an acceptance of the transformist theory, created a veritable storm in both religious and scientific circles.

So gratified was Madame Royer by the impression made by this preface and so pleased was she with the controversy which she had started, that she expanded her summary of the theory of evolution as therein given and published it in 1870 under the title of *Origine de l'Homme et de Societes.* This production was so revolutionary in character and so subversive of teachings long held sacred that it provoked an indignant protest from all quarters, and the author was at once ranked with such radical exponents of the new science as Voght, Büchner and Haeckel.

After the appearance of this production, she wrote numerous other works, several of them on subjects relating to natural science, especially in its connection with anthropology and prehistoric archaeology. And so great was her breadth of view and so exceptional was her grasp of all subjects discussed by her that Renan declared of her, *Elle est presque un homme de genie*—She is almost a man of genius.

Mme Royer was frequently spoken of as a candidate for the French Institute, but she was so well aware of the prejudices against the admission of women to membership in this learned body that she never allowed herself to consider the proposal seriously. She was certainly a brainy woman, and in her own department of intellectual effort she exhibited as much talent as did George Sand and Mme de Staël in literature and history.

An entirely different type of woman from the radical and disputatious Mme Royer was the charming and cultured lady, Miss Eleanor Ormerod, her contemporary, who, in her chosen department of science, won both fame and the lasting gratitude of her fellowmen.

Miss Ormerod, unlike Mme Royer, was pre-eminently a

specialist, and the branch of science in which she achieved distinction was entomology, or rather that branch of it known as economic entomology. From her childhood she manifested an unusual interest in all forms of insects, but particularly in those which are serviceable to mankind or are destructive to farms and gardens, orchards and forests.

Fortunately for the gratification of her peculiar bent of mind, nearly half of Miss Ormerod's life was spent in a locality which was specially favourable to the study of insects which are obnoxious to the gardener, the farmer and the forester. This was at the confluence of the Wye and the Severn, where her father owned a large landed estate, part of which was under cultivation and part wood and park land.

Here the young girl made her first collection of insects, and here she began her studies on the cause and nature of the parasitic attacks upon crops. Here she first realised the frightful ravages that were occasioned by the manifold insect pests that infest not only trees, shrubs, cereals and vegetables, but also flocks and herds as well. And here, too, she resolved to devote her life to devising preventive and remedial treatment for the evils which were robbing the husbandman of so great a part of the fruits of his toil.

After taking this generous resolution, the life of our young heroine was, like that of Liebig and Pasteur, devoted to the welfare of her fellowmen. And like these noble benefactors of their race, her thought was always how she might prevent the losses and increase the products of the tillers of the soil. Entomology with her was not mere nomenclature — a knowledge of strange and fantastic names, which, with the ignorant, constitutes a distinction — but one of the most practical and useful of the sciences.

Miss Ormerod might, had she so elected, have won fame as a systematic entomologist and as a distinguished contributor to the already long list of genera and species of insects. She might have devoted herself to theoretical work, or bent her energies towards the general advancement of the science, like Fabricius, Swammerdam, Westwood and Burnmeister; but she preferred to forego all the glory that might accrue from pursuing such a course, and to direct her efforts in such wise as to be of most service to humanity.

Like the great Pasteur, after his long and laborious experimental researches on silkworm diseases, Miss Ormerod could, at the end of her illustrious career, declare with truth: "The results which I have obtained are, perhaps, less brilliant than those which I might have anticipated from researches pursued in the field of pure science, but I have the satisfaction of having served my country in endeavouring, to the best of my ability, to discover the remedy for great misery. It is to the honour of a scientific man that he values discoveries which at their birth can only obtain the esteem of his equals, far above those which at once conquer the favour of the crowd by the immediate utility of their application; but, in the presence of misfortune, it is equally an honour to sacrifice everything in the endeavour to relieve it."[10]

Miss Ormerod's labours were not, it is true, instrumental in rescuing from destruction a nation's chief industries, as were Pasteur's in the case of his famous researches on the phyloxera of the grape vine or the pebrine of the silkworm. Nor had they to do with such frightful industrial disturbances as have frequently been occasioned by rinderpest or by the potato blight in Ireland in 1845.

This is true in so far as any one pest is concerned. But when one reflects on the scope of Miss Ormerod's investigations and considers how far-reaching were her researches and how many and diverse industries were embraced by the remedial and prophylactic measures which she proposed, one cannot but realise the immense importance of her life-work.

The fact that her activities were confined chiefly to old and well-known pests — insects from which the farmer and the gardener and the forester had suffered for centuries, and which they had come to regard as necessary and inevitable evils — does not detract from the merit and the value of her labours. That she should have taken up a work which affected so many people and have been so successful in abating, or in entirely removing evils which had so long afflicted agriculturists and stock-growers, shows that she was a woman of rare courage and determination as well as one of invincible persistence and of intellectual resources of a very high order.

During more than a quarter of a century Miss Ormerod devoted practically the whole of her time to the study of economic entomology and to spreading a knowledge of it among her countrymen. From 1877 to 1898 she published annual reports on injurious insects and sent them broadcast throughout Great Britain and her colonies. In addition to this she wrote a number of manuals and textbooks on insects injurious to food crops, forest trees, orchards and bush fruits.

Nor was this all. She also prepared for gratuitous distribution a large number of four-page leaflets on the most common farm pests. Of the leaflet, for instance, on the warble-fly, its life-history, methods of prevention and remedy, no less than a hundred and seventy thousand copies were printed. And so great was the demand for her leaflet on the gooseberry red spider that a single mail brought her an order for three thousand copies.

Miss Ormerod, it is proper to state here, received no remuneration whatever for her great services to the public. On the contrary, she gave not only all her time gratuitously, but bore a great part of the expense of printing and distributing her publications. The amount of good she thus did unaided and alone cannot be estimated.

In her leaflet on the warble-fly, also known as bot-fly, she estimates the annual damage to the stock-growers of the United Kingdom from this pest at from £3,000,000 to £4,000,000. The losses due to fruit, grain and vegetable insects of various kinds, before she began her insect crusade, were much greater. In Great Britain and her colonies they amounted to very many millions of pounds sterling every year.[11]

And most of these losses, as she demonstrated, were preventable by simple precautions which she eventually succeeded in inducing the people to adopt. How much she was instrumental in saving annually to the farmers and gardeners of England by her writings and lectures can only be imagined, but the sum must have been immense.

When we recollect that Miss Ormerod accomplished all her work before it occurred to the English Board of Agriculture to appoint a government entomologist, we shall realise what a pioneer she was in the career in which she achieved such distinction and through which she conferred such inestimable benefits upon her fellows.

Miss Ormerod's entomological publications, especially her annual reports, brought her into relations with people of all classes throughout the whole world. Her correspondence, in consequence, was enormous, and not infrequently amounted to from fifty to a hundred letters a day. The great entomologists of Europe and America held her in the highest esteem, and had implicit faith in her judgement in all matters pertaining to her specialty.

One day she would receive a letter from an English gardener begging for a remedy against the strawberry beetle. The next day she would have a similar letter regarding mite-galls on black currants, or pea-weevil larvae or clover-eel worms. Again there would be a communication from Norway requesting advice about the Hessian fly, or from Argentina asking information concerning a certain kind of destructive grass beetle, or from India appealing for help against a pernicious species of forest fly, or from South Africa seeking a relief from the boot-beetle. And still again, she was consulted by her foreign correspondents about termites, which were causing havoc among the young cocoa trees of Ceylon, or about certain peculiar species of Australian larvae, or about the devastating action of the pine beetle in the Scotch forests, or about the wheat midge and antler moth in Finland.

One day she had a communication from the Austrian Embassy regarding a beetle that was eating the oats about Constantinople, and not long afterwards she received a letter from the Chinese Minister in London begging for information as to how to prevent the ravages of certain noxious bugs in the lee-chee orchards of China.

In view of all these facts it is not surprising that Miss Ormerod became an active and valued colleague of some of England's most noted scientific men. Professor Huxley said of her in connection with certain work performed by her as a member of one of the committees to which he belonged that "she knew more about the business" than all the rest put together.

Miss Ormerod's services and attainments, it is gratifying to note, were not without recognition in high quarters. Besides being in constant correspondence with the most eminent entomologists of the world, consulting entomologist to the Royal Agricultural Society of England and examiner in agricultural entomology in the University of Edinburgh, she was a member of many learned societies in both the Old and the New World. She was also the recipient of many medals, two of which came from Russia.

The honour, however, which gave her the most pleasure was the degree of Doctor of Laws, which was conferred on her by the University of Edinburgh. It was the first time this old and conservative institution thus honoured a woman, but in honouring Miss Ormerod it honoured itself as well.[12]

But when one considers the magnitude of Miss Ormerod's services to her country and to the world, when one reflects on the tens of millions of pounds sterling which she saved to the British Empire by her researches and writings, these honours seem trivial and unworthy of the great nation which she so signally benefited. If any of her countrymen had laboured so long and so successfully and made so many sacrifices for the welfare of the nation as she had, he would have been knighted or ennobled. But age-long prejudices and traditions will not yet permit England to bestow the same honours on women as on men, no matter how brilliant their attainments or how distinguished their services to the crown and to humanity. Recognition of this kind may possibly come as one of the desirable innovations of the twentieth century. No lover of fair play can deny "'tis a consummation devoutly to be wished."[13]

The names of the women in the United States who have become prominent by their researches and writings in the various branches of the natural sciences would make a long list. And when one recalls the fact that it was only in the latter part of the nineteenth century that American women were afforded an opportunity to study science, it is a matter of surprise that the list is so extended. For practically no provision was made for the serious pursuit by them of the natural sciences until the opening of Vassar College in 1865, and it was not until the closing years of the century that the portals of many men's colleges were unlocked and thrown open to the hitherto proscribed sex. Considering all the obstacles they had to overcome, the ignorance, the prejudice, the opposition of all kinds they had to combat in the United States, women have already accomplished wonders and bid fair to achieve much more in the near future.

Now almost every educational institution in the land, private or state, has one or more women professors or associate professors. They teach all the branches of the natural sciences that are taught by their male colleagues, — botany, geology, mineralogy, zoölogy, anatomy, bacteriology and all the numerous subdivisions of these sciences, — and they teach them with success and eclat.

They also occupy responsible scientific positions in various state and federal institutions. Thus one woman has been the principal of the Denver School of Mines, while another has been the state entomologist for Missouri. Women are also found doing important work in the National Museum, in the Smithsonian Institution, and in the Agricultural Department in Washington, as well as in the various museums, botanical gardens and public laboratories of the country from the Atlantic to the Pacific.

Among those who have deserved well of science in the United States by their investigations and writings are Olive Thorne Miller and Florence Merriam in ornithology; Susanna Phelps Gage, Dr Ida H Hyde, Mary H Hinckley, Cornelia M Clapp, Edith J and Agnes M Claypole in biology; Rose S Eigenman in icthyology; Edith M Patch, Elizabeth W Peckham, Emily A Smith, Cora H Clarke, J M Arms Sheldon, Mary Treat, Mary E Murfeldt, Annie T Slosson in entomology; Elizabeth G Britton and Clara E Cummings in cryptogamic botany; Sarah A Plummer Lemmon, Katherine E Golden, Alice Eastman and Almira Lincoln Phelps in general botany; Ada D Davidson, Ella F Boyd and Florence Bascom in geology. Besides these, special mention should also be made of Dr Julia W Snow for her work on the microscopical forms of freshwater algae; Anna Botsford Comstock for her contributions to our knowledge of microscopic insects; Katherine J Bush for her monographs on shallow and deepwater molusca; Harriet Randolph and Fannie E Langdon for their studies on worms, and Katherine Foot for her papers on cellular morphology. Particularly notable, too, is the work that has been done on marine invertebrates by Mary J Rathbun in the United States National Museum and by Florence Wambaugh Patterson in vegetable physiology and pathology in the Department of Agriculture in Washington.

But much as the women just named deserve recognition for their achievements in the various branches of science to which they have severally devoted themselves, the one who will always be specially remembered, not only for her valuable contributions to divers branches of natural science, but also for her labours in behalf of higher female education — particularly as president of Radcliffe College — is Mrs Elizabeth Cary Agassiz, the wife of the celebrated Swiss-American naturalist, who gave such an impetus to the study of natural science in the United States, and whose influence on the general advancement of science in all its departments has proved so enduring and so far-reaching. As an inspirer of and collaborator with her gifted husband, Mrs Agassiz deserves a large page in the annals of science, while as an enthusiastic student of nature and as one who communicated her enthusiasm to her students, and at the same time held up before them the highest ideals of womanhood, she is sure of a portion of that immortality which has been decreed to her illustrious life-partner, Jean Louis Agassiz.

This chapter would not be complete without some reference to that large class of women travellers who, directly or indirectly, have contributed so much to the advancement of the natural sciences. The gifted Romanian writer and traveller, Princess Helena Kolzoff Massalsky — better known under her pseudonym, Doria d'Istria — somewhere expresses the opinion that a woman traveller admirably supplements the scientific work of the male explorer by bringing to it aptitudes that the latter does not possess. For she notes many things in nature, as well as in the national life and popular customs of the countries which she traverses, which escape the more hebetudinous perceptions of men, and thus a vast field, that would otherwise remain unknown, is opened to observation and critical study.

One of the most noted travellers of her sex in the nineteenth century was the famous Ida Pfeiffer, of Austria. During the years intervening between 1842 and 1858, the date of her death, she travelled nearly two hundred thousand miles and, in so doing, visited nearly every quarter of the globe. When one recalls the difficulties and discomforts of transportation in the early part of the last century, as compared with our present facilities and conveniences, and bears in mind the fact that her travelling expenses for an entire year were less than those of a Lamartine or a Chateaubriand for a single week, we must admit that her achievements were, indeed, extraordinary.

Besides being the author of numerous books which had for many years a great vogue — books which, by reason of the keen observations and the absolutely truthful narratives of their author, are still of special value to the student of geography and ethnology — she made collections illustrative of botany, mineralogy and entomology which were subsequently secured for the British Museum and other similar institutions in Europe.

No one more highly appreciated Frau Pfeiffer's efforts in behalf of science than did the illustrious Alexander von Humboldt, whose friendship was one of the greatest joys of this remarkable woman's life. Through his recommendation and that of the noted geographer, Karl Ritter, she was made an honorary member of the Geographical Society of Berlin. Besides this, the King of Prussia conferred on her the gold medal for arts and sciences.

Three other women, all representatives of Great Britain, likewise deserve notice for their extensive travels and the interesting and instructive accounts which they published of them. These are Constance Gordon Cumming, Isabella Bird Bishop and Amelia B Edwards.

More notable in many respects than these three distinguished women were Miss Mary H Kingsley and Madame Octavie Coudreau. For their contributions to science and for their daring adventures in savage lands, they have won for themselves an unique position among women explorers.

Miss Kingsley — the niece of the well-known writer and naturalist, Charles Kingsley — exhibited much of her uncle's literary ability and love of nature. So complete was her intellectual grasp of the most difficult problems, and so rare was her overflowing sympathy for all of God's creatures, that she was well described as possessing "the brain of a man and the heart of a woman."

In order to get at first-hand information that was necessary to complete a work which her father, George Kingsley, had, owing to his premature death, left unfinished, she determined to visit that part of West Africa "where all authorities agreed that the Africans were at their wildest and worst." Accompanied only by the natives, she travelled among cannibals, pushed her way through mangrove swamps and pestilential morasses. She spent months in a canoe exploring the territory watered by the Calabar and Ogowe rivers, often in imminent peril of death from wild animals or wilder men.

When not studying the manners and customs of the native tribes, she was hunting fishes and reptiles in streams and quagmires and collecting insects in the weird, grim twilight of the equatorial forest with its inextricable tangle of creepers, its great hanging tapestries of vines and flowers, its myriads of bush-ropes, suspended from the summits of tall buttressed trees, "some as straight as plumb lines, others coiled round and intertwined among each other until one could fancy one was looking on some mighty battle between armies of gigantic serpents that had been arrested at its height by some mighty spell."

The results of Miss Kingsley's wanderings in this dark and uncanny wilderness and among the savage tribes visited by her were her two instructive volumes entitled *Travels in West Africa* and *West African Studies*. In addition to these two works from her pen there are deposited in the British Museum an interesting collection of insects, fishes and reptiles — many of them new species and some of them named in her honour — which testifies to her activity as a collector and her enthusiasm as a naturalist.

Her brilliant and useful career was cut short in Cape Colony, whither she had gone as an army nurse during the Boer war. In view of her achievements one is not surprised to learn that her countrymen regarded her premature taking-off as a national misfortune. The noblest monument to her memory is "The Mary Kingsley Society of West Africa," whose object is to carry on, as far as may be, the beneficent work she began on the West African coast and to accomplish for English rule in this part of the world what the "Royal Asiatic Society" has achieved for British administration in India.

Madame Coudreau is designated in *Qui Etes-Vous* — the French Who's Who — as an *exploratrice.* This well characterises her; for, if not the first woman explorer by profession, she is certainly the most energetic and successful.

Her first work was in French Guiana, under instructions from the colonial minister of France. This was in 1894. The following year she began the scientific exploration of the province of Pará in northern Brazil, in collaboration with her husband, Henri Coudreau, who had previously distinguished himself by his achievements as a writer and as an explorer in French Guiana. The fruit of their joint work from 1895 to 1899 was six quarto volumes profusely illustrated by photographs which they had taken and by carefully executed charts of the various rivers which they had explored.

While engaged in the exploration of the Trombetas, a tributary of the Amazon, Henri Coudreau was taken seriously ill, and, after a few days' struggle against the disease with which he was stricken, he expired in the depths of the forest primeval, where he was buried by his desolate and disconsolate widow. After such a calamity any other woman would have left the tropics at once and returned to her home and friends. Not so Mme Coudreau. With matchless courage and determination she buried her grief in the work in which her husband had been so interested, and, after completing the unfinished survey, published the results of this expedition under the title *Voyage au Trombetas.*

Having completed this work, she was engaged by the states of Pará and Amazonas to explore a number of other rivers in the vast territory known as Amazonia. This commission involved the most arduous and dangerous kind of labour and was a task which few men would have been willing to undertake. It is doubtful if any other woman would have ventured on such an expedition, and it is quite certain that no other one could have been found that was so well equipped for this herculean undertaking or who would have carried it to a more successful issue.

Mme Coudreau was in the service of Amazonia, in the capacity of official explorer, from 1899 to 1906. Most of this time she spent in a canoe on the affluents of the Amazon, or in her tent in the dense forests under the equator. Her only companions were negroes, or Indians, or Brazilian half-breeds who served her as porters, cooks and boatmen. Frequently they were in the forest wilds for many months at a time and far away from every vestige of civilised life. As it was impossible to take sufficient provisions with them to last them during the whole of their journey, they had to depend on wild fruits and such fish and game as they were able to secure. Often they were forced to live for weeks at a time on an unchanging diet of manioc and tapir meat.

But their sufferings were not confined to hunger and disagreeable — often indigestible — food. There were the heavy steaming atmosphere and the broiling rays of a superheated sun, especially when reflected from the mirror-like surface of lake or river, which were so debilitating and exhausting that physical exertion of any kind was at times almost impossible. There were also the torrential and incessant rains — making it impossible for them to cook their food or dry their clothing — which added to their miseries whether in camp or in their canoe.

Great, however, as were their trials on the river, they were trifling in comparison with those in the woods. Here locomotion was impeded by tangled undergrowth which was bound together by strands of lianas and thorny vines which constituted an impenetrable barrier until a passage was hewn through it with a machete. Under foot was a yielding morass which threatened to absorb them. Overhead were countless chigoes, garapatas and fire-ants which infested the body or buried themselves in the flesh. Or there were clouds of mosquitoes which gave no rest day or night. And worst of all was the ever-present danger of fever and dysentery, not to speak of the dread diseases so common in certain sections of the equatorial regions. It was then that Mme Coudreau had to act the part of a physician, as well as of a leader, even though she was at the time such a sufferer herself that she was barely able to stand.

To make matters still more difficult for Mme Coudreau, her employees at times, especially when under the influence of liquor which they contrived to obtain some way or other, became mutinous and refused to accompany her to the end of her journey. At other times the expedition was halted by their fear of wild beasts or savage Indians, or by imaginary evils of many kinds, suggested to them by their superstitious minds. On such occasions Mme Coudreau never failed to show herself a born leader of men, for she invariably — alone as she was with a crew who were often half savages — was successful in suppressing incipient rebellion and in restoring obedience and order.[14]

Continually confronted, as she was, by such trials and difficulties, privations and dangers, one would imagine that the delicately reared Frenchwoman would have sought immediate release from an engagement that necessitated so much exposure and suffering and sought surcease of sorrow in the distractions and gaieties of pleasure-loving Paris.

Nothing, however, was farther from her thoughts. Intrepid and resourceful, she feared no danger and hesitated before no difficulty, however great. As an explorer she was as venturesome as Crevaux and as conscientious as La Condamine. Like them, who were both her countrymen, she spent many years of her life in the equinoctial regions, and, like them, she contributed immensely to our knowledge of the Land of the Southern Cross.

Never did the tropics have a greater fascination for anyone

than for Mme Coudreau. During the twelve years she spent there, exploring its rivers and traversing its interminable forests, the spell of Amazonia was ever upon her and was never broken, even for a moment.

"I have," she writes, "loved everything in Amazonia, the great majestic woodland and the mysterious virgin forest, the beautiful rivers with their traitorous waters and thundering cataracts, the suffocating air and the perfumed breeze, the burning sun and the sweet freshness of night, the impressive voice of the wind among the trees and the torrential rain. And, contrary to the usual custom of man of bringing everything under his domination, it is I who have become a captive of this savage life which I love, and have permitted it to take possession of all my soul and all my will."[15]

Elsewhere she declares: "In the solitude of the virgin forest I am calm, tranquil, experience no ennui and am almost merry. When I am obliged to leave the great woodland the power to struggle grows less in me. I become of an excessive sensibility. I feel more keenly life's blows. I am not armed for elbowing my way and making a place for myself in the sunshine. I neither love nor understand anything except my virgin forest. There, indeed, I suffer from the inclemency of the weather, from hunger, from sickness; but these are only physical sufferings and are soon forgotten, while moral and interior pains, on the contrary, are ineradicable."[16]

And still again she tells us: "The solitude of the virgin forest has become a necessity for me; it attracts me by its mysterious silence, and only in the great woods have I the impression of being at home."[17]

Can we wonder that such an ardent lover of Nature and such a strenuous votary of science was able to forget herself in her work and was able, notwithstanding her toils and her sufferings, to produce six quarto volumes of reports, in as many years, on the unexplored regions which she had so carefully surveyed and charted? Can we be surprised that her labours received due recognition from learned societies in both the New and the Old World, and that she was acclaimed as an explorer who had rendered distinct service to the cause of natural science, as well as to geography?[18]

When we recall the labours of this lone daughter of France in the wilds of the tropics, with no one to communicate with except her half-civilised servants and boatmen, we instinctively hark back to days not long past and estimate the enormous progress women have made in social and intellectual freedom within but a few decades.

Owing to the policy of repression which so long prevailed regarding the intellectual efforts of women, and the social obstacles which prevented them from publicly acknowledging the offspring of their genius, women like the Brontë sisters, George Sand and George Eliot were compelled to conceal their identity under male designations. Because it was considered immodest for a woman to appear before the public as an author, Lady Nairne, after Burns, the most popular song writer in Scotland, felt obliged to keep secret the authorship of her beautiful poems.

Similarly, family honour made it incumbent on Fanny Mendelssohn to refrain from publishing her musical compositions under her own name. Accordingly, they appeared along with those of her brother Felix, and so similar are they in colour and sentiment to his own productions that they are indistinguishable from them, unless the author's signature be attached. To satisfy an inane public opinion, they long contributed "to swell the volume of her brother's fame," and there is reason to believe that some of them still appear under his name at the present day.

Yes, truly, when one recalls these and similar facts, one cannot help exclaiming: "What a marvellous change in the attitude of the world toward women within the memories of those still living!" Women like Miss Ormerod, Miss Kingsley and Mme Coudreau would have been ostracised if they had dared to attempt, in the days of Lady Nairne, the Brontë sisters and Fanny Mendelssohn, what they may now do not only without censure but without exciting more than passing comment. The ban has been lifted from what was for ages taboo for women, and the sphere of their intellectual activities is now almost coextensive with that of the sterner sex. Not only does society no longer point the finger of scorn at the woman naturalist or the woman explorer, but it showers honours on her while living and erects monuments to her memory when dead. A great change, indeed, and one long and ardently desired. Verily, *tempora mutantur, nos et mutamur in illis.*

## Notes & References

1) A few days before Mme Curie's name was to come before the Academy of Sciences as a candidate for membership, the French Institute in its quarterly plenary meeting of the five academies, of which the Institute is composed, decided by a vote of ninety to fifty-two against the eligibility of women to membership, and put itself on record in favour of the "immutable tradition against the election of women, which it seemed eminently wise to respect."

Commenting on this decision of The Immortals, a writer in the well-known English magazine, *Nature*, under date of January 12, 1911, penned the following pertinent paragraph:

It remains to be seen what the Academy of Sciences will do in the face of such an expression of opinion. Mme Curie is deservedly popular in French scientific circles. It is everywhere recognised that her work is of transcendent merit, and that it has contributed enormously to the prestige of France as a home of experimental inquiry. Indeed, it is not too much to say that the discovery and isolation of the radio-active elements are among the most striking and fruitful results of a field of investigation pre-eminently French. If any prophet is to have honour in his own country — even if the country be only the land of his adoption — surely, that honour ought to belong to Mme Curie. At this moment, Mme Curie is without doubt, in the eyes of the world, the dominant figure in French chemistry. There is no question that any man who had contributed to the sum of human knowledge what she has made known, would years ago have gained that recognition at the hands of his colleagues, which Mme Curie's friends are now desirous of securing for her. It is incomprehensible, therefore, on any ethical principles of right and justice that, because she happens to be a woman, she should be denied the laurels which her pre-eminent scientific achievement has earned for her.

Compare this frank and honest statement with that of a contributor, about the same date, to *La Revue du Monde*, of Paris. Guided by his myopic vision and diseased imagination, this writer discerns in the admittance of women into the grand old institution of Richelieu and Napoleon the imminent triumph of what Prudhon called pornocracy and the eventual opening of the portals of the Palais Mazarin to representatives of the type of Lais and Phryne, on the Hellenic pretext that "Beauty is the supreme merit."

It is gratifying, however, to the friends of woman's cause to learn that Mme Curie's candidacy was defeated by only two votes. Her competitor, M Branly, received thirty votes against the Polish woman's twenty-eight. She thus fared far better than did Mme Pauline Savari, who aspired to the fauteuil made vacant by the death of Renan, regarding whose candidature the Academy curtly declared, "Considering that its traditions do not permit it to examine this question, the Academy passes to the order of the day." Thus, it will be seen that, in spite of the long-continued opposition to women members, the French Academy is more than likely to offer its next vacant chair to the pride and glory of Poland — the immortal discoverer of radium and polonium.

2) In his erudite work, *Geschichte der Botanik*, Vol III, p 517, Koenigsberg, 1856, Ernest H F Meyer gives in a few words his estimate of the excellence of Hildegard's *Physica*: "Aber als ehrwürdiges Denkmal des Alterthums und einer zu jener Zeit nicht gemeinen Naturkentniss empfehlen sich zumal deutschen Naturforschern ihre vier Bücher der *Physica*..... Denn nicht nur der deutsche Botaniker und Zoologe finden in ihrer Physik fast die ersten rohen Anfänge vaterländische Naturforshung, auch dem Artzt bietet sic für jene Zeit überraschende Erscheinung dar, eine nicht von Dioskorides abgeleitete, sondern unverkennbar aus der Volksüberlieferung geschöpfte Heilmittellehre; und der Sprachforscher stösst im lateinischen Text beinahe Zeile um Zeile auf deutsche Ausdrücke seltener Sprachformen."

3) Hildegardis *Opera Omnia*, p 1122, Migne's Edition, Paris, 1882.

4) "Constat permulta S Hildegardi nota jam fuisse, quae caeteri medii aevi scriptores nescierunt, quaeque sagaces demum recentiorum temporum indagatores reperierunt ac tamquam nova ventitarunt." *Ibid.* Dr Karl Jessen, in his thoughtful *Botanik der Gegenwart und Vorzeit in Culturhistorischer Entwickelung*, p 123, Leipzig, 1864, expresses himself on the extraordinary medical knowledge of the abbess of Bingen as follows: "Wer deutsche Volkarznei studieren will, der studiere Hildegard und er wird Respect davor bekommen."

5) *Compendio Storico della Scuola Anatomica di Bologna*, p 358, by Michele Medici, Bologna, 1857, and *Notizie degli Scrittori Bolognesi*, Vol VI, p 113, by Giovanni Fantuzzi, Bologna, 1788. Certain writers tell us of another woman who distinguished herself in anatomy in the early part of the fourteenth century. Her name

was Alessandra Giliani, who is said to have been a pupil and an assistant of the celebrated Mondino, father of modern anatomy. In addition to possessing great skill in dissection, she is reputed to have devised a means of drawing the blood from the veins and arteries — even the most minute — and then filling them with variously coloured liquids which quickly solidified. By this means, we are told, she was able to exhibit the circulatory system in all its details and complexity, and to have always on hand, for purposes of instruction, a model that was absolutely true to nature.

How much truth there may be in these statements regarding a young girl, who was only nineteen when she died, is difficult to determine. Medici, in concluding his account of her and referring to the inscription on her tomb, which seems to authenticate all the claims made for her, expresses himself as follows: "In quoting this document, I do not intend that my readers shall accord to it a credence that I myself abstain from giving it, but only that they may know of it, if for no other reason than to satisfy their curiosity." *Op cit*, pp 30 and 362, note I. Should the traditions regarding this precocious girl be verified, it would be most gratifying to the people of Bologna, for it would add one more to the long list of her illustrious women.

6) The titles of the two works of this remarkable woman are of sufficient interest to be given in full. They are as follows:

1. *Veritable Declaration de la Decouverte des Mines et Minieres par le Moyen desquelles Sa Majeste et Sujets se peuvent passer des Pays Etrangers*, Paris, 1632.
2. *La Restitution de Pluton à Mgr l'Eminent Card. de Richelieu, des Mines et Minieres de France, cachees jusqu'à present au Ventre de la Terre, par la Moyen desquelles les Finances de sa Majeste seront beaucoup plus Grandes que celles de tous les Princes Chrestiens et ses Sujets plus Heureux de tous les Peuples*, Paris, 1640.

7) *Die Verdienste der Frauen um Naturwissenschaft and Heilkunde*, p 169, von Dr C F Harless, Göttingen, 1830.

8) The Latin title of this interesting work is *De Generatione et Metamorphose Insectorum Surinamensium*, Amsterdam, 1705.

9) The Latin edition of this work is entitled *Erucarum Ortus, Alimenta et Paradoxa Metamorphosis*, Amsterdam, 1718. It was afterwards translated into French and published under the title *Histoire des Insectes de l'Europe*.

10) *Die Leistungen der deutschen Frau in den letzen vierhundert Jahren auf wissenschaftlichem Gebiebte*, p 85, von Elise Oelsner,

Guhrau, 1894.

11) In his preface to *Les Maladies des Vers à Soie.*

12) It is estimated that the loss to the United States from cattle ticks alone is $100,000,000 a year. According to the yearbook of the Agricultural Department for 1904, the annual losses to agriculture from destructive insects reach the enormous sum of $420,000,000.

13) The dean of the law faculty in presenting Miss Ormerod to the vice-chancellor on this occasion and speaking before an audience of three thousand people said, among other things: "The pre-eminent position which Miss Ormerod holds in the world of science is the reward of patient study and unwearying observation. Her investigations have been chiefly directed towards the discovery of methods for the prevention of the ravages of those insects which are injurious to orchard, field and forest. Her labours have been crowned with such success, that she is entitled to be hailed as the protectress of agriculture and the fruits of the earth — a beneficent Demeter of the nineteenth century." *Eleanor Ormerod, Economic Entomologist, Autobiography and Correspondence*, Edited by Robert Wallace, p 96, London, 1904.

14) *The Canadian Entomologist,* September, 1901, in an obituary notice of Miss Ormerod, well voiced the high appreciation in which she was held throughout the civilised world in the following paragraph: "Miss Ormerod was one of the most remarkable women of the latter half of the nineteenth century and did more than any one else in the British Isles to further the interests of farmers, fruit-growers and gardeners by making known to them methods for controlling and subduing their multiform insect pests. Her labours were unwearied and unselfish; she received no remuneration for her services, but cheerfully expended her private means in carrying out her investigations and publishing their results. We know not now by whom in England this work can be continued; it is not likely that anyone can follow in the unique path laid out by Miss Ormerod; we may, therefore, cherish the hope that the Government of the day will hold out a helping hand and establish an entomological bureau for the lasting benefit of the great agricultural interests of the country." Professor J Ritzema Bos, the distinguished entomologist of Holland, had no hesitation in proclaiming Miss Ormerod the first economic entomologist in England and one of the most famous economic entomologists in the world.

15) The following dialogue between Mme Coudreau and one of her boatmen, Joas-Felix, who was the spokesman of his companions, illustrates not only the bravery of the daring explorer, but also the

pusillanimity of her half-breed personnel when in the depths of the forest at night:

'Madam has no fear?'

'Fear of what?'

'Of tigers.'

'No, it is not of tigers that I have fear.'

'Of Indians?'

'Neither have I fear of Indians.'

'Then, madam, it is something which is in the woods, which we do not know, that can harm us.'

'You know very well what frightens me. I am afraid that the bats will attack my chickens during the night. If you hear them making a noise you must get up'

I laugh heartily in observing their astonished look and ask myself how men whose consciences are stained with many bloody crimes can have fear here. Joas-Felix gives me the explanation:

'Madam makes game of us. None the less, madam, I am a man in the city and in the savanna. With my poignard and machete I fear nothing, neither man nor beast. But here, madam, where everything is dark, even in the daytime; where an enemy may be lying in wait for us behind every tree; it is not the same thing. It would be impossible for me to live in the forest. One cannot see far enough in it.'

Now I understand better their terror. The mysterious depth of the virgin forest impresses them. The opaque obscurity of the night in the underwood contrasts too strongly with the moonlit savanna where they have been reared. The low and sombre vault of the woods oppresses them and they imagine they are going to be crushed. They lose their heads and see in every tree a phantom enemy. To reason with them is useless, for when fear takes possession of them, there is nothing to be done." *Voyage au Maycurú*, p 127.

16) *Voyage au Maycurú*, p 1, Paris, 1903.

17) *Voyage au Rio Curuá*, p 85, Paris, 1903.

18) *Ibid*, p 1.

# Women in Medicine and Surgery

AS WOMAN WAS the first nurse, so was she also the first practitioner of the healing art. Among savages the world over it is the women, in the great majority of cases, who have the care of the sick and wounded, and who, by reason of their superior knowledge of simples for the cure of diseases, occupy the position of doctors. In certain parts of the uncivilised world there are, it is true, shamans or medicine men; but these are conjurers or exorcists, who profess to expel disease, or rather the evil spirits causing the disease, by sorcery or incantation, rather than physicians who essay to cure ailments or relieve suffering by the use of substances which experience has showed to possess remedial properties. In a word, the shaman is a kind of a religious functionary who imposes on the ignorance of his tribe and who holds his position by the fear he excites, and not by any knowledge he possesses of the healing art. It was the same, we may believe, in the early history of our race — women, and not men, were the first physicians; and they were also most probably the first surgeons.

According to Greek mythology, the god of the medical art was Aesculapius, a male; but his six daughters, as antiquity beautifully expressed it, were not only goddesses but were also medical mistresses — *artifices medici* — of suffering humanity. Of these Hygiea was specially distinguished as the goddess of health, or, rather, as the conserver of good health, while Panacea was invoked as the restorer of health after it had been impaired or lost.

One of the most beautiful pictures in the Iliad is that representing the daughter of Augea, King of the Epei, caring for the wounded and suffering Greeks on the plain before Troy. She was:

> His eldest born, hight Agamede, with golden hair, A leech was she, and well she knew all herbs on ground that grew. Nothing deterred by the din of battle around her, she provided cordial potions for the disabled warrior and prepared.
> The gentle bath and washed their gory wounds.

What a beautiful prototype of another ministering angel in the same land nearly thirty centuries later, amid similar scenes of suffering — of one who, though unsung by immortal bard, the world will never let die — the courageous, the self-sacrificing Florence Nightingale.

That there were in Greece from the earliest times numerous women possessed of a high degree of medical skill is evidenced by many of the ancient writers. They were what we would call medical herbalists, and not a few of them exhibited a natural genius for determining the curative virtues of rare plants and a remarkable sagacity in preparing from them juices, infusions and soothing anodynes. Others there were who, in addition to evincing the cunning of leechcraft in the therapeutic art, were distinguished for nimble hands in treating painful lesions and festering sores, and who, when occasion required, were experts in "quickly drawing the barb from the flesh and healing the wound of the soldier."

In the Odyssey special mention is made of the surpassing expertness of the Egyptian female leech, Polydamna, whose name signifies the subduer of many diseases. The land of the Nile, the poet tells us, "teems with drugs," and

> There ev'ry man in skill medicinal
> Excels, for these are sons of Paeon all.

In this favoured cradle of civilisation, to which Greece owed so much of its knowledge and culture, there were many women who, like Polydamna, achieved distinction in the healing art, and many, too, we have reason to think, who communicated their knowledge to their sisters in the fair land of Hellas.

But not only were there in Greece women physicians like Agamede, who were noted for their general medicinal knowledge and practice, but there were also others who made a specialty of treating ailments peculiar to their own sex. This we learn from a passage in the *Hippolytus* of Euripides, wherein the nurse of Phaedra addressed the suffering queen in the following words:

> If under pains
> Thou labour, such as may not be revealed,
> To succor thee thy female friends are here.
> But if the other sex may know thy sufferings
> Let the physician try his healing art.

More positive information, however, is afforded us by the ancient Roman author Hyginus, who, in writing of the Greek maiden, Agnodice, tells us how the medical profession was legalised for all the free-born women of Athens. Instead of a literal translation of Hyginus, the version of his story is given in the quaint language of one Mrs Celleor, a noted midwife in the reign of James II.

"Among the subtle Athenians," writes Mrs Celleor, "a law at one time forbade women to study or practice medicine or physick on pain of death, which law continued some time, during which many women perished, both in child-bearing and by private diseases, their modesty not permitting them to admit of men either to deliver or cure them. But God finally stirred up the spirit of Agnodice, a noble maid, to pity the miserable condition of her own sex, and hazard her life to help them; which, to enable herself to do, she apparelled her like a man and became the scholar of Hierophilos, the most learned physician of the time; and, having learnt the art, she found out a woman that had long languished under private diseases, and made proffer of her service to cure her, which the sick person refused, thinking her to be a man; but, when Agnodice discovered that she was a maid, the woman committed herself into her hands, who cured her perfectly; and after her many others, with the like skill and industry, so that in a short time she became the successful and beloved physician of the whole sex."

When it became known that Agnodice was a woman "she was like to be condemned to death for transgressing the law — which, coming to the ears of the noble women, they ran before the Areopagites, and, the house being encompassed by most women of the city, the ladies entered before the judges and told them they would no longer account them for husbands and friends, but for cruel enemies, that condemned her to death who restored to them their health, protesting they would all die with her if she were put to death. This caused the magistrates to disannul the law and make another, which gave gentlewomen leave to study and practice all parts of physick to their own sex, giving large stipends to those that did it well and carefully. And there were many noble women who studied that practice and taught it publicly in their schools as long as Athens flourished in learning."[1]

After the time of Agnodice many Greek women won distinction in medicine, some as practitioners in the healing art, others as writers on medical subjects. Nor were their activities confined to the land of Hellas. They were also found succouring the infirm and instructing the poor and ignorant in Italy, Egypt and Asia Minor. Among these was Theano, the wife of Pythagoras, who, after her husband's death, assumed charge of his school of philosophy, and who, like her husband and teacher, was distinguished for her attainments in medicine. The names of many others occur in the pages of Hippocrates, Galen and Pliny; and frequent references are made to the works and prescriptions of women doctors who enjoyed more than ordinary celebrity during their time. Of these female practitioners many confined their practice to the diseases of women and children, while others excelled in surgery and pharmacy, as well as in general medical practice.

Among the medical women whom antiquity especially honoured, particularly during the Greco-Roman period, were Origenia, Aspasia — not the famous wife of Pericles — and

Cleopatra, who was not, however, as is often asserted, the ill-fated queen of Egypt. Likewise deserving of special mention was Metradora, of whom there is still preserved in Florence a manuscript work on the diseases of women,[2] and Antiochis, to whom her admiring countrymen erected a statue bearing the following inscription: "Antiochis, daughter of Diodotos of Tlos; the council and the commune of the city of Tlos, in appreciation of her medical ability, erected at their own expense this statue in her honour."

Pliny, the naturalist, felicitates the Romans on having been for nearly six hundred years free from the brood of doctors. These he does not hesitate to berate roundly. His statement regarding the non-existence of physicians, it must be observed, is somewhat exaggerated. It is true that during the first five centuries there were no professional doctors who lived entirely on their practice. There were, however, many men who had by long experience gained an extensive knowledge of drugs and simples, and who were able to dress wounds and treat diseases with considerable success.

The first Greek freeman to practice medicine in Rome was one Archagatos, about two centuries BC. He was soon followed by one of his countrymen named Asclepiades. These two soon built up a great reputation as successful practitioners, and were held in the highest esteem by the people of Rome. In consequence of this and of the favourable conditions offered foreigners for the practice of the healing art, there was soon a large influx of physicians and surgeons from Greece, not only into Rome but also into other parts of Italy.

Not long after the arrival of Greek doctors in the capital of the Roman world we learn of certain women physicians in Rome who were held in high repute. Among these were Victoria and Leoparda, both mentioned by the medical writer, Theodorus Priscianus. To Victoria, Priscianus dedicates the third book of his *Rerum Medicarum*, and in the preface to this book he refers to her as one who has not only an accurate knowledge of medicine, but also as one who is a keen observer and experienced practitioner.

The word *medica*, which occurs in Latin authors of the

classical period, testifies to the existence of the woman doctor as early as the age of Augustus.

But the most important documents bearing on women physicians, not only in the city of Rome but also in Italy, Gaul and the Iberian peninsula, are the large body of epigraphic monuments which have recently been brought to light, and which prove beyond all doubt that women were not only obstetricians, but that they were successful practitioners in the entire field of medical art. Thus a funeral tablet found in Portugal tells of a woman who was a most excellent physician — *medica optima* — while another describes the deceased not only as a woman incomparable for her virtues, but also as a mistress of medical science, *antistes disciplinae in medicina fuit.*

The Greek word for *medica* — *iatromaia* — occasionally found in some of the inscriptions, seems to refer specially to women of Greek origin or birth. This is particularly true of a monument erected to one Valiae, who is designated as *Kalista iatromaia* — the best doctor.[3]

Among the many women who became converts to Christianity during the early ages of the church a goodly number were physicians. Unfortunately, our information respecting these votaries of the healing art is not as complete as we could wish. One of the most noted of them is St Theodosia, whose name is given in the Roman martyrology for the twenty-ninth of May. She was the mother of the martyr, St Procopius, and was distinguished for her knowledge of medicine and surgery, both of which she practiced in Rome with the most signal success. She died a heroic death by the sword during the persecution of Diocletian.

Another woman who was as eminent for her knowledge of medicine as for her holiness of life was St Nicerata, who lived in Constantinople during the reign of the emperor Arcadius. She is said to have cured St John Chrysostom of an affection of the stomach from which he was a sufferer.

To the Roman lady Fabiola, remarkable as the daughter of one of the most illustrious patrician families of Rome, but more remarkable for her sanctity and her boundless charity toward the poor, was due the erection of the first hospital — a noble structure which she founded in Ostia, at the mouth of the Tiber, which was then the port of entry to the capital of the Roman empire. Here the noble matron received the poor and suffering from all parts, and did everything in her power to afford them succour in their wants and infirmities.

It is difficult for us now, when hospitals and charitable institutions of all kinds are so common, to understand what an innovation Fabiola's unheard-of institution was considered by her contemporaries. For her method of treating the needy and the suffering was as different from that which had hitherto obtained as were the debasing lessons of heathendom from the elevating precepts of the Gospels.

No wonder that the news of this godlike work was soon wafted to the uttermost bounds of the earth; that, in the words of St Jerome, "summer should announce in Britain what Egypt and Parthia had learned in the spring." No wonder that the same eloquent hermit of Bethlehem should proclaim the foundress of this home of the indigent and the afflicted to be "the glory of the church, the astonishment of the Gentiles, the mother of the poor and the consolation of the saints." No wonder that, in contemplating her countless acts of charity, he should ignore the fact that Fabiola was a daughter of the Fabii and a descendant of the renowned Quintus Maximus, who, by his sage counsel, had saved his country from her enemies, and that, recalling the words of Virgil, he should declare: "If I had a hundred tongues and a hundred mouths and iron lungs, I should not be able to enumerate all the maladies to which Fabiola gave the most prodigal care and tenderness — to the extent even of making the poor who were in health envy the good fortune of those who were sick."[4] No wonder that Fabiola's funeral, which brought together the whole of Rome, was more like an apotheosis than the transfer of the remains of the deceased to their last resting-place, and that Jerome should declare, "the glory of Furius and Papirius and Scipio and Pompey, when they triumphed over the Gauls, the Sammites, Numantia and Pontus" was less than that which was spontaneously accorded to Fabiola, the solace of the sick and the comforter of the distressed. For she had in her hospital at Ostia established a type of institution that was to effect more for ameliorating the condition of suffering humanity than anything that had before been dreamed of; something that was to contribute immensely to the efforts of physicians and surgeons in minimising the sad ravages of wounds and disease; something whose beneficent effects were to be felt through the centuries and in every part of the world down to the wards of the military hospital at Scutari, guarded by the watchful eyes of Florence Nightingale, and to the leper-tenanted lazarettos, blessed by the ministrations of Father Damien and the Sisters of Charity, on the desolate shores of plague-stricken Molokai.

After the fall of the Roman empire and through the long period of the Middle Ages, when the monasteries and convents were almost the only centres of learning and culture for the greater part of Europe, the practice of medicine was to a great extent in the hands of monks and nuns. For every religious house was then a hospital as well as a school, a place where drugs and ointments were compounded and distributed, as well as a place where manuscripts were transcribed and illuminated. At a time when there were but few professional physicians and when these few were widely separated from one another, the only places where the poor could always be sure to find free medical treatment as well as abundant alms were those sanctuaries of knowledge and charity where the love of one's neighbour was never lost sight of in the love of science and literature. And during this time, too, the care of the sick was regarded as a duty incumbent on everyone, but particularly on those devoted to the service of God in religion. It was considered, above all, as a duty devolving on women, especially on the lady in the castle and on the nun in the convent.

The old romance of *Sir Isumbras* gives us a charming picture of the nuns of long ago receiving the wounded knight and ministering unto him until he was made whole and strong, as witness the following verses:

The nonnes of him they were full fayne,
For that he had the Saracenes slayne
And those haythene houndes.
And of his paynnes sare ganne them rewe.
Ilke a day they made salves new
And laid them till his woundes;
They gave him metis and drynkis lythe,
And heled the knyghte wunder swythe.

So universally during medieval times was the healing art considered as pertaining to woman's calling that it became a part of the curriculum in convent schools; and no girl's education was considered complete unless she had an elementary knowledge of medicine and of that part of surgery which deals with the treatment of wounds. For during those troublous times a woman was liable to be called upon at any time to nurse the sick wayfarer or dress the wounds of those who had been maimed in battle or in the tourney.

Illustrations of these facts are found in many of the romances and fabliaux of the Middle Ages. Thus, when a sick or wounded man was given hospitality in a chateau or castle it was not the seigneur, but his wife and daughters, as being better versed in medicine and surgery, who acted as nurses and doctors and took entire charge of the patient until his recovery.

In the exquisite little story of *Aucassin et Nicolette*, the heroine is pictured as setting the dislocated shoulder of her lover in the following simple but touching language:

"Nicolette searched his hurt, and perceived that his shoulder was out of joint. She handled it so deftly with her white hands, and used such skilful surgery that, by the grace of God, who loveth all true lovers, the shoulder came back to its place. Then she plucked flowers and fresh grasses and green leafage, and bound them tightly about the setting with the hem torn from her shift, and he was altogether healed."

And in the medieval Latin poem, *Waltharius*, written by a

German monk, Ekkehard, reference is made to a sanguinary contest in which one of the combatants falls to the earth seriously wounded. Seeing this, Alpharides, in a loud voice, summons a young girl, who timidly comes forward and dresses the unfortunate man's wound.[5]

Still more to our purpose is a passage from the famous epic poem, *Tristan and Isolde,* written by *Godfrey of Strasburg,* in which Isolde, accompanied by her mother and cousin, is represented as administering restoratives to Tristan, who had fallen exhausted after his combat with the dragon. It shows that women, in accompanying an army to the field of battle, always went provided with bandages and medicaments for dressing wounds and fractured limbs. Similarly Angelica, in *Orlando Furioso,* and Ermina, in *Jerusalem Delivered,* are portrayed as surgeons with deftness of hand and leeches with rare knowledge and skill.

The frequent introduction of women doctors into the poems and romances of the Middle Ages would of itself, if other evidence were wanting, suffice to show what an important role women played in medicine and surgery at a time when, in many parts of Europe, women were far better educated and far more cultured than men — "when the knights and barons of France and Germany were inclined to look upon reading and writing as unmanly and almost degrading accomplishments, fit only for priests or monks, and especially for priests or monks not too well born."[6]

In the instances just quoted, as well as those mentioned by Homer and Euripides, the writers do no more than faithfully reflect conditions which then obtained, and truthfully report what were the occupations of women when their status was so different from what it is today. But, fortunately, we do not have to rely on works of the imagination for our knowledge respecting the women practitioners of the healing art, either during the Homeric period or during that which intervened between the downfall of Rome and the dawn of the Renaissance. For the history of medicine during medieval times affords too many examples of women who became famous for their knowledge of medicine, as well as for their success in surgical and medical practice, to leave any doubt about the matter. Besides this, we have still the writings of many of these women, and are thus able to judge of their competency in those branches of knowledge on which they shed so great lustre.

One of the most noted of them was the Benedictine abbess, St Hildegard, of Bingen on the Rhine, who was eminent not only as a theologian but also as a writer whose treatises on various branches of science are justly regarded as the most important productions of the kind during the Middle Ages prior to the time of Albertus Magnus. Besides this, she not only wrote many books on *materia medica*, on pathology, physiology and therapeutics, but, as a practitioner, she gloriously sustained the best traditions of her sex in both theoretical and practical medicine.

Her work entitled *Liber Simplicis Medicinae*, which deals with what in the Saint's time was called "simples" — for the belief was then current that each plant or herb was or provided a specific for some disease — contains accounts of many plants used in *materia medica*, as well as statements of their importance in therapeutics. Her descriptions often indicate an observer of exceptionally keen perception and one whose knowledge of science was far in advance of her epoch. The same observations may be made respecting Hildegard's work, *Liber Compositae Medicinae*, in which she treats of the causes, signs and treatment of diseases.[7]

Still more remarkable, in many respects, is a treatise in nine books, entitled *Physica* or *Liber Subtilitatum Diversarum Naturarum Creaturarum*, which, among other things, treats of the various elements, of plants, trees, minerals, fish, birds, quadrupeds, and of the manner in which they may be of service to man. Of so great importance was this book considered that several editions of it were printed as early as the sixteenth century. No less an authority than the late Rudolph Virchow, the founder of cellular pathology, characterises it as an early *materia medica*, curiously complete, considering the age to which it belongs.[8] And Haeser, in his history of medicine, directs attention to the historical value of the book, declaring it to be "an independent German treatise, based chiefly on popular experience."

Dr F A Reuss, of the University of Würtzburg, at the conclusion of his *Prolegomena* to the *Physica* published in Migne's *Patrologia*, expresses himself as follows regarding the writings and medical knowledge of the illustrious abbess of Bingen: "Among all the saintly *religieuses* who, during the Middle Ages, practiced medicine or wrote treatises on it, the first, without contradiction, is Hildegard. According to the monk Theodoric, who was an eye witness, she had to so high a degree the gift of healing that no sick person had recourse to her without being restored to health. There is among the books of this prophetic virgin a work which treats of physics and medicine. Its title is *De Natura Nominis Elementorum Diversarumque Creaturarum*, and it embodies, as the same Theodoric fully explains, the secrets of nature which were revealed to the saint by the prophetic spirit. All who wish to write the history of the medical and natural sciences should read this book, in which the holy virgin, initiated into all the secrets of nature which were then known, and having received special assistance from above, thoroughly examines and scrutinises all that which was, until then, buried in darkness and concealed from the eyes of mortals. It is certain that Hildegard was acquainted with many things of which the doctors of the Middle Ages were ignorant, and which the investigators of our own age, after rediscovering them, have announced as something entirely new."[9]

The life and works of St Hildegard throw a flood of light on many subjects that have long been veiled in mystery. It explains why the convents of the later Middle Ages were so famed as curative centres and why the sick flocked to them for relief from far and near. It reveals the real agencies employed in effecting the extraordinary cures that were reported in so many religious houses — cures so extraordinary that they were usually regarded by the multitude as miraculous — and discloses the secret of the success of so many nuns in the alleviation of physical and mental sufferings. It was not because they were thaumaturges, but because they were good nurses, and because of their thorough knowledge of the healing art, that they were able to diagnose and prescribe for diseases of all kinds with a success which, in the estimation of the multitude, savoured of the supernatural.

There was also another reason for the fame of convents as sanctuaries of health. They were usually situated in healthy locations where there was an abundance of pure water, fresh air and cheerful sunshine. Then there were likewise a wholesome diet, good sanitary conditions, and, above all, regularity of life.

The same can be said of the hospitals connected with the convents. They were not like some of the public hospitals of the eighteenth and nineteenth centuries in many of the large cities of Europe — repulsive, prison-like structures, with narrow windows and devoid of light and air and the most necessary hygienic appliances — institutions that were hospitals in name, but which were in reality too frequently breeding places of disease and death.[10]

Unlike these, the hospitals presided over by nuns of the type of Hildegard were splendid roomy structures with large windows and abundance of light, pure air, with special provisions for the privacy of the patients, and with sanitary arrangements that not only precluded the dissemination of disease but which contributed materially to those marvellous cures which the good people of the time attributed to supernatural agencies rather than to the medical knowledge and skill of the devoted nuns,[11] who were the real conquerors of disease and death.

But the inmates of the cloister were not the only women who, during the Middle Ages, achieved distinction by their writings on medical subjects and by their signal success in the practice of the healing art. In various parts of Europe, but especially in Italy and France, there were at this time among women, outside as well as inside convent walls, many daughters of Aasculapius and sisters of Hygeia who stood in such high repute among their contemporaries that they received the same honours and emoluments as were accorded to their masculine colleagues.

This was particularly the case in Salerno, which was the venerated mother of all Christian medical schools, and which, for nine centuries, was universally regarded as "the unquestioned fountain and archetype of orthodox medicine." Situated on the Gulf of Salerno, and laved by the cerulean waters of the Tyrrhenian sea, the *Civitas Hippocratica*, as it was called on its medals, rejoiced in a salubrious climate, and was celebrated throughout the world as the "City sacred to Phoebus, the sedulous nurse of Minerva, the fountain of physic, the votary of medicine, the handmaid of Nature, the destroyer of disease and the strong adversary of death."[12] For to this favoured city flocked from all quarters the lame and the halt and those afflicted with the tortures of disease and the disabilities of advancing years. The noble and the simple, crowned heads as well as the poorest of the poor, were found there, all of them in quest of life's most precious boon — health and strength.

Never did the far-famed sanctuary of the god of medicine in Epidaurus witness such an influx of invalids as gathered in the hospitals of Salerno and pressed through the streets of the Hippocratic city, seeking the aid of those doctors whose marvellous cures had given them a world-wide reputation. Small wonder, then, that the *Regimen Santatis Salernitanum* — that famous code of health of the school of Salerno — has been translated into almost all the languages of modern Europe, and that since 1480 no fewer than two hundred and fifty editions of it have been published. "Not to have been familiar with it from beginning to end, not to have been able to quote it orally as occasion might require, would, during the Middle Ages, have cast serious suspicion upon the professional culture of any physician."[13] But the noblest claims of the Hippocratic city to the gratitude of humanity yet remain to be told. A German traveller in the thirteenth century wrote:

> Laudibus Aeternum nullum negat esse Salernum
> Illuc pro morbis totus circumfluit orbis.[14]

This was because Salerno was universally recognised as the "day star" and "morning glory" of the best culture in the healing art, and, still more, because of the thorough instruction she gave in her schools of medicine and the pre-eminence she so long held in every department of medical lore.

The course of study in medicine was long and thorough, and the candidate applying for a degree had to pass a rigid examination and give proof not only of his proficiency in every branch of the healing art, but also of perfect acquaintance with the various branches of science and letters as well. At the time of Frederick II, who organised all the different schools of Salerno into a single university, a three years' course in philosophy and literature was required before one could present himself for entrance into the school of medicine. The courses in medicine lasted five years, at least, after which a year of practice with an old physician was required. In addition to this, if the candidate wished to practice surgery he was obliged to devote one year to the study of human anatomy and to the dissection of human bodies. Considering the progress of knowledge since the time of Frederick II, it must be admitted that the legal requirements enforced by the faculty of Salerno compare favourably with those of the best of our medical schools of today.

Still more to the credit of Salerno, long known as the Athens of the two Sicilies, was her boundless liberality toward scholarship and culture regardless of sex. For, with a chivalrous admiration for intellect, wherever found, and with a sense of intellectual justice that has put to shame all medical schools outside of Italy, until less than fifty years ago, the school of Salerno was the first to throw open its portals to women as well as men, and give to an admiring world a number of women — those celebrated *mulieres Salernitanae* — who were eminent not only as physicians, but also as professors of the theory and practice of medicine. For this reason, if for no other, it can be truly affirmed that "No school of medicine in any age or country, if only for this, can ever over-peer her in renown; and, even as formerly in the universities of Europe, at the bare mention of the name of the learned Cujacius, every scholar instinctively uncovered himself, so at the very name of Salernum, the fount and nurse of rational medicine, every physician should recall her memory 'with mute thanks and secret ecstasy' as among the most spotless and venerated chapters in the history of his art."[15]

The most noted professor and successful practitioner among the women of Salerno was Trotula, wife of the distinguished physician, John Platearius, and a member of the old noble family of the Ruggiero. She flourished during the eleventh century and enjoyed a reputation as a physician that was not inferior to that of the most noted doctors of her time. Besides occupying a chair in the school of medicine and having an extensive practice, she was the author of many works on medicine which had a great vogue among her contemporaries. Some of them, especially those relating to diseases of her own sex,[16] were published several times after the invention of printing, and many manuscript copies of her works are still found in various libraries of Europe. But she did not confine her practice to the diseases of women. She was also well versed in general medicine and exhibited, besides, as her works testify, marked skill as a surgeon in many cases that would even now be considered as peculiarly difficult of treatment.

One of her books was entitled *De Compositione Medicamentorum* — the Compounding of Medicaments — and it was this work, doubtless, that gave her much of the fame she enjoyed beyond the confines of Italy. Ruteboeuf, a noted French trouvere of the thirteenth century, gives us a quaint picture of a scene frequently witnessed in his day. Crowds were frequently attracted by herbalists — venders of simples — who, stationed at street corners or in other public places, near tables covered with a cloth of flaring colours, were wont to descant, somewhat after the style of certain of our patent-medicine hawkers and quack-salvers, upon the extraordinary curative properties of the various drugs and panaceas which they had for sale.

"Good people," one of these travelling herb doctors would begin, "I am not one of those poor preachers, nor one of those poor herbalists who carry boxes and sachets and spread them out on a carpet. No, I am a disciple of a great lady named Madame Trotte of Salerno, who performs such marvels of every kind. And know ye that she is the wisest woman in the four quarters of the world."

Ordericus Vitalis, an English Benedictine monk, in his *Historia*

*Ecclesiastica*, tells us of the impression made by Trotula on Rudolfo Malacorona, one of those famous itinerant scholars of the Middle Ages, who spent their lives in wandering from one university to another in pursuit of knowledge. He had been a student from his youth and was a man of remarkable attainments in every department of learning. After visiting and conferring with the learned men of the most celebrated universities of France and Italy, he finally arrived at Salerno, where, he informs us, he found no one who could cope with him in disputation except *quandam sapientem matronam* — a certain very learned woman.[17] This was Trotula, who, by reason of the extraordinary cures she effected, was known among her contemporaries as *magistra operis* — a consummate practitioner. When, however, we consider the thorough course of study that every one aspiring to a degree in medicine was obliged to complete, women as well as men, it is not so surprising that Trotula should be regarded both as a learned woman and as a successful physician.

Among other women doctors who did honour to Salerno and whose names have come down to us were three who are known in history as Abella, Rebeca de Guarna and Mercuriade. All of them achieved a great reputation by their writings on medical subjects, especially Mercuriade, who distinguished herself in surgery as well as in medicine. Still another woman deserving special mention is Francesca, wife of Matteo de Romana, of Salerno. After passing a very severe examination before a board composed of physicians and surgeons, she was accorded the doctorate in surgery. An official document of the time referring to this event reads as follows: "Whereas the laws permit women to practice medicine, and whereas, from the viewpoint of good morals, women are best adapted to the treatment of their own sex, we, after having received the oath of fidelity, permit the said Francesca to practice the said art of healing," etc.[18]

In view of the facts above mentioned regarding the University of Salerno — the excellence of its work, its liberality and breadth of view, its attitude toward the higher education of women, and its pre-eminence for so many centuries as a school of medicine — is it surprising that it was, until comparatively recent times, considered "the *mater et caput* of medical authority in ethical matters," and that, so late as 1748, the Medical Faculty of Paris should address an official letter to the faculty of Salerno requesting its judgement regarding the rights of precedence as between physicians and surgeons? But what is surprising, and what, too, passes all understanding, is that the University of London, after being empowered by royal charter to do all things that could be done by any university, was legally advised that it could not grant degrees to women without a fresh charter, because no university had ever granted such degrees.[19]

While women were winning such laurels in Salerno in every department of the healing art, their sisters north of the Alps were not idle. As early as 1292 there were in Paris no less than eight women doctors — called *miresses* or *mediciennes* — whose names have come down to us, not to speak of those who practiced in other parts of France. There was also a certain number of women who devoted themselves to surgery and called by the old Latin authors of the time *cyrurgiae*.

In Paris, however, conditions for studying and practicing medicine and surgery were far from being as favourable to women as they were in Salerno. As there were no schools open to them for the study of these branches, they had to depend entirely for such knowledge as they were able to acquire on the aid they could get from practicing doctors, the reading of medical books and their own experience. The consequence was that they were not at all so well equipped for their work as were the women who enjoyed all the exceptional advantages offered the students at Salerno. None of them was noted for scholarship, none of them was a writer of books, and only one of them — Jacobe Felicie, about whom more presently — rose above mediocrity.

The reason for the great difference between the conditions of the women doctors of Paris and those of Salerno is not far to seek. The Faculty of Medicine in Paris was, from the beginning of its existence, unalterably opposed to female medical practitioners. As early as 1220 it promulgated an edict prohibiting the practice of medicine by any one who did not belong to the faculty, and, according to its constitutions and by-laws, only unmarried men were eligible to membership.

For a long time the edict remained a dead letter. But eventually, as the faculty grew in power and influence, it was able to enforce the observance of its decrees. One of its first victims was Jacobe Felicie, just mentioned, who was hailed before court for practicing medicine in contravention of its edict issued many years before.

Jacobe Felicie was a woman of noble birth, and had won distinction by her success in the healing art. As the testimony at her trial revealed, she never treated the sick for the sake of gain. In nearly all cases the sick who had addressed themselves to her had been abandoned by their own physicians. All the witnesses who had been called testified that they had been cured by Jacobe Felicie, and all expressed their deepest gratitude to her for her care and devotion. But, in spite of all these facts, and in spite of the brilliant defence that this worthy woman made, she was condemned to pay a heavy fine — condemned because, as the indictment read, she had presumed to put her sickle into the harvest of others-*falcem in messem mittere alienam*— and this was a crime.[20] The faculty was a close corporation and insisted that its members should have a monopoly of all the honours and emoluments that were to accrue from the treatment of the sick and suffering. What a curious adumbration of similar proceedings within the memory of many still living!

The prosecution of Jacobe Felicie recalls that of Agnodice in Greece long ages before. And the plea urged for the necessity of a female physician — that many a woman would rather die than reveal the secrets of her infirmity to a man[21] — was the same as that offered by the women of Athens before the council of the Areopagus. It was the same agonising cry that had been heard thousands of times before and which has been heard thousands of times since. Isabella of Castile was not the first of the long list of victims who, for lack of a doctor of their own sex, have been sacrificed through womanly modesty, and, more's the pity, she will not be the last.

Unfortunately for the women of France, the result of the

prosecution of Mme Felicie was the very reverse of that instituted against Agnodice; for the latter came off victorious, while the former was condemned and punished. So crushing was the blow dealt to women practitioners, outside of obstetrics, that they did not recover from its effects for more than five hundred years. For it was not until 1868 that the École de Medicine of Paris opened its doors to women, and it was not until nearly twenty years later that female physicians were able to enter the hospitals of the French capital as *internes.*[22]

Until quite recent years there is very little to be said of women physicians in England and Germany. Their practice, outside of that of certain herb doctors, was confined chiefly to midwifery. There was no provision made in either of these countries for the education of women in medicine and surgery, and such a thing as a college where they could receive instruction in the healing art was unknown. It is true that an ecclesiastical law of Edgar, King of England, permitted women as well as men to practice medicine, but this law was subsequently abolished by Henry V.[23]

During the reign of Henry VIII a law was again enacted in favour of women physicians; for at that time an act was passed for the relief and protection of "Divers honest psones, as well men as women, whom God hathe endued with the knowledge of the nature, kind and operaçon of certeyne herbes, rotes and waters, and the using and ministering them to suche as be payned with customable diseases, for neighbourhode and Goddes sake and of pitie and charitie, because *that* 'The Companie and Fellowship of Surgeons of London, *mynding only their owne lucres and nothing the profit or case of the diseased or patient,* have sued, vexed and troubled' the aforesaid 'honest psones,' who were henceforth to be allowed 'to practyse, use and mynistre in and to any outwarde sore, swelling or disease, any herbes, oyntments, bathes, pultes or emplasters, according to their cooning, experience and knowledge — without sute, vexation, penaltie or loss of their goods.' "[24]

The italicised words in this quotation prove that the women doctors of England had the same difficulties as their sisters in France, and that the real reason of the opposition of the male practitioners was that they wished to monopolise the practice of medicine. They, like the medical faculty of Paris, strenuously objected to women "putting the sickle into their harvest," and they, accordingly, left nothing undone to circumvent the intrusion of those whom they always regarded as undesirable competitors.

It was argued by the men that women, to begin with, lacked the strength and capacity necessary for medical practice. It was also urged that it was indelicate and unwomanly for the gentler sex to engage in the healing art, and that, for their own good, they should be excluded from it at all costs. Those who were willing to waive these objections contended that women had not the knowledge necessary for the profession of medicine and should be excluded on the score of ignorance. When women sought to qualify themselves for medical practice by seeking instruction under licensed practitioners or in medical schools, they found a deaf ear turned to their requests. The doctors declined to teach them and the medical schools, one and all, closed their doors against them.

Thus it was that in England, France and Germany the practice of medicine and surgery was always practically in the hands of men until only a generation ago. Even the English midwives gradually "fell from their high estate," and were left far behind the female obstetricians of Germany and France. For these two countries can point to a number of midwives who, by their knowledge, successful practice, and the books they wrote, achieved a celebrity that still endures.

Chief among these in Germany were Regina Joseph von Siebold, her daughter Carlotta, and Frau Teresa Frei, all of whom, in the early part of the last century, enjoyed an enviable reputation in the Fatherland.

The first named, after following a course of lectures on physiology and the diseases of women and children, and passing a brilliant examination in the medical college of Darmstadt, devoted herself to the practice of obstetrics, and with so great success that the University of Giessen in 1819 conferred on her the degree of doctor of obstetrics. Her daughter, Carlotta, after studying obstetrics under her mother, went to the University of Gottingen, where she devoted herself to physiology, anatomy and pathology. After passing an examination and successfully defending a number of theses in the University of Giessen, she was also proclaimed a doctor of obstetrics. At a later date Frau Frei received a similar degree.[25]

More noted as *accoucheuses* and gynaecologists than the three distinguished women just mentioned were Mme Marie Louise La Chapelle and Mme Marie Bovin, who, shortly after the French Revolution, entered upon those wonderful careers in their chosen specialties which have given them so unique a place in the annals of medicine.

Mme La Chapelle was particularly celebrated for the numerous improvements she effected in lying-in hospitals, for the large number of skilled midwives whom she furnished, not only to France, but also to the whole of Europe, and, above all, for the excellent treatises which she wrote on obstetrics, which gave her a reputation second to none among her contemporaries, men or women. Her *Pratique des Accouchements,* in three volumes, based on the immense number of fifty thousand cases at which she presided, reveals an operator of rarest skill and genius. This production was long regarded as a standard work on the topics discussed, and for years exerted an immense influence in the medical world.

Less skillful as an operator, but of greater ability as a doctor than Mme La Chapelle, was her illustrious contemporary, Mme Bovin. Possessing extraordinary insight as an investigator and marvellous sagacity as a diagnostician, Mme Bovin achieved the distinction of being the first really great woman doctor of modern times. Her marvellous success as a practitioner — Dupuytren said she had an eye at the tip of her finger — her extended knowledge of the entire range of gynaecology, but above all her numerous treatises on the subject matter of her life work, gave her a prestige that none of her sex had ever before enjoyed, and commanded the admiration of the doctors of the world. Her *Memorial de l'Art des Accouchements* passed through many editions and was translated into several European languages. And so highly were her scientific attainments valued in Germany that the University of Marburg recognised them by conferring on her — *honouris causa* — the degree of doctor of medicine and, had its rules permitted the admission of women, the Royal Academy of Medicine would have honoured her with a place among its members. She was also the recipient of many other honours, besides being a member of several learned societies. But the greatest monument to her genius is a large illustrated treatise in two volumes, in which she exhibits a wonderful knowledge of anatomy, physiology, surgery, pathology and therapeutics. It gave her a large following in Germany as well as in France, and there were not wanting distinguished German *accoucheurs* who followed Mme Bovin's teachings to the letter.

The remarkable German and French women just named were all practically self-made women. They won fame as they had acquired knowledge — chiefly by courage, in spite of the countless obstacles that beset their paths. They owed nothing to schools or universities, nothing to government patronage or assistance, nothing to the medical fraternity as a whole. Universities would not admit them to their lecture rooms or laboratories, and the various medical faculties opposed them as intruders into their jealously guarded domain, and as competitors whose aspirations were to be frustrated, whatever the means employed. It is true that, when some of the women mentioned had won worldwide renown by their achievements, they were made the recipients of belated honours by certain universities and learned societies; but these societies and universities were then honouring themselves as much as the women who received their degrees and diplomas of membership.

How different it was in Italy, which, since the fall of the Roman Empire, has ever been in the van of civilisation, and which has always continued the best traditions of Greco-Roman learning and culture — Italy, which has been the home of such supreme masters of literature, science, art as Dante, Petrarch, Galileo, Leonardo da Vinci, Raphael, Michaelangelo, Brunelleschi — Italy, the mother of universities, the birthplace of the Renaissance, and the recognised leader of intellectual progress among the nations of the world. Here in the favoured land of the Muses and the Graces, women enjoyed all the rights and privileges accorded to men; here the doors of schools and universities were open to all regardless of sex; and art, science, literature, law, medicine, jurisprudence counted its votaries among women as well as among men; here, far from encountering jealousy and opposition in the pursuit of knowledge or in the practice of the professions, women never found aught but generous emulation and sympathetic cooperation.

For a thousand years women were welcomed into the arena of learning and culture on the same footing as men. In Salerno, Bologna, Padua, Pavia, they competed for the same honours and were contestants for the same prizes that stimulated the exertions of the sterner sex. Position and emolument were the guerdons of merit and ability, and the victor, whether man or woman, was equally acclaimed and showered with equal honour. Women asked for no favours in the intellectual arena and expected none. All they desired were the same opportunities and the same privileges as were granted the men, and these were never denied them. From the time when Trotula taught in Salerno to the present, when Giuseppina Catani is professor of general pathology in the medical faculty of Bologna, the women of Italy always had access to the universities and were at liberty to follow any course of study they might elect. We thus find them achieving distinction in civil and canon law, in medicine, in theology even, as well as in art, science, literature, philosophy and linguistics. No department of knowledge had any terrors for them, and there was none in which some of them did not win undying fame. They held chairs of language, jurisprudence, philosophy, physics, mathematics, medicine and anatomy, and filled these positions with such marked ability that they commanded the admiration and applause of all who heard them.

This is not the place to tell of the triumphs of the women professors in the Italian universities, or to recount the achievements of those who were honoured with degrees within their classic walls. Let it suffice to recall the names of a few of those who won renown in medicine and surgery and whose names are still in their own land pronounced with respect and veneration.

One of the most noted practitioners in Southern Italy, after the death of Trotula and her compeers, was one Margarita, who had studied medicine in Salerno. One of her patients was no less a personage than Ladislaus, King of Naples. Among those that had diplomas for the practice of surgery were Maria Incarnata, of Naples, and Thomasia de Matteo, of Castro Isiae.

That women enjoyed in Rome the same privileges in the practice of medicine and surgery as their sisters in the southern part of the peninsula is manifest from an edict issued by Pope Sixtus IV in confirmation of a law promulgated by the Medical Faculty of Rome, which reads as follows: "No man or woman, whether Christian or Jew, unless he be a master or a licentiate in medicine, shall presume to treat the human body either as a physician or as a surgeon."[26]

In central and northern Italy — in Florence, Turin, Padua, Venice — as well as in the southern part, we find constantly recurring instances of women practicing medicine and surgery and winning for themselves an enviable reputation as successful practitioners.

But after the decline of Salerno, consequent on the establishment by Frederick II of a school of medicine in Naples, the great centre of medicine and surgery, as of civil and canon law, was Bologna.[27] So renowned did it become as a teaching and intellectual centre that it was, as Sarti informs us, known throughout Europe as *Civitas Docta* — the learned city — and *Mater Studiorum* — the mother of studies. On its coins were stamped the words *Bononia Docet* — Bologna teaches — and on the city seal, which is still used for certain public documents, were the words *Legum Bononia Mater* — Bologna, the Mother of Laws.

Here, more than in Salerno, more than in any other city in the world, was, for long centuries, witnessed a blooming of female genius that has, since the time of Gratian and Irnerius, given the University of Bologna pre-eminence in the estimation of all friends of woman's education and woman's culture. For here, within the walls of what was for centuries the most celebrated university in Christendom, women had, for the first time, an opportunity of devoting themselves at will to the study of any and all branches of knowledge. And it can be truthfully affirmed that no seat of learning can point to such a long list of eminent scholars and teachers among the gentler sex as is to be found on the register of Bologna's famous university. For here, to name only a few, achieved distinction, either as students or as professors, such noted women as Bitisia Gozzadina, Bettina and Novella Calendrini, Dorotea Bocchi, Giovanna and Maddalena Bianchetti, Virginia Malvezzi, Maria Vittoria Dosi, Elizabetta Sirani, Ippolita Grassi, Properzia de Rossi, Maria Mastellagri, Laura Bassi, Maddelena Noe-Candedi, Clotilda Tambroni and Anna Manzolini. In this honour list we have a group of savantes that were famed throughout Europe for their attainments in law, philosophy, science, ancient and modern languages, medicine, and surgery — the rivals, and sometimes the superiors, in scholarship of the ablest men among their distinguished colleagues.

It would be a pleasure to recount the achievements of these justly celebrated daughters of Italy; but lack of space precludes the mention of more than one of them. This was Maria dalle Donne, who was born of poor peasants near Bologna, and who at an early age exhibited intelligence of a superior order. After pursuing her studies under the ablest masters, she obtained from the University of Bologna, *maxima cum laude,* the degree of doctor in philosophy and medicine. On account of her knowledge of surgery, as well as of medicine, she was soon afterward put in charge of the city's school for midwives. When Napoleon, in 1802, passed through Bologna he was so struck by the exceptional ability of the young *dottoressa* that, on the recommendation of the savant Catersani, he had instituted for her in the university a chair of obstetrics — a position which she held until the time of her death, in 1842, with the greatest credit to herself and to the institution with which she was identified.

Maria dalle Donne is a worthy link between that long line of women doctors, beginning with Trotula, who have so honoured their sex in Italy, and those still more numerous practitioners in the healing art who, shortly after her death, began to spring up in all parts of the civilised world.[28]

For it was about this time that the movement which had long been agitated in behalf of the higher education of women began suddenly to assume extraordinary vitality, not only throughout Europe but in America as well. And to no women did this movement appeal so strongly as to those who had long been looking forward to an opportunity to qualify themselves for the learned professions, especially medicine. No sooner did they descry the first flush of dawn on their long-deferred hopes than they began to consider ways and means for putting their fondly nurtured projects into execution.

Seven years, almost to the day, after the death of Maria dalle Donne, Miss Elizabeth Blackwell, a young woman in America, of English birth, decided to enter college with a view of studying medicine and surgery. But, at the very outset, she encountered all kinds of unforeseen difficulties — difficulties that would have caused a less courageous and determined woman to give up her plans in despair. She was told, in the first place, that it was highly improper for a woman to study medicine and that no decent woman would think of becoming a medical practitioner. As to a lady studying or practicing surgery that, of course, was out of the question.

But a more serious obstacle than the conventionalities in the case was the difficulty of finding a medical college that was willing to admit a woman to its lecture rooms and laboratories. Miss Blackwell applied to more than a dozen of the leading institutions of America, and received a positive refusal to her request. Finally, when hope had almost vanished, she received word from a small college in Geneva, New York, announcing that her application had been favourably considered and that she would be admitted as a student whenever she presented herself.

The truth is that the faculty of the college was opposed to the young woman's admission, but wished to escape the odium incident to a direct refusal by referring the question to the class with a proviso which, it was believed, would necessarily exclude her. "But in this it was greatly surprised and disappointed. For the entire medical class, to the number of about one hundred and fifty, decided unanimously in favour of the fair applicant's admission. And they did more than this. They put themselves on record regarding the equality of educational opportunities for women and men in a way that must have put their timid professors to shame. Their resolution, accompanying an invitation to the young woman to become a member of the student body, was worded as follows:

" 'Resolved, That one of the radical principles of a republican government is the universal education of both sexes; that to every branch of scientific education the door should be equally open to all; that the application of Elizabeth Blackwell to become a member of our class meets our entire approbation, and, in extending our unanimous invitation, we pledge ourselves that no conduct of ours shall cause her to regret her attendance at this institution.' "

The students were as good as their word. Their conduct, as Miss Blackwell wrote years afterward, was always admirable and that of "true Christian gentlemen." But the women of Geneva were shocked at the female medical student. They stared at her as a curious animal; and the theory was fully established that she was "either a bad woman, whose designs would gradually become evident, or that, being insane, an outbreak of insanity would soon be apparent."[29]

In due time Miss Blackwell finished her course in medicine and surgery, and graduated at the head of her class. The orator of the day, who was a member of the faculty, naturally referred to the new departure that had been made — the admission of a woman for the first time to a complete medical education — and among other things declared that the experiment, of which every member of the faculty was proud, "had proved that the strongest intellect and nerve and the most untiring perseverance were compatible with the softest attributes of feminine delicacy and grace."[30]

The awarding of the degree of MD for the first time to a woman in America excited general comment and widespread interest, not only in the United States, but in Europe as well. The public press was not unfavourable in its opinion of the new departure, and even *Punch* could not resist writing some verses, sympathetic, albeit humorous, in honour of the fair MD.[31]

After spending some time abroad studying in the great hospitals of Europe, Miss Blackwell started the practice of medicine in New York City. At first, as she declares in her autobiographical sketches, it was "very difficult, though steady, uphill work. I had," she tells us, "no medical companionship, the profession stood aloof, and society was distrustful of the innovation."

The aloofness of the profession arose from a dread of successful rivalry, and the men did not wish to encourage "the invasion by women of their own preserves." "You cannot expect us," one of them frankly admitted to her, "to furnish you with a stick to break our heads with."

But, undeterred by opposition, Miss Blackwell continued her work, daily making converts to the new movement and receiving substantial aid, as well as sympathetic cooperation, from many people, both men and women, prominent in society and public life. In 1854 she started a free dispensary for poor women. Three years later she founded a hospital for women and children, where young women physicians as well as patients could be received. These were the humble beginnings of the present flourishing institutions known as the New York Infirmary and the College for Women. And in less than ten years after her graduation, Miss Blackwell saw the new departure in medical practice successfully established, not only in New York, but also in other large cities of the United States. In 1869 the early pioneer medical work by women in America was completed.

"During the twenty years which followed the graduation of the first woman physician, the public recognition of the justice and advantage of such a measure had steadily grown. Throughout the northern states the free and equal entrance of women into the profession of medicine was secured. In Boston, New York and Philadelphia special medical schools for women were sanctioned by the legislatures, and in some long-established colleges women were received as students in the ordinary classes."[32]

Meanwhile, the women in Europe were not idle nor heedless of the example set by their brave sisters in America. The University of Zurich threw open its portals to women, and was soon followed by those of Bern and Geneva. The first woman to obtain a degree in medicine in Zurich — it was in 1867 — was Nadejda Suslowa, a Russian. She was soon followed by scores of others from Europe and America, who found greater advantages and more sympathy in Swiss universities than elsewhere.

In 1869 the Medico-Chirurgical Academy of St Petersburg conferred the degree of MD upon Madame Kaschewarow, the first female candidate for this honour. When her name was mentioned by the dean it was received with an immense storm of applause which lasted several minutes. The ceremony of investing her with the insignia of her dignity being over, her fellow students and colleagues lifted her on a chair and carried her with triumphant shouts throughout the halls.

The first woman graduate from the University of France was Miss Elizabeth Garrett, of England. She received her degree in medicine in 1870, and the following year the same institution conferred the doctor's degree on Miss Mary C Putnam, of New York.

After these precedents had been established, the universities of the various countries on the continent, following the examples set by those in the United States and Switzerland, opened one after the other their doors to women, and in most of them accorded them all the privileges of *cives academici* enjoyed by the men.

Great Britain held out against the new movement long after most of the continental countries had fallen into line, nor did she surrender until after a protracted and bitter fight, during which the men leading the opposition exhibited evidences of selfishness and obscurantism that now seem incredible.

The leader in Great Britain of pioneer medical work for women was Miss Sophia Jex-Blake, whose academic pathway was beset with difficulties far sterner than had in the United States confronted her friend and colleague, Miss Blackwell.

Hearing much of the tolerance and liberality of the University of London, she applied to it for admission as a student, but was informed at once that the charter of the institution "had purposely been so worded as to exclude the possibility of examining women for medical degrees."

After this rebuff she made application to the University of Edinburgh, which, like the other Scotch universities, had always boasted of its broad-mindedness and freedom from educational trammels. She was received provisionally, and was, after a while, joined by six other women who had in view the same object as herself. For a time, notwithstanding opposition from certain quarters, everything was quiet and apparently satisfactory. But the gathering storm soon broke, and the seven young women, as they were one day entering the university gates, were actually mobbed by a ruffianly band of students who had all along been opposed to the presence of women in the class and lecture rooms. They pelted the helpless females with street mud and hurled at them all the vile epithets and heaped upon them all the abuse that their foul tongues could command. These outrageous proceedings on the part of the rabble of rowdies were allowed to continue for several days, and, had it not been for a brave band of chivalrous young Irishmen among the students, who formed themselves into a bodyguard for the protection of their fair classmates, and were, in consequence, known as "The Irish Brigade," the hapless women students would not have escaped bodily harm. What a marked contrast between the conduct toward Miss Blackwell of the gallant students of the modest little American town and that of the cowardly ruffians of the vaunted "Athens of the North!"

But this was not all. The seven young women in question had matriculated as students of the university with the understanding that they were to have all the rights and privileges of the male students. But after the disgraceful conduct of the mob just referred to, they discovered that the authorities of the university were prepared to break faith with them, and prevent them from getting their coveted degrees, and thus debar them from all chance of medical practice.

The reason why the university was induced to annul its contract, after the women on their part had fully complied with all its stipulations, soon became apparent. It was purely and simply to make it impossible for women to secure a license as medical practitioners. Both in and outside of Edinburgh the conviction daily grew stronger that women doctors were a menace to the monopoly so long enjoyed by the medical fraternity, and that the movement in their favour should be crushed by fair means or foul before it got beyond control. The *Spectator* made this clear by stating at the time of the controversy that "every profession in this country" — England — "is more or less of a trades union," and yet the members of these professions "would shake their heads and prate about the necessity of stamping out trades unionism among workmen." "Women," whined one of the doctors, "would snatch the bread from the mouths of poor practitioners." Another doctor who had championed the cause of women physicians, when commenting on the hypocritical objection that it was unbecoming for women to practice medicine or surgery, expressed the same idea in other words. "It appears," he declared, "that it is most becoming and proper for a woman to discharge all the duties which are incidental to our profession for thirty shillings a week; but, if she is to have three or four guineas a day for discharging the same duties, then they are immoral and immodest and unsuited to the soft nature that should characterise a lady."

After Miss Jex-Blake and her companions learned that the university was determined to refuse them the degrees to which they were entitled, they brought suit against it for breach of contract. But, after a long and expensive trial, the judge rendered a decision against them. They then appealed to Parliament, and, after a protracted and strenuous campaign on the part of friends whom they had enlisted in their cause, they saw their opponents not only dragged at the chariot wheels of progress but forced to help to turn them; for, in 1878, after nearly ten years of a persistent, continuous struggle such as had rarely been witnessed in woman's long battle for things of the mind — a struggle in which the intrepid, dauntless Miss Jex-Blake "made the greatest of all the contributions to the end attained" — the women of Great Britain had the supreme satisfaction of winning what was probably the most glorious victory which their sex had ever won.[33] The war was over and henceforward they were free — as were their sisters in other parts of the world — as the women in Italy had been for a thousand years — to devote themselves at will to the study and practice of the healing art without let or hindrance.

What a wonderful change has taken place in the medical world almost within the space of a single generation! The tiny grain of mustard that was sown by two lone women, the Misses Blackwell and Jex-Blake, in their chosen field of effort has grown and "waxed a great tree." Women doctors are now found in all parts of the civilised world and are numbered by thousands. And so great has been their professional success, so widespread is the desire to secure their services, especially in countries like America and England, where opposition was in the beginning especially bitter, that the proportion of women practitioners in medicine and surgery is now regarded as the best index of a nation's enlightenment.

The healing art of Greece and Rome has broadened out into the noble sciences of medicine and surgery of today. For, based as they now are on the sciences of chemistry, botany, biology, hygiene, physiology, anatomy and bacteriology, which have all witnessed such extraordinary developments during the last half century, they both deserve a pre-eminent place in the history of the sciences. And the success which has crowned woman's efforts in surgery and medicine is not only a conclusive indication of her capacity, so long denied by her self-interested opponents, but also the most convincing indication that she is at last properly occupied in a field of activity from which she was too long excluded. Her contributions as writer and investigator toward the progress of both sciences, even during the short time in which she has been able to give proof of her ability, have been notable and augur well for the share she will have in their future advancement. But more important still is the refining influence she has already exerted on both professions, and the relief she has been able to afford to countless thousands of her own sex who would otherwise have been the voluntary victims of untold misery. Women doctors are, indeed, not only worthy representatives of Aesculapia Victrix and of the two sciences which they have so elevated and so ennobled, but are also ministering angels to poor, suffering humanity comparable only with the heroic Sisters of Charity and the devoted nurses of the Red Cross.

## Notes & References

1) In order that the reader may realise the immense extent of territory that was covered by this strenuous woman's explorations, during the twelve years she spent in Amazonia, it suffices to give the titles of her books, all of which are profusely illustrated by photographs taken by herself and by accurate charts of rivers, whose courses were previously almost unknown.
The books written in collaboration with her husband are *Voyage au Tapajos*, *Voyage au Xingu*, *Voyage au Tocantins-Araguaya*, *Voyage au Itaboca et à l'Etacayuna*, *Voyage entre Tocantins et Xingu*, *et Voyage au Yamunda*.
The books written by Mme Coudreau after her husband's death are *Voyage au Trombetas*, *Voyage au Cuminá*, *Voyage au Rio Curuá*, *Voyage a la Mapuerá* and *Voyage au Maycurú*.
When one remembers that many of the watercourses here named would be considered large rivers outside of South America; that, notwithstanding their countless rapids and waterfalls, necessitating numberless portages, Mme Coudreau explored all these rivers from their embouchures to as near their sources as the water would carry her rude dugouts, we can form some idea of the miles she travelled and of the stupendous labour that was involved in making these long journeys in the sweltering and debilitating and insect-laden atmosphere of the Amazon basin.

2) Quoted in *Medical Women*, p 11, by Sophia Jex-Blake, MD, Edinburgh, 1886. Cf Hyginus, *Fabularum Liber*, No 274.

3) *Charles Daremberg, who, at the time of his death in 1872, was professor of the history of medicine in the Faculty of Medicine in Paris, had the intention of publishing this work Ðåñé ôùí ãõíáéùí ðáèùí. — On the Diseases* of Women — but his premature death prevented him from executing his project. It is to be hoped that some one else, interested in woman's medical work, may at an early date give this production to the public with an appropriate commentary.

4) Cf Hertzen et Rossi *Inscriptiones Urbis Romae Latinae*, p 1245, No 9478, Berlin, 1882.

5) "Non mihi si linguae centum, oraque centum, ferrea vox..... omnia morborum percurrere nomina possim quae Fabiola in tanta

miserorum refregeria commutavit ut multi pauperum sani languentibus inviderent." *Epistola ad Oceanum.*

6) Haec inter timidam revocat clamore puellam Alpharides, veniens quae saucia quaeque ligavit.
— Ekkehardi Primi *Waltharius*, Berlin, 1873.

7) That the Germans, at the time under discussion, regarded learning as having an effeminating effect on men is well illustrated by the following characteristic anecdote: "when Amasvintha, a very learned woman who was a daughter of the Ostrogoth King, Theodoric, selected three masters for the instruction of her son, the people became indignant. 'Theodoric,' they exclaimed, 'never sent the children of the Goths to school, learning making a woman of a man and rendering him timorous. The sabre and the lance are sufficient for him.'" Procopius, *De Bello Gothico*, I, 2, Leipzig, 1905.
If we may judge by a letter from Pace to Dean Colet, the noted classical scholar and founder of St Paul's school in London, such views found acceptance in England as late as the time of More and Erasmus. For we are told of a British parent who expressed his opinion on the education of men in these words: "I swear by God's body I'd rather that my son should hang than study letters. The study of letters should be left to rustics."

8) This work was for a long time regarded as lost, but a manuscript copy was recently found in Copenhagen, and it has since been published by Teubner of Leipsic, under the title of *Hildegard's Causae et Curae.*

9) *Archiv für Pathologische Anatomie und Physiologie und für Klinische Medicin*, Vol 18, p 286, Berlin.

10) *S Hildegardis Opera Omnia*, Editor: Migne, p 1122, Paris, 1882.

11) "In the municipal and state institutions of this period the beautiful gardens, roomy halls and springs of water of the old cloistral hospital of the Middle Ages were not heard of, still less the comforts of their friendly interiors." *A History of Nursing*, Vol I, p 500, M Adelaide Nutting and Lavinia L Dock, New York, 1907.
The mortality in some of the state hospitals from the latter part of the seventeenth to the middle of the nineteenth century was appalling, often as high as fifty and sixty per cent. This was due not only to shockingly unsanitary conditions, but also to inordinate overcrowding. A large proportion of the beds, incredible as it may seem, were purposely made for four patients, and six were frequently crowded into them. "The extraordinary spectacle was then to be seen of two or three small-pox cases, or several surgical cases, lying on one bed." John Howard, in his *Prisons and Hospitals*, pp

176-77. Warrington, 1874, tells us of two hospitals that were so crowded that he had "often seen five or six patients in one bed, and some of them dying."

It is gratifying to learn that the chief agents in changing this revolting condition, due to faulty construction and management of hospitals, were women. Prominent among these benefactors of humanity were Mme Necker, Florence Nightingale, and the wise and alert superiors of the various nursing sisterhoods.

12) How like Chaucer's prioress who
Was so charitable and so piteous,
And al was conscience and tender herte.

13) Cf *Lib de Virtutibus et Laudibus*, by Aegidius, head physician to Philip Augustus of France, in which occur the following verses:
Urbs Phoebo sacrata, Minervae sedula nutrix,
Fons physicae, pugil eucrasiae, cultrix medicinae,
Assecla Naturae, vitae paranympha, salutis
Promula fida; magis Lachesis soror, Atropos hostis.
Morbi pernicies, gravis adversaria mortis.
quoted in the appendix, p xxxii, to S de Renzi's, *Storia Documentata della Scuola Medica di Salerno*, Naples, 1857.

14) Cf The introduction to the English translation of the *Regimen Sanitatis Salernitanum*, p 28, by J Ordronaux, Philadelphia, 1870.

15) "Immortal praise adorns Salerno's name
To seek whose shrine the world once came.

16) See *Storia Documentata della Scuola Medica di Salerno*, ut sup, p 474 *et seq*, and p lxxvi *et seq* of Appendix; also Ordronaux, ut sup, p 16.

17) Probably her most noted work is the one which bears the title *De Morbis Mulierum et Eorum Cura* — The Diseases of Women and Their Cure.

18) "Physicae quoque scientiam tam copiose habuit ut in urbe Psaleritana, ubi maxime medicorum scholae ab antiquo tempore habentur, neminem in medicinali arte, praeter quandam sapientem matronam, sibi parem inveniret." Migne, Patrologiae Latinae, Vol 188, Col 260.

19) As this decree is of singular interest and importance, a copy of the original is here given in full:

Karolus, etc, Universis per Justitieratum Principatus citra Serras Montorii constitutis presentes litteras inspecturis fidelibus paternis et suis salutem, etc. In actionibus nostris utilitati puplice libenter oportune perspicimus et honestatem morum in quantum suadet modestia conservamus. Sane Francisca uxor Mathei de Romana de

Salerno in Regia Curia presens exposuit quod ipsa circa principale exercitium cirurgie sufficiens circumspecto in talibus judicio reputatur. Propter quod excellentie nostre supplicavit attentius ut licentiam sibi dignaremus concedere in arte hujusmodi practicandi. Quia igitur per scriptum puplicum universitatis terre Salerni presentatum eidem Regie Curie, inventum est lucide quod Francisca prefata fidelis est et genere orta fidelium ac examinata per medicos Regios paternos nostrosque cirurgicos, in eadem arte cirurgie tamquam ydiota sufficiens est inventa, licet alienum sit feminis conventibus interrese virorum, ne in matronalis pudoris contumelia irruant et primum culpam vetite transgressionis incurrant. Quia tamen de juris indicto medicine officium mulieribus est concessum expedienter attento quod ad mulieres curandas egrotas de honestate morum viris sunt femine aptiores, not recepto prius ab eadem Francisca solito fidelitatis et quod iuxta tradiciones ipsius artis curabit fideliter corporaliter Juramento, licentiam curandi et practicandi sibi in eadem arte per Justitieratum jam dictum auctoritate presentium impartimus. Quare fidelitati vestre precipimus quatenus eandem Franciscam curare et practicari in prefata arte per Justitieratum predictum ad honourem et fidelitatem paternam et nostram ac utilitatem fidelium presentium earumdam libere permittatis, nullum sibi in hoc impedimentum vel obstaculum interentes. Datum Neapoli per dominum Bartholomeum de Capua, etc, Anno domini mcccxxi, die x Septembris v, indictionis Regnorum dicti domini patris nostri anno xiii.

*Collectio Salernitana*, Vol III, p 338, by G Henschel, C Daremberg, and S de Renzi, Naples, 1852-59.

20) *Universities in the Middle Ages*, Vol II, Part II, p 712, by H Rashdall, Oxford, 1895. The most exhaustive work on the University of Salerno and its famous doctors, men and women, is a joint work in five volumes entitled *Collectio Salernitana; ossia Documenti Inediti e Trattati di Medicina appartenenti alla scuola Salernitana, raccolti e illustrati*, by G Henschel, C Daremberg e S Renzi, Naples, 1852-59. Cf also, *Storia Documentata della Scuola Medica di Salerno*, by S de Renzi, Naples, 1857; *L'École de Salerne*, by C Meaux, with introduction by C Daremberg, Paris, 1880, and Piero Giacosa's *Magistri Salernitani Nondum Editi*, Turin, 1891.

21) *Chartularium Universitatis Parisiensis*, Vol II, p 150, and pp 255 and 267, by Denifle and Chatelain, Paris, 1889-91.

22) "Mulier antea permitteret se mori, quam secreta infirmitatis sui homini revelare propter honestatem sexus muliebris et propter verecundiam quam revelando pateretur." *Chartularium Universitatis Parisiensis*,

Vol II, p 264, Paris, 1891.

23) It may interest the reader to know that the first two women to get the doctorate in the Paris School of Medicine were Miss Elizabeth Garret, an English woman, and Miss Mary Putnam, an American. The first woman permitted to practice in the Paris hospitals was likewise an American, Miss Augusta Klumpke, of San Francisco.

24) "Possunt et vir et foemina medici esse." Cf Chiappelli, *Medicina negli Ultimi Tre Secoli del Medio Evo*, Milan, 1885.

25) Quoted in *Woman's Work and Woman's Culture*, p 87, Josephine E Butler, London, 1869. Dom Gasquet in his *English Monastic Life*, p 175, tells us that in the Wiltshire convents "the young maids learned needlework, the art of confectionery, surgery — for anciently there were no apothecaries or surgeons; the gentlewomen did cure their poor neighbours — physic, drawing, etc."

26) The first woman to receive the doctorate of medicine in Germany was Frau Dorothea Christin Erxleben. Hers, however, was a wholly exceptional case, and required the intervention of no less a personage than Frederick the Great. In 1754, Frau Erxleben, who had made a thorough course of humanities under her father, presented herself before the faculty of the University of Halle, where she passed an oral examination in Latin which lasted two hours. So impressed were the examiners by her knowledge and eloquence that they did not hesitate to adjudge her worthy of the coveted degree, which was accorded her by virtue of a royal edict.

Her reception of the doctorate was made the occasion of a most enthusiastic demonstration in her honour. Felicitations poured in upon her from all quarters in both prose and verse. One of them, in lapidary style, runs as follows:

Stupete nova litteraria,
In Italia nonnumquam,
In Germania nunquam
Visa vel audita
At quo rarius eo carius.

This, freely translated, adverts to the fact that an event, which before had been witnessed only in Italy, was then being celebrated in Germany for the first time, and was, for that very reason, specially deserving of commemoration.

27) "Nemo masculus aut foemina, seu Christianus vel Judaeus, nisi Magister vel Licentiatus in Medicina foret, auderet humano corpori mederi in physica vel in chyrurgia." Marini, *Archiatri Pontifici*, Vol I, p 199, Roma, 1784.

28) Thomas Aquinas, the Angel of the Schools, who had taught in

Salerno, and was well acquainted with the leading universities of Europe, was wont to say "Quattuor sunt urbes caeteris praeeminentes, Parisius in Scientiis, Salernum in Medicinis, Bononia in legibus, Aurelianis in actoribus….." there are four preëminent cities: Paris, in the sciences; Salerno, in medicine; Bologna, in law; Orleans, in actors Op 17. *De Virtutibus et Vitiis*, Cap ult.

The medieval poet, Galfrido, expressed the same idea in verse when he wrote:

In morbis sanat medici virtute Salernum
Aegros: in causis Bononia legibus armat
Nudos: Parisius dispensat in artibus illos
Panes, unde cibat robustos: Aurelianis
Educat in cunis actorum lacte tenellos.

29) It may be remarked that it was a woman, Lady Mary Montagu, who introduced inoculation with small-pox virus into Western Europe, and that it was also a woman — a simple English milkmaid — who communicated to Jenner the information which led to his discovery of a prophylactic against small-pox. But of far greater importance was the introduction into Europe of that priceless febrifuge and antiperiodic — chinchona bark. This was due to the Countess of Chinchon, vicereine of Peru. Having been cured by its virtues of an aggravated case of tertian fever in 1638, while living in Lima, she lost no time, on her return to Spain, in making known to the world the marvellous curative properties of the precious quinine-producing bark. The powder made from the bark was most appropriately called *Pulvis Comitessae* — the countess's powder — and by this name it was long known to druggists and in commerce. Thanks to Linnaeus, the memory of the gracious lady will always be kept green, because her name is now borne by nearly eight score species of the beautiful trees which constitute the great and incomparable genus Chinchona. See *A Memoir of the Lady Ana de Osorio, Countess of Chinchon, and Vice-Queen of Peru*, by Clements R Markham, London, 1874.

30) *Pioneer Work in Opening the Medical Profession to Women*, p 70, by Dr Elizabeth Blackwell, London, 1895.

31) *Ibid*, p 91.

32) "Young ladies all, of every clime,
Especially of Britain
Who wholly occupy your time
In novels or in knitting,
Whose highest skill is but to play,
Sing, dance or French to clack well,

Reflect on the example, pray,
Of excellent Miss Blackwell.

* * * *

For Doctrix Blackwell, that's the way
To dub in rightful gender —
In her profession, ever may
Prosperity attend her.
Punch a gold-headed parasol
Suggests for presentation
To one so well deserving all
Esteem and Admiration.

33) *Op cit*, p 241.

# Women in Archaeology

ARCHAEOLOGY, IN ITS broadest sense, is one of the most recent of the sciences, and may be said to be a creation of the nineteenth century. In its restricted sense, however, it dates back to the beginning of the Italian Renaissance. For it was at this period that the collector's zeal began to manifest itself, and that were brought together those priceless treasures of ancient art which are today the pride of the museums of Rome and Florence. It was then that Pope Sixtus IV and Julius II, his nephew, laid the foundations of the great museums of the Capitol and the Vatican, and enriched them with such famous masterpieces as the Ariadne, the Nile, the Tiber, the Laocoon and the Apollo Belvidere. Their example was quickly followed by such cardinals as Ippolito d'Este, Fernando de' Medici, and by representatives of the leading princely houses of the Italian peninsula. In rapid succession the palaces of the Borghese, Chigi, Pamphili, Ludovisi, Barbarini and Aldobrandini became filled with the choicest Greek and Roman antiques. In the course of time many of these treasures found their way to the museums of Venice, Madrid, Paris, Munich and Dresden, while still others were purchased by wealthy art connoisseurs in various parts of Europe and Great Britain.

In the beginning these antiques in marble and bronze were used chiefly for decorative purposes. "Courts, stairs, fountains, galleries and palaces were adorned with statues, busts, reliefs and sarcophagi applied in such a manner as to become incorporated in contemporary art and thereby to gain fresh life."[1]

These treasures of antiquity, statues, bas-reliefs, mosaics, coins, medals, busts, sarcophagi, and productions of ceramic art, although at first used almost exclusively for decorating palaces and villas and enriching museums, were eventually to become of inestimable value in the study of the history of art and the civilisation of Greece and Rome, as well as of the various nations of antiquity with which they had come into contact. Besides this, they supplied the necessary raw material not only for classical archaeology, but also for that more comprehensive science of archaeology which deals with the art, the architecture, the language, the literature, the inscriptions, the manners, customs and development of our race from prehistoric times until the present day.

Among the women who took a prominent part in collecting material toward the advancement of archaeologic science were those illustrious ladies — as celebrated for their knowledge and culture as for their noble lineage and their patronage of men of letters — who presided over the brilliant courts of Urbino, Mantua, Milan and Ferrara.

Pre-eminent among these were Elizabetta Gonzaga, Duchess of Urbino, and Isabella d'Este, Marchioness of Mantua. The palace of the former — "that peerless lady who excelled all others in excellence" — was famous for its precious antiques in bronze and marble, but above all for its superb collection of rare old books and manuscripts in Greek, Latin and Hebrew.

Isabella d'Este, who was through life the most intimate friend of Elizabetta Gonzaga, was acclaimed by her contemporaries as "the first lady in the world." She was a true daughter of the Renaissance, in the heart of which she was brought up; and "the small, passing incidents of her everyday life are to us memorials of the classic age when the gods of Parnassus walked with men."[2] She was an even more enthusiastic collector than the Duchess of Urbino, and her magnificent palace in Mantua was filled with the choicest works of Greek and Roman art that were then procurable.

She has been described as one who secured everything to which she took a fancy. She had but to hear of the discovery of a beautiful antique, a rare work in bronze or marble uncovered by the spade of the excavator, when she forthwith made an effort to procure it for her priceless collection. If that was not possible, she would not rest until she could secure something else even more precious. She aimed at supremacy in everything artistic and intellectual, and would be content with nothing short of perfection. Hence it is that her collection of antiques, like those of her friend, the Duchess of Urbino, is rightly regarded as having been of singular value in preparing the way for the foundation of scientific archaeology — a foundation that was laid by the eminent German scholar, Winckelmann, in the eighteenth century by the publication of his masterly work — *History of the Art of Antiquity*.

The first woman of eminence to take an active part in archaeologic excavation was the youngest sister of Napoleon Bonaparte, "the beautiful, clever and ambitious Caroline." When Joachim Murat became king of Naples, after his brother-in-law, Joseph Bonaparte, had in 1808 been transferred to the throne of Spain, his wife, Queen Caroline, gave at once a new impetus to the work of the excavation of Pompeii along the lines planned a few years before by the eminent Neapolitan scholar, Michele Arditi. She exhibited the keenest interest in the work, and the notable discoveries which were made under her inspiring supervision of this important undertaking show how much classical archaeology owes to her intelligent and munificent patronage.

Queen Caroline proved her interest in the excavations that were to contribute so much to our knowledge of antiquity "by appearing frequently at Pompeii and stimulating the workmen to greater efforts. She frequently spent entire days, during the great heat of summer, at the excavations, to encourage the lazy workmen and to reward them in the event of success. The funds were increased so as to make the employment of six hundred men possible. The Street of Tombs was next uncovered, forming a complete and solemn picture, greatly impressing the beholder even today. For the first time a complete outline of an ancient marketplace and its surroundings could be obtained. The market, closed and inaccessible to wheeled traffic, was surrounded by a colonnade filled with monuments, with the great temple in the background, and beyond the arcades were other temples or public buildings, among the principal being the stately Basilica. Constant and increased efforts were thus crowned by important results. The Queen did not withhold generous assistance. The French architect, Fr Mazois, received from her fifteen hundred francs while preparing his monumental work at Pompeii."[3]

It is not too much to say that Queen Caroline's archaeological work at Pompeii was as far-reaching in its results as was that of her illustrious brother in the land of the Pharaohs. It drew in the most impressive manner the attention of the world to the vast treasures of art which lay concealed under the earth-covered ruins of the once noted cities of the ancient world, and stimulated scholars and learned societies to undertake similar researches in Sicily, Greece, Mesopotamia, Asia Minor, and the almost forgotten islands of the Aegean Seas.

While this energetic sister of the great Napoleon was occupied in bringing to light those priceless treasures of art which had for seventeen centuries lain beneath the ashes of Vesuvius, a bright, refined, *spirituelle* young girl, born in Dublin and bred in England, was unconsciously preparing herself for a brilliant career in the branch of archaeology known as Christian iconography. Her name was Anna Murphy, better known to the world as Mrs Jameson. At an early age she gave evidence of unusual intelligence, and she had hardly attained to womanhood when she was noted for her knowledge of languages and for her remarkable attainments in art and literature. Numerous journeys to France, Italy and Germany and a systematic study in the great museums and art galleries of these countries, but, above all, her association with the most distinguished scholars of Europe, completed her education and prepared her for those splendid works on Christian art which have made her name a household word throughout the world.

Mrs Jameson was a prolific writer, but those of her works on which her fame chiefly rests are the ones which are classed under the general title, *Sacred and Legendary Art.* They treat of God the Father and Son, of the Madonna and the Saints, as illustrated in art from the earliest ages to modern times. So masterly and exhaustive was her treatment of the difficult subjects discussed in this *chef d'oeuvre* of hers that no less an authority than the eminent German archaeologist, F X Kraus, writes of this elaborate production as follows:

> Neither before nor since has the subject matter of this work been handled with such skill and thoroughness. The older iconographic works were mere dilettanteism. For the first time since classical archaeology had applied the principles of modern criticism to Greek and Roman iconography, and had presented an example of scientific treatment free from such reproach, was a serious iconography of our early Christian monuments possible. Mrs Jameson was the first to attempt this on a large scale. It was clear to her — and here lay the advance which her work reveals — that in order to accomplish her colossal task two things must be realised. She must not build on a foundation of material that is imperfect or brought together in a haphazard way. She must not only see and test everything available in the way of monuments, but she must likewise place the productions of literature and poetry beside those of the plastic arts. It was clear to her, also, that, in this case, one would throw light on the other, and that the investigator who would lay claim to the name of archaeologist must, moreover, study the spirit of a people in all its monumental and literary manifestations.
>
> Mrs Jameson strove to learn the mind and the mode of early Christian times from the works of the Fathers. She saw in the hymns of the Middle Ages and in the writings of the mystics the sources of the art ideas which disclose themselves in the wall and glass paintings of our cathedrals and in the entrancing creation of a Fiesole. She had also the special advantage of being thoroughly imbued with Dante's ideas of the plastic arts of the Middle Ages.

> And all this is evidenced in a form which exhibits neither dry dissertation nor wearisome nomenclature. Each of her articles is a little essay. It teaches us what place the Madonna, or St Catherine, or some other saint has held in the memory and in the imagination of past centuries. We behold the sainted forms flitting before our eyes in all the charm of poetic perfection which was given them by the childlike fantasy of the Middle Ages, and in all the power which they exercised over men's minds, and which, however we may view the religious side of the question, certainly had the effect of creating forms of infinite beauty and pictures of unspeakable reality.[4]

When we recollect that Mrs Jameson achieved so much before the foundations of Christian archaeology had been fully laid; before de Rossi's monumental publications had supplied the means of interpreting early Christian sculpture; before critics and archaeologists were at one regarding the significance of early Christian and Middle Age symbolism, or agreed on the principles that were to guide to a correct understanding of the pictures of Roman and Gothic art, and while students were yet in ignorance as to the real influence of Byzantine art on that of western Europe, we cannot but wonder at the courage and the energy of this gifted woman in undertaking and in bringing to a happy issue a work which, even today, with all our increased facilities and greater array of facts, would be considered a herculean task.

As we read her admirable volumes on *Sacred and Legendary Art* we can, as did a close friend of hers, see the enraptured author "kindle into enthusiasm amidst the gorgeous natural beauty, the antique memorials and the sacred Christian relics of Italy," and we are prepared to believe, with the same friend, that there was not "a cypress on the Roman hills, or a sunny vine overhanging the southern gardens, or a picture in those vast sombre galleries of foreign palaces, or a catacomb spread out, vast and dark, under the martyr churches of the City of the Seven Hills, which was not associated with some vivid flashes of her intellect and imagination." And we can also understand how "the strange, mystic symbolism of the early mosaics was a familiar language to her," and why she should experience special delight when she found herself "on the polished marble of the Lateran floor or under the gorgeously sombre tribune of the Basilica of Santa Maria Maggiore, reading off the quaint emblems or expounding the pious thoughts of more than a thousand years ago."[5]

It is gratifying to know that Queen Victoria recognised the surpassing merits of this noble woman by placing her on the civil list, and that our own Longfellow was able to say of her masterpiece, *Sacred and Legendary Art*, "It most amply supplies the cravings of the religious sentiment of the spiritual nature within."

A countrywoman of Mrs Jameson and her contemporary, who also deserves an honourable place in the literature of archaeology, is Louise Twining. Although inferior in intellectual attainments and literary activity to the accomplished author of *Sacred and Legendary Art*, her two works on *Types and Figures of the Bible Illustrated by Art* and *Symbols and Emblems of Early Medieval Christian Art* have given her a well-deserved reputation on the Continent as well as in the British Isles. The latter volume Mrs Jameson herself declares in her *Legends of the Madonna* to be "certainly the most complete and useful book of the kind which I know of."

A third woman who has won fame for her sex in the island kingdom in the domain of archaeology is Miss Margaret Stotes. Her activities, however, have been chiefly confined to the antiquities of Ireland, on which she is a recognised authority.

The notable part she took in editing Lord Dunraven's great work, *Notes on Irish Architecture*, established her reputation on a firm basis. Among her other important works are *Early Christian Art in Ireland* and *Christian Inscriptions in the Irish Language*, chiefly collected and drawn by George Petrie, one of the annual volumes of the Royal Historical and Archaeological Association of Ireland. This work has justly been described as an epoch-making contribution to Christian epigraphy and to our rapidly developing knowledge of Keltic language and literature. The learned Dr. Krauss, than whom there is no more competent judge, in referring to this splendid performance, does not hesitate to affirm, "No man could have done better than this brave college girl, whom I would wish to greet across the Channel with a cordial *Macte virtute*."

The women archaeologists so far mentioned, with the exception of Queen Caroline Murat, were conspicuous as writers rather than active investigators in the field. There have been, however, quite a number who have won distinction as "archaeologists of the spade" — women who, either alone or with their husbands, have superintended excavations in different lands, which have yielded results of untold scientific value. Among the most conspicuous of these are Mme Sophia Schliemann, Mme Dieulafoy and the enterprising Yankee girl, Miss Harriet A Boyd.

Of these the first named is the wife of the late Dr Henry Schliemann, who immortalised himself by his famous excavations at Troy, Tiryns and Mycenae — enterprises which solved for us the great problem of nearly thirty centuries and demonstrated in the most startling manner "the truth of the foundations on which was framed the poetical conception that has for thousands of years called forth the enchanted delight of the educated world." During his meteoric career as an archaeologist, Schliemann was able to realise the dreams of his youth, and succeeded in unveiling the mystery that had so long hung over Sacred Ilios, and to give the heroes of the Iliad a local habitation on the rediscovered Plain of Troy. And his glorious achievements we must credit largely to that brave and devoted woman — his wife — who was ever at his side to share in his trials and labours and to raise his drooping spirits in hours of depression, or when hostile criticism treated him as a visionary in the pursuit of a chimera.

Mrs Schliemann is a Greek lady who was born and bred under the shadow of the Acropolis and a worthy descendant of those proud Athenian women who wore the golden grasshopper in their hair as a sign that they were natives of the City of the Violet Crown. She was not only dowered with intellectual gifts of a high order, but she was also her husband's most congenial companion and sympathetic friend in all his literary work, while she was his very right hand in those glorious enterprises at Hissarlik and Mycenae, which secured for both of them undying fame.

Dr Schliemann was the first to attest the never-failing assistance which he received from this noble woman who, as he informs us, was "a warm admirer of Homer" and "with glad enthusiasm" joined her husband in executing the great work which he had conceived in his early boyhood. Usually they worked together, but at times Mrs Schliemann superintended a gang of labourers at one spot while the Doctor was occupied at another in the immediate vicinity. Thus it was she who excavated the heroic tumulus of Batieia in the Troad — that Batieia who, according to Homer, was a queen of the Amazons and undertook a campaign against Troy.[6]

Mme Jane Dieulafoy is noted as the collaborator of her husband, Marcel Dieulafoy, in the important archaeological mission to Persia that was entrusted to him by the French government. The results of this mission, in which Mme Dieulafoy had a conspicuous part, were published in Paris in 1884 in five octavo volumes.

It was during this expedition to the ancient empire of Cyrus and Artaxerxes that this indefatigable couple became interested in the ruins of Susa, the ancient capital of the Persian kings. On their return to France they succeeded in securing money and supplies for conducting excavations among these ruins which, in the end, yielded results which were, in some respects, as important as those which rewarded the labours of the Schliemanns in Greece and Asia Minor.

So completely had Susa — the City of the Lilies — been buried and forgotten for nearly two thousand years that even its site was almost as much a matter of dispute as was that of ancient Troy. And yet it was one of the greatest and richest cities of antiquity — the city of Esther and Daniel, the city of the mighty Assuerus who reigned from India even unto Ethiopia, over a hundred and twenty-seven provinces — the city where the great Alexander celebrated his nuptials with Statira, the daughter of Darius, with a magnificent festival at which, according to Plutarch, "there were no fewer than nine thousand guests, to each of which he gave a golden cup for the libations."

In December 1884, the two brave and venturesome explorers were on their way to Susa with high hopes, but not without a full knowledge of the difficulties and dangers that they would have to confront among the fanatical nomads of Arabistan, where the very name of Christian inspires rage and horror. It meant, as Mme Dieulafoy herself tells us, "to cross the Mediterranean, the Red Sea, the Indian Ocean, the Persian Gulf and the deserts of Elam three times in less than a year; to pass whole weeks without undressing; to sleep on the bare ground; to struggle nights and days against robbers and thieves; to cross rivers without a bridge; to suffer heat, rain, cold, mists, fever, fatigue, hunger, thirst, the stings of divers insects; to lead this hard and perilous existence without being guided by any interest other than the glory of one's country."[7]

In spite, however, of all the opposition which they encountered among the fanatical Mussulmans of Arabistan and of the dreadful sufferings incident to living in a desert where it was at times impossible to secure the necessaries of life, their mission was successful, and their account of their finds in the ancient capital of Elam was as thrilling in its way as anything reported of the excavations at Troy or Pompeii. Their splendid collection of specimens of ancient Persian art and architecture, now on exhibition in the Museum of the Louvre, testifies to the successful issue of their expedition and to their indomitable energy in conducting researches under the most untoward conditions.[8] So highly did the French government value the part Mme Dieulafoy had taken in this arduous enterprise that it conferred on her a distinction rarely awarded to a woman for scientific work — that of Chevalier of the Legion of Honour.

As an archaeologist, the gifted and energetic American woman, Miss Harriet Boyd — now Mrs C H Hawes — has achieved an international reputation for her remarkable excavations in the island of Crete. She is a frequent contributor to archaeological journals; but it is upon her splendid work in the field that her fame will ultimately rest.

Her first work of importance was undertaken as Fellow of the

American School of Classical Studies at Athens. This was in 1900, and the field of her investigations was the Isthmus of Hierapetra in Crete. Here she excavated numerous tombs and houses of the early Geometric Period, *circa* 900 BC, and paved the way for those brilliant discoveries which rewarded her labours during the following three years.

The investigations conducted during these three years under Miss Boyd's directions yielded results of transcendent value. Assisted by three young American women — the Misses B E Wheeler, Blanche E Williams, and Edith H Hall — she superintended the work of more than a hundred native employees whom she had on her payroll. By good fortune in the choice of a site for excavation and by well-directed efforts she was soon able to unearth one of the oldest of Cretan cities and to expose to view the ruins of what was probably one of the ninety cities which Homer tells us in his Odyssey graced the land of Crete — "a fair land and a rich, in the midst of a wine-dark sea."

So remarkable were the finds in this long-buried Minoan town and so well preserved are its general features that it has justly been called the Cretan Pompeii. It antedates by long centuries the oldest cities of Greece and was a flourishing centre of commerce ages before the heroes of the Iliad battled on the plains of Troy.

It is not too much to say that the extraordinary discoveries made by this enterprising Yankee girl at Gournia, no less than those made by British and Italian archaeologists at Knossos and Phaestos, have completely revolutionised our ideas respecting the state of culture of the inhabitants of Crete during the second and third millennia before the Christian era. They have thrown a flood of light on the origins of Mediterranean culture, and have, at the same time, supplied material for a study of European civilisation that was before entirely wanting.

An enduring monument to Miss Boyd's ability as an archaeologist is her notable volume containing an account of her excavations at Gournia, Vasilike and other prehistoric sites on the Isthmus of Hierapetra. It will bear comparison with any similar productions by the Schliemanns or the Dieulafoys. A later work on

*Crete, the Forerunner of Greece,* which she wrote in collaboration with her husband, Mr C H Hawes, is also a production of recognised merit. As a study on the origin of Greek civilisation it opens up many new vistas in pre-history and illumines many questions that were before involved in mystery.

Besides Mrs Hawes, three other American women have achieved marked distinction by their archaeological researches. These are Mrs Sarah Yorke Stevenson, Miss Alice C Fletcher and Mrs Zelia Nuttall.

Mrs Stevenson has long been identified with the progress of archaeological research, especially with that in Egypt and the Mediterranean. A prominent member of many learned societies, she is likewise a writer and lecturer of note. She enjoys the distinction of being the first woman whose name appears as a lecturer on the calendar of the University of Harvard. In acknowledgment of her scholarly ability and eminent services in the development of its Department of Archaeology, the University of Pennsylvania has conferred upon her the honorary degree of Doctor of Science.

That American women have not been behind their sisters in Europe in their enthusiasm for archaeological investigation is evinced by the researches and writings of Miss Alice C Fletcher and Mrs Zelia Nuttall, both of whom enjoy an international reputation in the learned world.

Miss Fletcher's chosen field of labour has been in ethnology and anthropology. Her studies of the folk lore and the manners and customs of various tribes of North American Indians have a distinct and permanent value, while those of her contributions which have been published by the Smithsonian Institution and the Bureau of Ethnology — contributions based on personal knowledge of a long residence among the tribes she writes about — show that she has exceptional talent for the branches of archaeology to which she has devoted many years of earnest and successful study.

Mrs Nuttall is the daughter of an American mother and an English father. Thanks to the care that was bestowed on her education by her parents and to her long residence in the different countries of Europe, she is proficient in seven languages. This knowledge of tongues has been of inestimable advantage to her in her researches in European libraries and in those historical and archaeological investigations which have rendered her famous. She has devoted special attention to the early history, languages, religions and calendar systems of the primitive inhabitants of Mexico and Central America, in all of which she is a recognised authority.

When, some years ago, the mysterious ruins of Mexico began to attract the special attention of archaeologists, Mrs Nuttall was selected by the University of California as the field director of the commission which it sent to pursue archaeological researches in this Egypt of the New World. A more competent or a more enthusiastic director could not have been chosen. Her finds in the Pyramids of the Sun and Moon at Teotihuacan and elsewhere in our sister republic were especially important. In recognition of her achievements President Porfirio Diaz nominated Mrs Nuttall honorary professor in the Mexican National Museum. She was also offered the position of curator of the archaeological Museum of Mexico; but this office she declined. She holds membership in a large number of learned societies in America and Europe and is a frequent contributor to numerous magazines on historical and archaeological subjects. She has had the good fortune to discover a number of important manuscripts illustrating the early history of Mexico. Chief among these are a Hispano-American manuscript which she dug out of one of the libraries of Madrid and another which was found in a private collection in England and reproduced in facsimile in this country. In honour of its fair discoverer it is now known as the Codex Nuttall, and is regarded by experts as one of the most precious records of ancient Mexico.

What is probably Mrs Nuttall's most valuable contribution to archaeological science is her erudite work entitled *The Fundamental Principles of Old and New World Civilisations.* It is a comparative research based on a study of the ancient Mexican, religious, sociological and calendar systems, and represents thirteen years of assiduous labour. It is a worthy monument to the scientific ability of this gifted Americanist, and one which brilliantly illumines some of the most controverted points of comparative archaeology.

The Nestor of women archaeologists is Donna Ersilia Caetani-Bovatelli — the daughter of the famous Dante scholar, the late Duke Don Michel Angelo Caetani-Sermonetta. Since the days of Boniface VIII, whom Dante scornfully denounced as *lo principe de' Pharisei,* the family of the Caetani has been one of the most illustrious of the Roman nobility, and is today ranked with those of the Colonna and Orsini.

Besides his thorough knowledge of Dante, whose *Divina Commedia* he regarded as the great artistic production of the human mind — a work which he knew by heart — the Duke of Sermonetta was deeply versed in philology and archaeology. No one was more familiar with the history and antiquities of Rome than he was, nor a greater friend and patron of scholars of every nationality. The Palazzo Caetani was the resort of not only the savants of Rome, but also and especially of those who gathered from all quarters of the world to study the rich collections of antiquities for which the Eternal City is so famous. Here the ablest authorities in history and archaeology discussed the latest discoveries among the ruins of Greece and Asia Minor, and the most recent finds in the Forum or amidst the crumbling ruins of the palaces of the Caesars.

Having such a father and brought up in such an environment it is not surprising that Donna Ersilia acquired at an early age that taste for archaeology which was, as events proved, to constitute the chief occupation of her long and busy life. Having enjoyed and studied literature and the languages under the best masters in Rome, she was thoroughly prepared for the work of deciphering Greek and Latin inscriptions and for an intelligent study of the ancient monuments of Italy and Hellas.

Her learned countryman, A de Gubernatis, assures us that she has such a thorough knowledge of Latin and Greek that she writes both with ease and elegance, and that she is endowed with an admirable memory for philology and archaeology. Besides being a mistress of several modern languages, she is also familiar with Sanskrit.

Since the death of her husband, in 1879, she has devoted all her time, outside of that given to the care and education of her children, to the pursuit of classical archaeology, in which she has long been regarded as an authority of the first order. Her salon, unlike those of the frivolous leaders of high life, has for many years been the favourite rendezvous in Rome of learned men and women from every clime. Here were seen the noted historians Gregorovius, Theodore Mommsen, and Giovanni Battista de Rossi, the illustrious founder of Christian archaeology. Here the representatives of the French, German and American schools of archaeology meet to exchange views on their favourite science and to find inspiration in the knowledge and enthusiasm of their gifted hostess, who always takes an active part in their recondite discussions, and never fails to contribute her share to these meetings, which have contributed so much toward the advancement of science and the history of antiquity. Whether the discussion turn on the deciphering of an ancient text, the inscription of a monument or a recently excavated sarcophagus, Donna Ersilia's opinion is eagerly sought, and her judgement is generally unerring.

This cultured and erudite daughter of sunny Italy has been a prolific writer on her favourite branch of research. Besides contributing to such publications as the *Nuova Antologia* and the bulletins of the archaeological commissions in Rome, she has found time to prepare for the press a number of volumes of the highest value on divers questions of Roman and Greek archaeology.

It is interesting, in this connection, to note the fact that, after Mme Curie had been refused admittance into the French Academy, one of the members of this institution, who had voted against her on the ground that she was a woman, had occasion to attend a meeting of the Academy of the Lincei in Rome, an association which plays the same role in Italy as does the French Academy in France, and found, to his astonishment, that the dean of the department of archaeology, as well as the presiding officer of some of the most important meetings of the academy, was a woman. She was no other than Donna Ersilia Caetani-Bovatelli, the learned and gracious scion of an honoured race. So taken aback was the Gallic opponent of *feminisme* that he could but exclaim: "*Diable!* they order things differently in Italy from what we do in *la belle France*."

Considering their attainments and achievements, the two women who occupy the highest place as archaeologists in the English-speaking world are Mrs Agnes Smith Lewis and Margaret Dunlop Gibson. They are the twin daughters of the Rev John Smith, an English clergyman, and have long enjoyed an enviable reputation among Scriptural scholars and Orientalists.

During their youth they had the advantage of instruction under the best masters, and, among other things, acquired a wide knowledge of the modern and classical languages. Subsequent study and frequent visits to Greece and the Orient made them proficient in modern Greek, Arabic, Hebrew and Syriac. Becoming interested in the search for ancient manuscripts, they resolved to make the long and arduous journey to the Greek convent of St Catherine on Mt Sinai.

In the latter part of January 1892, these two brave and enterprising women left Suez for their destination in the heart of the Arabian desert. They were accompanied only by their dragoman and Bedouin servants. Eleven camels carried the two travellers, their baggage, tents and provisions for fifty days. They had laid in supplies not only for the two or three weeks they were to spend on the way to and from Sinai, but also for the month they expected to remain at the Convent of St Catherine.

Arriving at the end of their journey, they were most cordially received by the monks, who afforded them every facility for examining the treasures of their unique and venerable library. They immediately set to work, and before they left the room in which the manuscripts were preserved they had made one of the most remarkable finds of the century. For, in closely inspecting a dirty, forbidding old manuscript whose leaves had probably not been turned for centuries, they discovered a palimpsest, of which the upper writing contained the biographies of women saints, while that beneath proved to be one of the earliest copies of the Syriac

Gospels, if not the very earliest in existence.

No find since the celebrated discovery by Tischendorf of the Sinaitic Codex, in the same convent nearly fifty years before, ever excited such interest among Scriptural scholars or was hailed with greater rejoicings. It was by all Biblical students regarded as an invaluable contribution to Scriptural literature, and as a find which "has doubled our sources of knowledge of the darkest corner of New Testament criticism." To distinguish it from the *Codex Sinaiticus*, the precious manuscript brought to light by Mrs Lewis has been very appropriately named after the fortunate discoverer, and will hereafter be known as the Codex Ludovicus.[9]

Another find of rare importance made by the gifted twin sisters was a Palestinian Syriac lectionary similar to the hitherto unique copy in the Library of the Vatican. A special interest attaches to this lectionary from the fact that it is written in the language that was most probably spoken by our Lord.

Among other notable discoveries of Mrs Lewis and her sister during the four visits[10] which they made to Mt Sinai and Palestine between the years 1892 and 1897 were a number of manuscripts in Arabic and a portion of the original Hebrew manuscript of Ecclesiastes which was written about 200 BC. Previously the oldest copies of this book of the Old Testament were the Greek and Syriac versions.

What is specially remarkable about the discoveries made by Mrs Lewis and Mrs Gibson is that they were able to make so many valuable finds after the convent library at Mt Sinai had been so frequently examined by previous scholars. The indefatigable Tischendorf made three visits to this library and had but one phenomenal success. But neither "he nor any of the other wandering scholars who have visited the convent attained," as has been well said, "to a tithe of the acquaintance with its treasures which these energetic ladies possess."

But more remarkable than the mere discovery of so many invaluable manuscripts, which was, of course, an extraordinary achievement, is the fact that these manuscripts, whether in Syriac,

Arabic or Hebrew, have been translated, annotated and edited by these same scholarly women. Already more than a score of volumes have come from their prolific pens, all evincing the keenest critical acumen and the highest order of Biblical and archaeological scholarship. The reader who desires a popular account of their famous discoveries should by all means read Mrs Gibson's entertaining volume, *How the Codex Was Found*, and Mrs Lewis' charming little work entitled, *In the Shadow of Sinai*. As to those men — and the species is yet far from extinct — who still doubt the capacity of women for the higher kinds of intellectual effort, let them glance at the pages of the numerous volumes given to the press by these richly dowered women under the captions of *Studia Sinaitica* and *Horae Semiticae*; and, if they are able to comprehend the evidence before them, they will be forced to admit that the long-imagined difference between the intellectual powers of men and women is one of fancy and not one of reality.[11]

And yet, strange to relate, while Mrs Lewis and Mrs Gibson were electrifying the learned world by their achievements in the highest form of scholarship, the slow-moving University of Cambridge was gravely debating "whether it was a proper thing to confer degrees upon women," and preparing to answer the question in the negative. The fact that there were "representatives of the unenfranchised sex at their gates who had gathered more laurels in the field of scholarship than most of those who belong to the privileged sex" did not appeal to the university dons or prevent them from putting themselves on record as favouring a condition of things which, at this late age of the world, should be expected only among the women-enslaving followers of Mohammed.

The saying that "a prophet hath no honour in his own country" was fulfilled to the letter in the case of the two women who had shed such lustre on the land of their birth. While foreign institutions were vying with one another in showering honours on the two brilliant Englishwomen, with whose praises the whole world was resounding, the University of Cambridge was silent. The University of St Andrews conferred on them the degree of LLD, while conservative old Heidelberg, casting aside its age-old traditions, made haste to honour them with the degree of Doctor of Divinity. In addition to this, Halle made Mrs Lewis a Doctor of Philosophy. One would have thought that sheer shame, if not patriotic spirit, would have compelled the university in whose shadows the two women had their home, and in which Mrs Lewis' husband had held for years an official appointment, to show itself equally appreciative of superlative merit and equally ready to reward rare scholarship, regardless of the sex of the beneficiaries. But no. The illustrious archaeologists and Biblical scholars were women, and this fact alone was in the estimation of the Cambridge authorities enough to withhold from them that recognition which was so spontaneously accorded them by the great universities of the Continent.

Nor was this the only instance of the kind. While the celebrated twin sisters just referred to were so materially contributing to our knowledge of Biblical lore, another Englishwoman, Jane E Harrison, who lived within hearing of the church bells of Cambridge, was lecturing to delighted audiences in Newnham College on the history, mythology and monuments of ancient Athens, and writing those learned works on the religion and antiquities of Greece which have given her so conspicuous a place among modern archaeologists.[12] But, as in the case of her distinguished neighbours, the discoverers of the *Codex Ludovicus*, the degrees she was honoured with came not from Cambridge, with which, through her fellowship in Newnham, she was so closely connected.

And while this gifted lady was deserving so well of science and literature, the undergraduate students of Cambridge, following the cue given by the twenty-four hundred graduates who had just rejected the proposal to give honorary degrees to women who could pass the required examinations, were giving an exhibition of rowdyism which far surpassed that which, a few years before, had so disgraced the University of Edinburgh, when the same question of degrees for women was under consideration.

According to the report of an eye witness of the turbulent scene at Cambridge, "The undergraduate students appeared to be, as a body, viciously opposed to the proposal to give degrees to women, and became fairly riotous. They hooted those who supported the reform and fired crackers even in the Senate House and made the night lurid with bonfires and powder. They put up insulting effigies of girl students, and such mottoes as 'Get you to Girton, Beatrice. Get you to Newnham. Here is no place for maids!' "

Verily, when such scenes are possible in one of the world's great intellectual centres — a place where, above all others, women should receive due recognition for their contributions toward the progress of knowledge — one is constrained to declare that what we call civilisation is still far from the ideal. And, when one witnesses the total indifference of institutions like Cambridge and the French Academy to the splendid achievements of women like Mrs Lewis, Mrs Gibson and Mme Curie, one cannot but exclaim in words Apocalyptic: "How long, O Lord, holy and true," is this iniquitous discrimination against one-half of our race to endure? O Lord, how long?

## Notes & References


1) For an interesting account of the long campaign for the admission of women to medical schools and practice, see *Medical Women — A Thesis and a History*, by Dr Sophia Jex-Blake, Edinburgh, 1886. For a more elaborate work on women in medicine, the reader may consult with profit, *Histoire des Femmes Medecins*, by Mlle Melanie Lepinska, Paris, 1900.

2) A Michaelis, *A Century of Archaeological Discoveries*, p 6, New York, 1908.

3) *The Most Illustrious Ladies of the Renaissance*, p 152, by Christopher Hare, London, 1904.

4) Michaelis, *Op cit*, p 20, Cf also Fiorelli's *Pompeinarum Antiquitatum Historia*, Vol I, Par III, Naples, 1860. Arditi characterised Queen Caroline's interest in the excavations as "*entusiasmo veramente ammirabile*."

5) *Frauenarbeit in der Archaeologie in Deutsche Rundschau*, March 1890, page 396.

6) *Memoirs of the Life of Anna Jameson*, pp 296-97, by her niece, Geraldine Macpherson, London, 1878.

7) *Ilios, the City and Country of the Trojans*, pp 657-58, by Dr Henry

Schliemann, New York, 1881.

As an illustration of Mrs Schliemann's devotion to the work which has rendered her, as well as her husband, immortal, a single passage from the volume just quoted, p 261, is pertinent. Referring to the sufferings and privations which they endured during their third year's work at Hissarlik, Dr. Schliemann writes as follows:

My poor wife and myself, therefore, suffered very much since the icy north wind, which recalls Homer's frequent mention of the blasts of Boreas, blew with such violence through the chinks of our house-walls, which were made of planks, that we were not even able to light our lamps in the evening, while the water which stood near the hearth froze into solid masses. During the day we could, to some degree, bear the cold by working in the excavations; but, in the evenings, we had nothing to keep us warm except our enthusiasm for the great work of discovering Troy.

So high was Dr Schliemann's opinion of his wife's ability as an archaeologist that he entrusted to her — as well as to their daughter, Andromache, and son, Agamemnon — the continuation of the work which death prevented him from completing.

8) See Mme Dieulafoy's graphic account of the expedition in a work which has been translated into English under the title, *At Susa, the Ancient Capital of the Kings of Persia, Narrative of Travel Through Western Persia and Excavations Made at the Site of the Lost City of the Lilies, 1884-1886*, Philadelphia, 1890.

See also her other related work — crowned by the French Academy — entitled, *La Perse, La Chaldee et la Susiane*, Paris, 1887.

9) Among the specimens secured were two of extraordinary beauty and interest. One of them is a beautiful enamelled frieze of a lion and the other, likewise a work in enamel, represents a number of polychrome figures of the Immortals — the name given to the guards of the Great Kings of Persia. Both are truly magnificent specimens of ceramic art, and compare favourably with anything of the kind which antiquity has bequeathed to us. Commenting on the pictures of the Persian guards, Mme Dieulafoy writes: "Whatever their race may be, our Immortals appear fine in line, fine in form, fine in colour and constitute a ceramic work infinitely superior to the bas-reliefs, so justly celebrated, of Lucca della Robbia." *Op cit*, p 222.

10) One passage in this codex bears so strongly on a leading argument of this work that I cannot resist the temptation to give it with Mrs Lewis' own comment:

"The piece of my work," she writes, *In the Shadow of Sinai*, p 98 *et seq*, "which has given me the greatest satisfaction, consists in the decipherment of two words in John IV, 27. They were well worth all our visits to Sinai, for they illustrate an action of our Lord which seems to be recorded nowhere else, and which has some degree of inherent probability from what we know of His character. The passage is 'His disciples came and wondered that with the women he was *standing and talking*'.....

"Why was our Lord standing? He had been sitting on the wall when the disciples left Him; and, we know that He was tired. Moreover, sitting is the proper attitude for an Easterner when engaged in teaching. And an ordinary Oriental would never rise of his own natural free will out of politeness to a woman. It may be that He rose in His enthusiasm for the great truths He was uttering; but, I like to think that His great heart, which embraced the lowest of humanity, lifted Him above the restrictions of His race and age, and made Him show that courtesy to our sex, even in the person of a degraded specimen, which is considered among all really progressive peoples to be a mark of true and noble manhood. To shed even a faint light upon that wondrous story of His tabernacling amongst us is an inestimable privilege and worthy of all the trouble we can possibly take."

11) Mrs Gibson, unaccompanied by her sister, has since made two more visits to Mt Sinai in order to complete the work so auspiciously begun.

12) The following partial list of the works of these erudite twins on subjects connected with Scripture and Oriental literature gives some idea of their extraordinary attainments and of their prodigious activity in researches that are usually considered entirely foreign to the tastes and aptitudes of women.

*Some Pages of the Four Gospels Retranscribed from the Sinaitic Palimpsest*, with a translation of the whole text by Agnes Smith Lewis.

*An Arabic Version of St Paul's Epistles to the Romans, Corinthians, Galatians and part of Ephesians*, edited from a ninth century MS by Margaret Dunlop Gibson.

*Apocrypha Sinaitica.* Containing the Anaphora Pilati in Syriac and Arabic: the Syriac transcribed by J Rendel Harris, and the Arabic by Margaret Dunlop Gibson; also two recensions of the *Recognitions of Clement*, in Arabic, transcribed and translated by Margaret Dunlop Gibson.

*An Arabic Version of the Acts of the Apostles and the Seven Catholic Epistles*, from an eighth or ninth century MS, with a treatise on the Triune Nature of God and translation; edited by Margaret

Dunlop Gibson.

Apocrypha Arabica, edited by Margaret D Gibson, containing 1, *Kitab al Magall* or the *Book of the Rolls*; 2, *The Story of the Aphikia Wife of Jesus Ben Sira* (Carshuni); 3, *Cyprian and Justa*, in Arabic and Greek.

*Select Narratives of Holy Women*, from the Syro-Antiochene or Sinai Palimpsest, as written above the Old Syriac Gospels in AD 778. Translation by Agnes Smith Lewis.

*Apocrypha Syriaca Sinaitica*, being the *Protevangelium Jacobi* and *Transitus Mariae*, from a Palimpsest of the fifth or sixth century, edited by Agnes Smith Lewis.

*Forty-One Facsimiles of Dated Christian Arabic Manuscripts*, with Text and English Translation, arranged by Agnes Smith Lewis and Margaret Dunlop Gibson, with introductory observations in Arabic calligraphy by the Rev David S Margoliouth.

*The Didascalia Apostolorum in Syriac*, edited from a Mesopotamian MS, with various readings and collations of other MS, by Margaret Dunlop Gibson.

*The Arabic Version of the Acta Apocrypha Apostolorum*, edited and translated by Agnes Smith Lewis, with fifth century fragments of the Acta Thomae, in Syriac.

*The Gospel of Isbodad in Syriac and English*, by Margaret D Gibson.

*Acta Mythologica Apostolorum in Arabic*, with translation by Agnes Smith Lewis.

For an elaborate and sympathetic account of the labours and discoveries of Mrs Lewis and her sister, the reader is referred to an article from the pen of the learned Professor V Ryssel, in the *Schweiserische Theologische Zeitschrift*, XVI, Jahrgang, 1899.

# Women as Inventors

"THERE HAVE BEEN very learned women as there have been women warriors, but there have never been women inventors."[1] Thus wrote Voltaire with that flippancy and cocksureness which was so characteristic of the author of the *Dictionnaire Philosophique* — a man who was ever ready to give, offhand, a categorical answer to any question that came before him for discussion. His countryman, Proudhon, expressed the same opinion in other words when he wrote, *Les femmes n'ont rien invente, pas meme leur quenouille* — women have invented nothing, not even their distaff.

Had these two writers thoroughly sifted the evidence available, even in their day, for a proper consideration of this interesting subject, they would, both of them, have reached a very different conclusion from that which is expressed in the sentences just quoted. Had they consulted the records of antiquity, they would have learned that most of the earliest and most important inventions were attributed to women; and, had they studied the reports of explorers among the savage tribes of the modern world, they would have found that these early legends and traditions regarding the inventions of women were fully confirmed by what was being done in their own time. Man's first needs were food, shelter and clothing; and tradition in all parts of the world is unanimous in ascribing to woman the invention, in essentially their present forms, of all the arts most conducive to the preservation and well-being of our race.

In Egypt, as Diodorus Siculus informs us, the inventors of specially useful things were, as a reward of their deserts, enrolled among the gods, as were certain heroes among the ancient Greeks and Romans. Foremost among these was Isis, who laid the foundation of agriculture by the introduction of the culture of wheat and other cereals. Before her time the Egyptians lived on roots and herbs. In lieu of these crude articles of food, Isis gave them bread and other more wholesome aliments. She invented the process of making linen and was the first to apply a sail to the propulsion of a boat. To her also was attributed the art of embalming, the discovery of many medicines and the beginnings of Egyptian literature.

Even more prominent was Pallas Athene, one of the greatest divinities of the Greeks. Virgil, in his *Georgics,* invokes her as

> Inventor, Pallas, of the fatt'ning oil,
> Thou founder of the plow and the plowman's toil.

But not only was she regarded as the *oleae inventrix* — inventress of the olive — as Virgil phrases it, but also as the inventor of all handicrafts, whether of women or men. Like Isis, she was deemed the originator of agriculture and many of the mechanic arts. But, above all, she was the inventor of musical instruments and those plastic and graphic arts which have for ages placed Greece in the forefront of civilisation and culture.

From the beginning it was woman who first made use of wool and flax for textile fabrics; and of this prehistoric woman one can affirm what Solomon, in his *Book of Proverbs,* said of the virtuous woman of his day:

> She seeketh wool and flax and worketh diligently with her hands;
> She layeth her hands to the spindle and her hands hold the distaff.

She was also the first one to weave cotton and silk. It was Mama Oclo, the wife of Manco Capac, as the Inca historian, Garcilasso de la Vega, tells us, who taught the women of ancient Peru "to sew and weave cotton and wool and to make clothes for themselves, their husbands and children."

And it was a woman, Se-ling-she, the wife of the emperor, Hwang-te, who lived nearly three thousand years before Christ, to whom the most ancient Chinese writers assign the discovery of silk. Her name is perpetuated in the name China, the goddess of silkworms, and under this appellation she still receives divine honours.

The preparation and weaving of silk were introduced into Japan by four Chinese girls, and the new industry soon became there, as in China, one of the chief sources, as it is today, of the country's wealth. To perpetuate the memory of these four pioneer silk weavers the grateful Japanese erected a temple in their honour in the province of Setsu.

According to tradition, the eggs of the silk moth and the seed of the mulberry tree were conveyed to India, concealed in the lining of her headdress, by a Chinese princess. She was thus instrumental in establishing in the region watered by the Indus and the Ganges the same industry which her countrywomen had introduced into the Land of the Rising Sun.

Cashmere shawls and attar of roses, the costliest of perfumes, are attributed to an Indian empress, Nur Mahal, whom her husband, in view of her achievements, as well as on account of his passionate love for her, called "The Light of the World."[2]

And what shall we say of those exquisite creations of woman's brain and hand — needle-point and pillow lace? These two inventions, like the manufacture of silk, have given employment to tens of thousands of women throughout the world; and, in such countries as Italy, Belgium and France, where lace-making has received special attention, they have for centuries been most prolific sources of revenue. Silk fabrics in ancient Rome were worth their weight in gold. The finest specimens of point lace are, even today, as highly prised as precious stones, and, like the great masterpieces of plastic art, are handed down as heirlooms from generation to generation. In no other instance, except possibly in the hairspring of a watch, is there such an extraordinary difference in value between the raw material and the finished product as there is in the case of the finest thread lace.

A great sensation was caused in Italy a few decades ago when a humble workwoman, Signora Bassani, succeeded in rediscovering the peculiar stitch of the celebrated Venetian point, which had been lost for centuries. She was at once granted a patent for her invention, which was by her countrymen regarded as an event of national importance.

After painting and sculpture, probably no art has contributed more to the development of the aesthetic sense among the nations of the world than has the art whose chief tools are the needle and the bobbin in the deft hands of a beauty-loving woman. If the name of the first lace-maker had not been lost in the mists of antiquity, it is reasonable to suppose that she, too, would long since have had a monument erected to her memory, as well as the weavers of silk and makers of attar of roses and cashmere shawls. She was surely as deserving of such an honour.

More conclusive information respecting woman as an inventor is, strange as it may appear, afforded by a systematic study of the various races of mankind which are still in a state of savagery. Such a study discloses the interesting fact that woman, contrary to the declaration of Proudhon, has not only been the inventor of the distaff, but that she has furthermore — pace Voltaire — been the inventor of all the peaceful arts of life, and the inventor, too, of the earliest forms of nearly all the mechanical devices now in use in the world of industry.

Architecture, as well as many other things, was credited by the ancient Greeks to Minerva. This was a poetical way of stating the fact — now generally accepted by men of science — that women were the first homemakers. But the first home was a very simple and a very humble structure. When not a cave, it was a simple shelter made of bark or skins, sufficient to afford protection to the mother and her child. Subsequently it was a lodge made of earth, of stone or wattle work or adobe.

Women were, in the light of anthropology, as well as in that of mythology and tradition, the first to discover the nutritive and medicinal values of fruits, seeds, nuts, roots and vegetables. They were consequently the first gardeners and agriculturists and the first to build up a materia medica. While men were engaged in the chase or in warfare, women were gradually perfecting those divers domestic arts which, in the course of time, became their recognised specialties. They soon found that it was better to cultivate certain food plants and trees than to depend on them for nourishment in the wild state. This was particularly true in the case of such useful and widely distributed species as wheat, rice, maize, the yam, potato, banana and cassava.

At first most of these food products were used in the raw state, but woman's quick inventive genius was not long in making one of the most important and far-reaching discoveries — a method for producing fire. In a certain sense this was the greatest discovery ever made, and the Greeks showed their appreciation of the value of it by asserting that fire was stolen from heaven. Considering its multifarious uses in heating and cooking, thereby immensely adding to the comfort and well-being of primitive man, we are not surprised that in certain parts of the world fire has always been considered something sacred, and that the old Romans instituted Vestal Virgins, and the ancient Peruvians Virgins of the Sun, to preserve this precious element and have it ever ready when required for sacrifice or for any of their various liturgical functions. If any one ever deserved a "monument more durable than bronze," it was the woman who, "on the edge of time," first drew the Promethean spark from a piece of pyrites by striking it with flint or produced it by the friction of two pieces of wood.

After building a home and establishing in it a fireplace for the preparation of food, woman's next concern was to secure more raiment than was afforded by the traditional fig leaf. This she found in the bark of certain trees, in the fibre of hemp and cotton and in the wool of sheep and goats. With these and her distaff she spun thread, and from the thread thus obtained she was by means of her primitive loom — likewise her invention — able to provide all kinds of textile fabrics for clothing for herself and family.

But there was much more to invent before the home of primitive man, or rather primitive woman, could be considered as fairly equipped. Furniture and culinary utensils were required, and these, too, were provided by the deft and cunning fingers of woman. She was the first potter and the first basket-maker; and anyone who has lived among the savages of any land, especially among the aborigines in the interior of South America, knows what an important part is played in domestic economy by native basketry and ceramic ware. Both of these articles were at first of the simplest character, but woman's innate aesthetic sense soon enabled her to produce those highly ornate specimens of pottery and basketry that are so highly prised in the public and private collections of this country and Europe.

The first device for converting grain into flour was, like the many other articles already named, the invention of woman. Whether the simple mortar and pestle of the North American Indian, or the Mexican metate and muller, or the Irish quern, it was, in every case, the product of woman's brain and handiwork, as it was also the basal prototype of our most improved types of flouring mills. And so was the soapstone pot — the predecessor of the iron or brass kettle — a woman's invention, as well as many similar contrivances for preparing food.

But what is probably the most remarkable culinary invention of woman in the state of savagery is her unique contrivance for converting the poisonous root of the *manihot utilissima* — the staple food of tropical America — into a wholesome and nutritious aliment. It is a bag, called *matapi,* which serves both as a press and as a sieve. For the inhabitants of the vast basins of the Amazon and the Orinoco, where the chief articles of diet are derived from the manihot and the plantain, this invention of woman is the most important ever made and ranks in importance with the discovery by the same skilled food purveyor of the dietetic value of manihot itself.

The first knife was a woman's invention, as the arrow-head and the spear point were the inventions of her hunter husband. It was in the beginning a most primitive implement; but, whether in the form of a simple flake of flint of obsidian, or in that of an Eskimo ulu — the woman's knife — it was the archetype of all the forms of cutlery now in use. With this rude knife the primitive housewife skinned and carved the game brought to her by her male companion. With it she scraped the interior of the hide and cut it up into articles of clothing. She was thus the first furrier and tailor. With it she made the first sandals and moccasins, and, in doing so, became the first shoemaker and the original St Crispin.

To woman, the originator of the first home, is due also the invention of the oven and the chimney. She was also the first maker of salt — that all-important condiment and sanitary agent — and the first to obtain nitre from wood ashes. She was the first engineer, as is evinced in her invention of the parbuckle and in the bamboo conduit, which was the predecessor of the great canals of Babylonia[3] and the imposing aqueducts of ancient Rome.

Important, however, as are all the foregoing inventions, we must not forget what was an equally important contribution by woman to the welfare and progress of our race — the domestication of animals. No discovery after that of artificially producing fire has contributed more toward the development of our race than the taming of milk- and fleece-bearing animals, like the cow, the sheep, the goat and the llama, or of burden-bearing animals, like the horse, the ass, the camel and the reindeer, or of hunting and watching animals like the faithful, ubiquitous dog. For, in the first place, the domestication of these supremely useful animals diminished man's labour as burden bearers. It likewise supplemented the fecundity of women and facilitated the multiplication of the race, because it supplied to the child a nourishment that previously could be obtained only from the mother, who had been obliged to suckle her young several years longer than was necessary after the friendly goat and cow came to her aid. Still another consequence of the domestication of animals was that it immensely diminished the amount of woman's care and labour, afforded her the necessary leisure to develop the arts of refinement, and stimulated intellectual growth in a way that otherwise would have been impossible.

It is often stated by certain writers who love to indulge in fanciful speculations that women inventors got their ideas as home builders and weavers and potters from nest-building birds, from web-weaving spiders, and from clay workers like termites and mud wasps. Be this as it may, the fact remains in all its inspiring truth that, in the matter of industrialism, as opposed to the militancy of man, we can unhesitatingly declare, with Virgil, *Dux femina facti*— woman was the leader in all the arts of peace — arts which have been slowly perfected through the ages until they present the extraordinary development which we now witness.

When we contemplate the splendid porcelain wares of Meissen and Sevres, or the countless varieties of cutlery produced in the factories of Sheffield, or the beautiful textile fabrics from the looms of Lowell and Manchester, or the delicate silks woven in the famous establishments of Lombardy and Southern France, or the countless forms of footwear made in Lynn and Chicago, or the exquisite furs brought from Siberia and the Pribyloff Islands, and dyed in Leipsic and London, or the astonishing output of food products from the factories of Pittsburgh and the immense roller mills of Minneapolis, we little think that the colossal wheels of these vast and varied industries were set in motion by the inventive genius of woman in the dim and distant prehistoric past.

And yet such is the case. Her handiwork from the earliest pottery may be traced through its manifold stages from its first rude beginnings to the most gorgeous creations of ceramic art. The primeval knife of flint or obsidian has become the keen tool of tempered steel; the simple distaff has issued in the intricate Jacquard loom; the metate and pestle actuated by a woman's arm have, by a long process of evolution, developed into our mammoth roller mills impelled by water power, steam or electricity.[4]

But these extraordinary changes from the rude implements of prehistoric time to the complicated machinery of the present is but a change of kind, not one of principle. It is a change due to specialisation of work which became possible only when men, liberated from the avocations of hunting and warfare, were able to take up the occupations of women, and develop them in the manner with which we are now familiar.

Why men, rather than women, should have achieved this work of specialisation; whether it was due to social causes or to woman's physical and mental organisation, or to these various factors combined, we need not inquire; but such is the fact. Whereas in primitive times every woman having a home was a cook, a butcher, a baker, a potter, a weaver, a cutler, a miller, a tanner, a furrier, an engineer, man, in assuming the work which was originally exclusively feminine and performed by one and the same person, has subdivided and specialised by improved forms of machinery and otherwise, so that what is now done is accomplished more rapidly and to better purpose, and with correspondingly greater results in the development of industry and in the progress of civilisation.

And the remarkable fact is that many of the most important of these improvements due to specialisation have been made within the memory of those yet living, while still others have been originated in quite recent years. Nevertheless, great as has been the work of specialisation and coordination in every department of human industry during the last few decades, it is, to judge by the reports of the Patent Office, as yet in little more than its initial stage.

We are now prepared for the consideration of the part woman has taken in this specialising movement and for a discussion of her share in modern inventions and in the improvements of those manifold inventions which were due to her genius and industry untold ages ago. Considering the short time during which her inventive mind has been specially active, and the many handicaps which have been imposed on her, the wonder is not that she has achieved so little in comparison with man, but rather that she has accomplished so much.

The first woman to receive a patent in the United States was Mary Kies. It was issued May 5, 1809, for a process of straw-weaving with silk or thread. Six years later Mary Brush was granted a patent for a corset. It seems to have been quite satisfactory, for no other patent for this article of feminine attire was issued to a woman until 1841, when one was granted to Elizabeth Adams. During the thirty-two years which elapsed between the issuing of a patent to Mary Kies and Elizabeth Adams, but twenty other patents were granted to women. The chief of these were for weaving hats from grass, manufacturing moccasins, whitening leghorn straw, for a sheet-iron shovel, a cook stove and a machine for cutting straw and fodder.

During the decade following 1841, fourteen patents were issued to as many different women. Among the articles patented by them were an ice-cream freezer, a weighing scale and a fan attachment for a rocking chair. It was not recorded, however, that this last invention, valuable as it was apparently, ever became particularly popular. But by far the most remarkable of woman's inventions during this period was a submarine telescope and lamp, for which a patent was awarded in 1845 to Sarah Mather.

From 1851 to 1861, twenty-eight patents were issued to women —just twice the number awarded them during the preceding decade. Most of these patents were for articles of domestic use or feminine apparel. Four of them, however, comprised a scale for instrumental music, for mounting fluid lenses, a fountain pen and an improvement in reaping and mowing machines.

The following decade is remarkable for the wonderful increase in the number of inventions due to women, for there was a sudden jump from twenty-eight to four hundred and forty-one patents awarded them between the years 1861 and 1871. Women now began to have confidence in their inventive faculties, and, no longer content with exercising their genius on articles of clothing and culinary utensils, sewing, washing and churning machines, they began to devote their attention to objects that were entirely foreign to their ordinary home activities. This is clearly evinced by the patents they obtained for such inventions as improvements in locomotive wheels, devices for reducing straw and other fibrous substances for the manufacture of paper pulp, improvements in corn huskers, low-water indicators, steam and other whistles, corn ploughs, a method of constructing screw propellers, improvements in materials for packing journals and bearings, in fire alarms, thermometers, railroad car heaters, improvements in lubricating railway journals, in conveyors of smoke and cinders for locomotives, in pyrotechnic night signals, burglar alarms, railway car safety apparatus, in apparatus for punching corrugated metals, desulphurising ores and other similar inventions in the domain of mechanical engineering, inventions that, at first blush, would seem to be quite alien to the genius and capacity of woman.

From now on women's inventions in the United States increased at an extraordinary rate, for from 1871 until July 1, 1888, when the first government report was made on the patents issued to women inventors, she had to her credit nearly two thousand inventions, many of which were of prime importance.[5]

During the seven years following 1888 she was awarded twenty-five hundred and twenty-six patents — more than the total number that had been granted her during the preceding seventy-nine years. Between 1895 and 1910, three thousand six hundred and fifteen more patents were placed to her credit, making a grand total for her first century of inventive achievement of eight thousand five hundred and ninety-six patents. No Patent Office reports are available since 1910, but the number of inventions for which women have received patents since Mary Kies was awarded hers on May 5, 1809, for "straw-weaving with silk or thread," cannot be far from ten thousand. This fact will, doubtless, be a revelation to that large class of men who still seem to share the views of Voltaire and Proudhon that women are incapable of inventing even the simplest article of domestic use.

The following story well illustrates the prevailing ignorance regarding the part women have taken in the invention of certain articles that are so common that most people think they were never patented.

"I was out driving once with an old farmer in Vermont," writes Mrs Ada C Bowles, "and he told me, 'You women may talk about your rights, but why don't you invent something?' I answered, 'Your horse's feed bag and the shade over his head were both of them invented by women.' The old fellow was so taken aback that he was barely able to gasp, 'Do tell!' "

Had he investigated further he would have found that the flynet on his horse's back, the tugs and other harness trimmings, the shoes on his horse's feet[6] and the buggy seat he then occupied were all the inventions of women. He would, doubtless, also have discovered that the currycomb he had used before starting out on his drive, as well as the snap hook of the halter and the checkrein and the stall unhitching device were likewise the inventions of members of that sex whose capacity he was so disposed to depreciate; for women have been awarded patents — in some instances several of them — for all the articles that have been mentioned. He might furthermore have learned that the fellies in his buggy wheels and his daughter's side saddle had been made under women's patents; and that, to complete his surprise and confusion, the leather used in his harness had been sewn by a machine patented by a woman who was not only an inventor but who was also for many years the manager and proprietor of a large harness factory in New York City.

What particularly arrests one's attention in reading the Patent Office reports is not only the large number of inventions by women, but also the very wide range of the devices which they embrace. It is not surprising to find them inventing and improving culinary utensils, house furniture and furnishings, toilet articles, wearing apparel and stationery, trunks and bags, toys and games, designs for printed and textile fabrics, for boxes and baskets, screens, awnings, baby carriers, musical instruments, appliances for washing and cleaning, attachments for bicycles and type-writing machines, art, educational and medical appliances; for these things are in keeping with their proper *metier*; but it is surprising for those who are not familiar with the history of modern inventions to learn of the share women have had in inventing and improving agricultural implements, building appurtenances, motors of various kinds, plumbing apparatus, theatrical stage mechanisms, and, above all, countless railway appliances from a coupling or fender to an apparatus for sanding railroad tracks, or a device for unloading boxcars.

Those who are still of the opinion of Voltaire and Proudhon — and their name is legion — respecting woman's inventive powers, might be willing to accord to her the capacity to design a new form of clothes pin, or hair crimper, or rouge pad, or complexion mask, or powder puff, or baby jumper; but they would limit her ability to contrivances of this character. But what would these same people say if they were told that over and above the things just mentioned for which many women have actually received patents, the much depreciated female sex had been granted patents for locomotive wheels, stuffing boxes, railway car safety apparatus, life rafts, cut-offs for hydraulic and other engines, street cars, mining machines, furnaces for smelting ores, sound-deadening attachments for railway cars, feed pumps and transfer apparatus for traction cars, machines for driving hoops on to barrels, apparatus for destroying vegetation on and removing snow from railroads, coke crushers, artificial stone compositions, elevated railways, new forms of cattle cars, dams and reservoirs, welding seams of pipes and hardening iron, alloys for bell metal and alloys to resemble silver, methods of refining and hardening copper, processes for concentrating ores, improvement in elevators and designs for raising sunken vessels? And yet, incredible as it may appear to these scoffers at woman's genius, patents for all these inventions, methods and processes — many of them of exceeding value — and for hundreds of others of a similar nature, have been issued to women during recent years. And the activity of the fair inventors, far from abating, is becoming daily more pronounced, and promises to reward their efforts with far greater triumphs. Indeed, women are becoming so active in the numerous fields of invention — even in such unlikely ones as metallurgy and civil, mechanical and electrical engineering — that they bid fair to rival men in what they have long regarded as their peculiar specialty.

In 1892 a woman in New York was granted two patents, one for a process of malting beer and the other for hooping malt liquors. These inventions, however, are not so foreign to the avocation of woman as they at first appear. For, if we may believe the teachings of ethnology and prehistoric archaeology in this matter, women were the first brewers. The one, therefore, who two decades ago secured the two patents just mentioned was but taking up anew an occupation in which her sex furnished the first invention many thousand years ago.

An instructive fact touching woman's inventive achievements is that her fullest success is coincident with her enlarged opportunities for education, and began with the breaking down of the prejudices which so long existed against her having anything to do with the development of the mechanical or industrial arts. When one recollects that the public schools of Boston, established in 1642, were not open to girls until a century and a half later, and then only for the most elementary branches and for but one-half the year; and that girls did not have the benefit of a high school education in the centre of New England culture until 1852; and when one furthermore recalls the attitude of the general public toward women and girls extending their activities beyond the nursery and the kitchen, it is easy to understand that there was not much encouragement for them to exercise their inventive talent, even if they had felt an inclination to do so.

The experience of Miss Margaret Knight, of Boston, who in 1871 was awarded a valuable patent for making a paper-bag machine is a case in point and well illustrates some of the difficulties that women inventors had to contend with only a few decades ago.

"As a child," she writes to a friend, "I never cared for the things that girls usually do; dolls never had any charms for me. I couldn't see the sense of coddling bits of porcelain with senseless faces; the only things I wanted were a jack-knife, a gimlet and pieces of wood. My friends were horrified. I was called a tomboy, but that made very little impression on me. I sighed sometimes because I was not like other girls, but wisely concluded that I couldn't help it, and sought further consolation from my tools. I was always making things for my brothers. Did they want anything in the line of playthings, they always said, 'Mattie will make them for us.' I was famous for my kites, and my sleds were the envy and admiration of all the boys in town. I'm not surprised at what I've done; I'm only sorry I couldn't have had as good a chance as a boy, and have been put to my trade regularly."

Even after she had demonstrated her skill as an inventor, Miss Knight had to encounter the scepticism of the workmen to whom she entrusted the manufacture of her machines. They questioned her ability to superintend her own work, and it was only her persistency and remarkable competency that ultimately converted their incredulity into respect and admiration.

Since women have come into the possession of greater freedom than they formerly enjoyed, and have been afforded better opportunities of developing their inventive faculties, many of them have taken to invention as an occupation, and with marked success. They find it the easiest and most congenial way of earning a livelihood, and not a few of them have been able thereby to accumulate comfortable fortunes, besides developing industries that have given employment to thousands of both sexes.

Thus the straw industry in the United States is due to Miss Betsy Metcalf, who, more than a century ago, produced the first straw bonnet ever manufactured in this country. Since then the industry which this woman originated has assumed immense proportions. The number of straw hats now made in Massachusetts alone, not to speak of those annually manufactured elsewhere, runs into the millions.

Scarcely less wonderful is the industry developed by Miss Knight, already mentioned, through her marvellous invention for manufacturing satchel-bottom paper bags. Many men had previously essayed to solve the problem which she attacked with such signal success, but all to no purpose. So valuable was her invention considered by experts that she refused fifty thousand dollars for it shortly after taking out her patent.

Often what are apparently the most trivial inventions prove the most lucrative. Thus, a Chicago woman receives a handsome income for her invention of a paper pail. A woman in San Francisco invented a baby carriage, and received fourteen thousand dollars for her patent. The gimlet-pointed screw, which was the idea of a little girl, has realised to its patentee an independent fortune. Still more remarkable is the Burden horseshoe machine, the invention of a woman, which turns out a complete horseshoe every three seconds and which is said to have effected a saving to the public of tens of millions of dollars.

The cotton gin, one of the most useful and important of American inventions — a machine that effected a complete revolution in the cotton industry throughout the world — is due to a woman, Catherine L Greene, the wife of General Nathaniel Greene, of Revolutionary fame. After she had fully developed in her own mind a method for separating the cotton from its seed, which was after her husband's death, she entrusted the making of the machine to Eli Whitney, who was then boarding with her, and who had a Yankee's skill in the use of tools. Whitney was several times on the point of abandoning as impossible the task which had been assigned to him, but Mrs Greene's faith in ultimate success never wavered, and, thanks to her persistence in the work and the putting into execution of her ideas, her great undertaking was finally crowned with success. She did not apply for a patent for her invention in her own name, because so opposed was public opinion to woman's having part in mechanical occupation that she would have exposed herself to general ridicule and to a loss of position in society. The consequence was that Whitney — her employee — got credit for an invention which, in reality, belonged to her. She was, however, subsequently able to retain a subordinate interest in it through her second husband, Mr Miller.

This is only one of many instances in which patents, taken out in the name of some man, are really due to women. The earliest development of the mower and reaper, as well as the clover cleaner, belongs to Mrs A H Manning, of Plainfield, New Jersey. The patent on the clover cleaner was issued in the name of her husband; but, as he failed to apply for a patent for the mower and reaper, his wife was, after his death, robbed of the fruit of her brain by a neighbour, whose name appears on the list of patentees of an invention which originated with Mrs Manning.

A few years ago men of science awoke to the startling fact that the earth's supply of nitrates was being rapidly exhausted. It was then realised that, unless some new store of this essential fertiliser could be found, it would soon be impossible to provide the food requisite for the world's teeming millions. What was to be done? Never was a more important problem presented to science for solution, and never did science more quickly and efficaciously respond. It was soon recognised that the earth's atmosphere was the only available storehouse for the much-needed nitrogen. Forthwith scientists and inventors the world over proceeded to tap this source of supply and to convert its vast stores of nitrogen into the nitrates which are so indispensable to vegetable life.

To form some idea of the importance of the problem and the urgency of its solution, it may be stated that the amount of fertiliser required for the cotton crop alone in the Southern States in 1911 was no less than three million tons. What, then, must have been the total amount used through the world for cereals and other crops that need constant fertilising? The famous nitrate deposits of Chile could supply only a small fraction of the stupendous amount required, and they, according to recent calculations, cannot continue to meet the present demands on them for more than a hundred years longer, at most.

The process involved, when once conceived, was simple enough, for it merely required the conversion of the nitrogen of the air into nitric acid, which in turn was employed in the production of nitrate of lime. But, simple as it was, mankind had to wait a long time for its origination, and action was taken only when necessity compelled. At present there are numerous nitrate factories in France, Germany, Austria, Sweden, Norway and the United States, and the output is already enormous and constantly increasing. Electricity, that mysterious force which has so frequently come to man's assistance during the last few decades, is the agent employed.

But who was the originator of the idea of utilising the atmosphere for the production of nitrates? Who took out the first patent for a process for making nitrates by using the nitrogen of the air? It was a Frenchwoman — Mme Lefebre, of Paris — long since forgotten. As early as 1859 she obtained a patent in England for her invention, but, as the need of fertilisers was not so urgent then as it is now, it was allowed to drop into oblivion, and the matter was not again taken up until a half-century later, when others secured the credit for an idea which was first conceived by a woman who happened to have the misfortune to live fifty years in advance of her time.

It were easy to extend the list of important inventions due to women and of patents which were issued in the name of their husbands or other men; to tell of inventions, too, of whose fruits, because they happened to be helpless or inexperienced women, the real patentees were often robbed; but the foregoing instances are quite sufficient to show what woman's keen inventive genius is capable of achieving in spite of all the restrictions put on her sex, and in spite of her lack of training in the mechanic arts.

Had women, since the organisation of our Patent Office, enjoyed all the educational opportunities possessed by men; had they received the same encouragement as the lordly sex to develop their inventive faculties; had the laws of the country accorded them the rewards to which their labour and genius entitled them, they would now have far more inventions to their credit than those indicated in our government reports; and they would, furthermore, be able to point to far more brilliant achievements than have heretofore, under the unfavourable conditions under which they were obliged to work, been possible. But when we recall all the obstacles they have had to overcome and remember also the fact that most of the patents referred to in the preceding pages have been secured by women living in the United States — little being said of the modern inventions of women in foreign countries — we can see that their record is indeed a splendid one, that their achievements are not only worthy of all praise, but also a happy augury for the future. When they shall have the same freedom of action as men in all departments of activity in which they exhibit special aptitude, when they shall have the same advantages of training and equipment and the prospect of the same emoluments as the sterner sex for the products of their brainwork and craftsmanship, then may we expect them to achieve the same distinction in the mechanic arts as has rewarded their efforts in science and literature; and then, too, may we hope to see them once more regain something of that supremacy in invention which was theirs in the early history of our race.

## Notes & References

1) For an evidence of this learned lady's competency to deal with the most recondite stores of history and archaeology, the reader is referred to two of her later works, viz., *Primitive Athens as Described by Thucydides*, Cambridge, 1906, and *Prolegomena to the Study of Greek Religion*, Cambridge University Press, 1903.

2) "On a vu des femmes tres savantes, comme en fût des guerrieres, mais il n'y en eut jamais d'inventrices." *Dictionnaire Philosophique, sub voce Femmes.* Condorcet, in commenting on this statement, remarks that "if men capable of invention were alone to have a place in the world, there would be many a vacant one, even in the academies."

3) That marvelous structure known as the Taj Mahal — India's noblest tribute to the grace and goodness of Indian womanhood — is sometimes said to be a monument to the memory of Nur Mahal. This is not the case. This matchless gem of architecture —

..... The proud passion of an emperor's love
Wrought into living stone, which gleams and soars
With body of beauty shrining soul and thought.

is a monument to Nur Mahal's niece and successor as empress, Mumtaz-Mahal — The Crown of the Palace — who, like her aunt, was a woman of rare beauty and talent and endeared herself to her people by her splendid qualities of mind and heart.

4) The inventor of canals as well as of bridges over rivers and causeways over morasses was, according to Greek historians, the famous Assyrian queen, Semiramis, the builder of Babylon with its wonderful hanging gardens.

5) Among the works which treat of the subject-matter of the foregoing pages the reader may consult with profit, *Woman's Share in*

*Primitive Culture*, by O T Mason, London, 1895; *Man and Woman*, the introductory chapter, by Havelock Ellis, London, 1898; and *Histoire Nouvelle des Arts et des Sciences*, by A Renaud, Paris, 1878.

6) Cf *Women Inventors to whom patents have been granted by the United States Government, Compiled under the Direction of the Commissioner of Patents*, Washington, 1888. See also subsequent reports of the Patent Office.

# Women as Inspirers and Collaborators in Science

ONE OF THE most interesting literary figures of the fifth century was Caius Apollinaris Sidonius, who, after holding a number of important civil offices, became the bishop of Clermont. The most valuable of his extant works are his nine books of letters which are a mine of information respecting the history of his age and the manners, customs and ideals of his contemporaries.

In one of these letters, addressed to Hesperius, a young friend of his who exhibited special talent in polite literature, he expresses a sentiment which applies as well to the votary of science as to the man of letters. Referring to the assistance which women had given to their husbands and friends in their studies, he conjures him to remember that in days of old it was the wont of Martia, Terentia, Calpurnia, Pudentilla and Rusticana to hold the lamp while their husbands, Hortensius, Cicero, Pliny, Apuleius and Symmachus, were reading and meditating.[1]

This picture of women as light-bearers to the great orators and philosophers just named is symbolic of them as the helpmates and inspirers of men in every field of human activity and in every age of the world's history. Always and everywhere, when permitted to occupy the same social plane as men, women have been not only as lamps unto the feet and as lights unto the paths of their male compeers in the ordinary affairs of life, but have also been their guiding stars and ministering angels in the highest spheres of intellectual effort.

For nearly fifteen centuries St Jerome has had the gratitude of the church for his masterly translation, known as the Vulgate, of the Hebrew Scriptures. But, had it not been for his two noble friends, Paula and Eustochium, who were as eminent for their intellectual attainments as they were for their descent from the most distinguished families of Rome and Greece, there would have been no Vulgate. For they were not only his inspirers in this colossal undertaking, but they were his active and zealous collaborators as well.

Dante and Petrarch are acclaimed as the morning stars of modern literature, but both of them owed their immortality to the inspiration of two pure-minded and noble-hearted women.

In the concluding paragraph of his Vita Nuova — the most beautiful love story ever written — Dante records his purpose to say of his inspirer, the gentle, gracious Beatrice Portinari, "what was never said of any woman." The outcome of this exalted purpose was the Divina Commedia, the world's greatest literary masterpiece.

Petrarch, the father of humanism, is the first to give Laura de Noves credit for his attainments as a poet. In one of his poems he sings:

> Blest be the year, the month, the hour, the day,
> The season and the time, and point of space,
> And blest the beauteous country and the place
> Where first of two eyes I felt the sway.

Elsewhere in one of his prose dialogues with St Augustine he declares, "Whatever you see in me, be it little or much, is due to her; nor would I ever have attained to this measure of name and fame unless she had cherished by those most noble influences that my feeble implanting of virtues which nature had placed in this breast."[2]

A no less remarkable inspirer, but in an entirely different sphere of activity, was the devout and spotless Italian maiden, Chiara Schiffi, better known as St Clara. She was, as is well known, the ardent cooperator of St Francis Assisi in his great work of social and religious reform which has contributed so much toward the welfare of humanity. But it is not generally known what an important part she had in this great undertaking, and how she sustained the Poverello during long hours of trial and hardship. It was during these periods of care and struggle that we see how courageous and intrepid was "this woman who has always been represented as frail, emaciated, blanched like a flower of the cloister."

"She defended Francis not only against others but also against himself. In those hours of dark discouragement which so often and so profoundly disturb the noblest souls and sterilise the grandest efforts, she was beside him to show the way. When he doubted his mission and thought of fleeing to the heights of repose and solitary prayer, it was she who showed him the ripening harvest with no reapers to gather it in, men going astray with no shepherd to herd them, and drew him once again into the train of the Galilean, into the number of those who give their lives as a ransom for many."[3]

It is under the shade of the olive trees of St Damian, with his sister-friend Clara caring for him, "that he composes his finest work, that which Ernest Renan called the most perfect utterance of modern religions sentiment, *The Canticle of the Sun*."[4]

This canticle, however, beautiful as it is, lacks, as has well been remarked, one strophe. "If it was not upon Francis' lips, it was surely in his heart:"

> Be praised, Lord, for Sister Clara;
> Thou hast made her silent, active, and sagacious,
> And, by her, thy light shines in our hearts.[5]

It was through the inspiration and influence of Theodora that the famous Church of St Sophia, that matchless poem in marble and gold, that imperishable monument to the glory of the true God, came into existence. It was through her that Justinian conceived the idea of those *Pandects* and *Institutes* which constitute the greatest glory of his reign, and which are the basis of the *Code Napoleon* and of all modern jurisprudence.

It was to Vittoria Colonna that Michaelangelo dedicated many of the most exquisite productions of his peerless genius. "He saw," as has been said, "with her eyes and acted by her inspiration."

Almost every one of Chopin's compositions was inspired by women, and a large proportion of them are dedicated to them. The same may be said of Mozart, Mendelssohn, Schubert, Beethoven, Weber, Schumann and other illustrious composers. All these sons of genius believed with Castiglione that "all inspiration must come from woman;" that she had been expressly created and sent into the world to inspire them with intelligence and creative power.

M Claviere declares that "There is hardly a philosopher or a poet of the sixteenth century whose pages are not illuminated or gladdened by the smile of some high-born lady."[6]

What the brilliant Frenchman says of the influence of woman on the poets and philosophers of a single century could with equal truth be said of the poets and philosophers of every century from Anacreon and Plato to the present day. And, still more, it can be predicated of woman's inspiration and influence in every department of intellectual effort, in art and architecture, in music and literature, in science in all its departments, whether deductive or inductive.

It has been well said, "Were history to be rewritten, with due regard to women's share in it, many small causes, heretofore disregarded, would be found fully to explain great and unlooked-for results..... For it is not in outward facts, nor great names, nor noisy deeds, nor genealogies of crowned heads, nor in tragic loves, nor ambitious or striking heroism, nor crime, that we find proofs of the constant and secret working whereby woman most effectually asserts herself. Certainly she has played her part in the outward and visible history of the world, but in that history which is told and written, which is buried in archives and revivified in books, woman's part is always small when set beside that of her companion, man. She contributes but little, and at this she may surely rejoice, to the tales of battles and treaties of successions and alliances, of violence, fraud, suspicions and hatreds. But if the inward history of human affairs could be described as fully as the outward facts; if the story of the family could be told together with the story of the nation; if human thoughts could with certainty be divined from human deeds, then the chief figure in this history of sentiment and morals would certainly be that of Woman the Inspirer."[7]

This same statement would hold equally good if applied to the part taken by women in the history of science. Their achievements have, in most cases, been so overshadowed by those of men that their work has been usually regarded as a negligible quantity. But when one considers the mainsprings of actions, and examines the silent undercurrents which escape the notice of the superficial observer, one finds, as in social and political history, that the most important scientific investigations are often conducted, and the most momentous discoveries are made, in consequence of the promptings of some devoted woman friend, or in virtue of the still, small voice of a cherished wife, or sister, who prefers to remain in the background in order that all the glory of achievement may redound to the man.

There have been, it may safely be asserted, few really eminent men in science, as there have been few really eminent men in art or letters, or in the great reform and religious movements of the world, who have not been assisted by some woman light-bearer, as were Hortensius by Martia, Tully by Terentia and Pliny by Calpurnia. There have been few that have not, during hours of doubt and discouragement, been sustained and stimulated as was Francis by Clara, and Jerome by Paula and Eustochium. And there have been still fewer who have not had, like Petrarch and Dante, their Laura or their Beatrice of whom each could say:

> This is the beacon guides to deeds of worth,
> And urges me to see the glorious goal:
> This bids me leave behind the vulgar throng.

In the preceding chapters we have had notable examples of women whose beneficent influence and cooperation have enabled distinguished men of science to achieve results that would otherwise have been impossible. Among these — to mention only a few — were Mme Lavoisier and Mme Curie in chemistry, Mme Lapaute and Miss Herschel in astronomy, Mrs Agassiz and Mme Coudreau in natural science and exploration, Mme Schliemann and Mme Dieulafoy in archaeology.

One of the most illustrious women inspirers of France was Catherine de Parthenay, who, after attaining womanhood, became the brilliant Princess de Rohan, and was recognised as one of the most learned and most remarkable women of the sixteenth century. As a young girl she exhibited rare intelligence and displayed special aptitude for the exact sciences. For this reason her mother saw to it that her child had the benefit of instruction under the ablest masters that could be secured.

The most noted of these was François Viete, the learned French mathematician, who is justly regarded as the father of modern algebra. In his day, especially in the higher classes of society, the education given to women was often more thorough than that afforded to men. For this reason, too, women not infrequently became distinguished in astronomy, which was then usually known under the name of astrology.

Viete, in initiating his gifted pupil into the principles of this science, became himself so enthusiastic a student of astronomy that he determined to prepare an elaborate work on the subject — something on the plan of the *Almagest* of Ptolemy — a work which he designated *Harmonicum Celeste*.

In order that the instruction given his pupil might not be lacking in precision, Viete wrote out, with the most scrupulous care, the lessons designed for her benefit. The manuscripts containing these lessons were long preserved among the family archives, but nearly all of them were unfortunately consigned to the flames during the French Revolution in 1793.

No one was more interested in Viete's mathematical researches — those researches which have rendered him so famous in the history of science — than was the Princess de Rohan. The former pupil was the first to receive notice of her distinguished master's discoveries and the first to congratulate him on his success.

It was to this cherished pupil, who always remained his friend and benefactress, that Viete dedicated his important work on mathematical analysis entitled *In Artem Analyticam Isagoge*. The words of the dedication are a tribute to the learning and the genius of the pupil as well as an expression of the gratitude of the teacher. It reads as follows:

> It is to you especially, august daughter of Melusine, that I am indebted for my proficiency in mathematics, to attain which I was encouraged by your love for this science, as well as your great knowledge of it, and by your mastery of all other sciences, which one cannot too much admire in a person of your noble lineage.[8]

More interesting, and at the same time more pathetic, were the relations of an Italian nun, Sister Maria Celeste, and the man whom Byron so happily designates as

> The starry Galileo, with his woes.

Sister Celeste, who was a Franciscan nun in the convent of St Matthew, in Arcetri, was the great astronomer's eldest and favourite daughter. They were greatly attached to each other, and the gentle religieuse was not only her father's confidante and consoler in the hours of trial and affliction, but was also his inspirer and ever-vigilant guardian angel. She watched over him, not as a daughter over a father, but as a mother watches over an only son.[9]

All this is beautifully exhibited in her one hundred and twenty-four letters which were published in 1891 for the first time. A few of these letters, it is true, were published as early as 1852 by Alberi, in his edition of the complete works of Galileo, and others were given to the press at subsequent dates; but the world had to wait more than two and a half centuries for a complete collection of all the known letters of this remarkable daughter of an illustrious sire.

These documents are precious for the insight they give into the sterling character of a noble woman, but they are beyond price as sources of information respecting the tenderly affectionate relations which existed between her and one of the foremost men of science, not only of his own age, but of all time. They show how he made her his confidante in all his undertakings, and how she was his amanuensis, his counsellor, his inspirer; how her love was an incentive to the work that won for him undying fame; how she was his support and comfort when suffering from the jealousy of rivals or the enmity of those who were opposed to his teachings.

These letters cover a period of nearly eleven years — the most momentous years of her father's busy and troubled life. Now playful, quaint, elfish, then serious, vivid, confidential, they show that the writer's intelligence was as rare as her nature was loyal and affectionate. At times she half-apologises for the length of a letter, "but you must remember," she adds in excuse, "that I must put into this paper everything that I should chatter to you in a week."

No daughter was ever prouder of her father or loved him with a more abounding love. "I pride myself," she says, "that I love and revere my dearest father more, by far, than others love their fathers, and I clearly perceive that, in return, he far surpasses the greater part of other fathers in the love which he has for me, his loved daughter."

When he was ill she prepared dishes and confections that she knew would tempt his appetite. But she was not satisfied with looking after the welfare of his body, for she took occasion to send with the cakes and preserved fruits a sermonette for the benefit of his soul.

An extract from one of her letters gives an insight into the character of this devoted daughter, who, Galileo says in a letter to his friend, Elia Diodati, "was a woman of exquisite mind, singular goodness and most tenderly attached to me."

"Of the preserved citron you ordered," she writes him on the nineteenth of December, 1625, "I have only been able to do a small quantity. I feared the citrons were too shrivelled for preserving, and so they proved. I send two baked pears for these days of vigil. But the greatest treat of all I send you is a rose, which ought to please you extremely, seeing what a rarity it is at this season. And with the rose you must accept its thorns, which represent the bitter passion of Our Lord, while the green leaves represent the hope we may entertain that, through the same sacred passion, we, having passed through the darkness of this short winter of our mortal life, may attain to the brightness and felicity of an eternal spring in heaven, which may our gracious God grant us through His mercy."[10]

She always insists upon his keeping her fully informed about his studies and discoveries. She is particular, also, about receiving without delay copies of his latest publications. "I beg you," she writes in one of her letters, "to be so kind as to send me that book of yours which has just been published, *Il Saggiatore*, so that I may read it; for I have a great desire to see it."

On another occasion, after his difficulties with the Holy Office, when she fancies her father is not keeping her fully informed about the subject matter of his writings, she implores him to tell her on what topic he is engaged, "if," she archly adds, "it be something I can understand and you are not afraid that I will blab."

And on still another occasion Sister Celeste reminds her father of a promise of his to send her a small telescope. From this we should infer that she desired to repeat the observations on the heavenly bodies that had created such a sensation in the learned world, and which had given occasion for such acrimonious controversy.

In one of her earlier letters Sister Celeste calls her father's attention to a promise of his to spend an afternoon with her and her sister Arcangela, also a nun in the same convent. And, referring to one of the regulations of the Franciscan cloister, she playfully observes: "You will be able to sup in the parlour, since the excommunication is for the table cloth" — O Sister Celeste! — "and not for the meats thereon."

What would one not give for a stenographic report of the conversations held that afternoon in the convent garden of Arcetri, as father and daughters leisurely strolled through the peaceful enclosure, all quite oblivious of the fleeting hours? How interesting would be a faithful record of the confidences exchanged at the frugal meal in the evening in the humble parlour of S Matteo! We would willingly exchange many of the famous *Dialoghi di Galileo Galilei* for a verbatim report of what passed between Sister Celeste and the father whom she so idolised.[11]

Judging from her letters, she had many questions to ask him about his studies, his experiments, his discoveries, his books, as well as about more personal and domestic matters.

Although there is no documentary proof of the fact, yet there is every reason to believe that Galileo had taken personal charge of the education of this, his favourite daughter. She shared his taste for science and inherited not a little of his genius. Such being the case, we may well believe that a faithful account of their conversations of that day would be not only of surpassing interest, but would also throw a flood of light on many questions now ill understood. They would certainly tend to fill up the numerous lacunae caused by the disappearance of the letters of Galileo, which he wrote in answer to those of his ever-cherished daughter.[12]

They would also show more clearly than any facts now available what an unbounded influence the gentle nun had over the greatest intellect of his time, and would, more clearly than anything in her correspondence, exhibit Sister Celeste as the efficient co-worker and the abiding inspirer of the father of modern physics and astronomy.

But, although we have no record of this soul-communion between father and daughter on the occasion in question; although we are deprived of the invaluable letters which he wrote in reply to hers, we are, nevertheless, from the evidence at hand, justified in regarding this unique pair as being ever one in heart, aspirations and ideals, and comparable in their mutual influence on each other with any of those famous men and women who, through achievement on the one side and inspiration and collaboration on the other, have ever been recognised as the greatest benefactors of their race.

One of Galileo's countrymen, G B Clemente de Nelli, was right when he declared that, had it not been for the assistance and consolation which he received from Sister Celeste, Galileo would have succumbed to the blows that were showered upon him during the most trying part of his career. An indication of this is given in one of the letters written by Sister Celeste in the last year of her life.

While in a fit of despondency and imagining his friends had forgotten him, Galileo, in a moment of bitterness, wrote in a letter to his daughter: "My name is erased from the book of the living." "Nay," came at once Sister Celeste's cheering reply, "say not that your name is struck *de libro viventium*, for it is not so; neither in the greater part of the world nor in your own country. Indeed, it seems to me that, if for a brief moment your name and fame were clouded, they are now restored to greater brightness, at which I am much astonished, for I know that generally *Nemo propheta acceptus est in patria sua*. I am afraid, however, if I begin quoting Latin, I shall fall into some barbarism. But, of a truth, you are loved and esteemed here more than ever."[13]

How much Sister Celeste was to her father in every way was not known until after her premature death in her thirty-fourth year. He was never the same man afterward. Disconsolate and broken, he fancied he heard the voice of the daughter he so fondly loved resounding through the house. Brooding over his great loss, the heart-broken old man writes to a friend in words of infinite pathos, "*Mi sento continuamente chiamare della mia diletta figlioula* — I continually hear myself called by my dearly beloved daughter." The eighth of January, 1642, he answered her call and went to join her in a better world.

Two other noted investigators, one of them a contemporary of Galileo, owed much to the inspiration and encouragement which they received from women. These were Descartes and Leibnitz. And the women that had the most influence on them were representatives of royal families, who were famous in their day for their love and knowledge and the extent of their intellectual attainments.

One of the most noted of these was Elizabeth of Bohemia, Princess Palatine. She was the favourite pupil of Descartes, and it was to her that he dedicated his great work, *Principia Philosophiae*. She, he declared, understood him better than any one else he had ever met, for "in her alone were united those generally separated talents for metaphysics and for mathematics which are so characteristically operative in the Cartesian system."[14]

To this earnest student who was always absorbed in the mysteries of metaphysics and the problems of geometry, Descartes could refuse nothing. When distance separated them he continued his instructions by correspondence. One of the results of this correspondence was his treatise on *Passions de l'Ame*, in which he develops certain ethical views suggested by the *Vita Beata* of Seneca.

Another distinguished pupil of Descartes who exercised a marked influence over him was the celebrated daughter of Gustavus Adolphus, Queen Christine of Sweden. A mistress of many languages and an ardent votary of science, she was a munificent patron of scientific men, a great number of whom she had attracted to her court. The most distinguished of these was Descartes, to whom she was deeply attached, and with whom she had planned great things for science in Sweden, when his career was cut short by a premature death.

Not the least influence on the intellectual life of Leibnitz was Sophia Charlotte, Queen of Prussia and mother of Frederick the Great. She was the niece of Descartes' illustrious friend, Elizabeth of Bohemia, and, as the pupil of Leibnitz, quite as gloriously associated as had been her aunt with the father of Cartesianism.

Leibnitz was as distinguished by genius as his royal pupil was by birth. Besides being eminent as a philosopher and a statesman, he shared with Newton the honour of discovering the calculus. Huxley pronounced him "a man of science, in the modern sense, of the first rank," while the King of Prussia declared of him, "He represents in himself a whole academy." Through the cooperation of Sophia Charlotte he founded the Berlin Academy of Sciences. For her he wrote one of the most notable of his productions —his famed *Theodicy*.

It would be difficult to estimate the influence of this learned queen on Leibnitz, but it was undoubtedly greater than any other single influence whatever. Her death was the greatest loss he ever suffered, and when she was no more, the beautiful Berlin suburb,

Charlottenburg — named after her — where he had been so happy in reading and philosophising with his illustrious pupil, lost all attraction for him.

A more striking illustration of woman's helpfulness is afforded in the case of François Huber, the celebrated Swiss naturalist. Although blind from his seventeenth year, he was able to carry on researches requiring the keenest eyesight and the closest observation. This he was able to do through the affectionate cooperation of his devoted wife, Marie Aimee.

When her friends tried to dissuade her from marrying Huber, to whom she had been engaged for some time, saying he had become blind, her reply was worthy of her generous and noble nature: "He then needs me more than ever."

During the forty years of their married life her tenderness and devotion to her husband were as unfailing as they were inspiring. He worked through the eyes and hands of his wife as if they were his own. She was his reader, his observer, his secretary, his enthusiastic collaborator in all those investigations that have rendered him so famous. The blind man devised the experiments to be made, and the quick-witted wife executed them and recorded the observations which supplied the material for his epoch-making work on bees, entitled *Nouvelles Observations sur les Abeilles.* So accurate are his descriptions of the habits of the winged creatures, to the study of which he devoted the best years of his life, that one would think his great work was the production, not of a man who had been blind for a quarter of a century, when he wrote it, but of one who was gifted with exceptional keenness of vision and powers of observation.

"As long as she lived," exclaimed the great naturalist after his trusty Aimee's death, "I was not sensible of the misfortune of being blind." Nay, more. During her lifetime, when, though sightless, he was always so happy in his work, he went so far as to aver that he would be miserable were he to recover his eyesight. "I should not know," he declared, "to what an extent a person in my condition could be beloved. Besides, to me, my wife is always young, fresh and pretty, which is no light matter." He could truly say of her, as Wordsworth said of his sister Dorothy,

> She gave me eyes, she gave me ears,
>
> * * * *
>
> And love and thought and joy.

We hear much of the achievements of Galvani and Faraday in the domain of electricity and electromagnetism, but little is said of the women to whom they were so greatly indebted for their success and fame.

It was Galvani's wife who first directed his attention to the convulsions of a frog's leg when placed near an electrical machine. This induced him to make those celebrated investigations which led to the foundation of a new science which has ever since been identified with his name.

It was Mrs Marcet's works on science — especially her *Conversations on Chemistry* — that inspired Faraday with a love of science and blazed for him that road in chemical and physical experimentation which led to such marvellous results. He was always proud to call her his first teacher, and never hesitated to attribute to her that taste for scientific research for which he became so pre-eminent. And it was his devoted wife who was not only a helpmate but a soulmate as well for nearly half a century, that had very much to do with the splendid development of the germ which had been placed in his youthful mind by Mrs Marcet.

The same may likewise be asserted of the wives of two distinguished geologists — Charles Lyell and Xavier Hommaire de Hell. Mrs Lyell was intimately associated with her husband in all his scientific undertakings, and her ready intellect contributed immensely toward securing for him that enviable position which he attained of being the premier geologist of his century. Mme Hommaire de Hell deserves special mention in the history of geology for the invaluable assistance which she gave her husband in the scientific exploration of the basin of the Caspian Sea. Not only did she share his labours and perils in this then wild part of the world, and collaborate with him in the preparation of the report for which the French government conferred on him the Cross of the Legion of Honour, but she also wrote unaided the two descriptive volumes of their great work, *Steppes de la Mer Caspienne*. Her part of this great undertaking received the special commendation of M Villemain, who was the minister of public instruction, and had she not belonged to the disenfranchised sex, she, too, would have been decorated with the Cross of the Legion of Honour.

All the world has heard of the daring explorations of Baker and Livingstone in the Dark Continent, but how few are aware of the important part taken in their great enterprises by their devoted and heroic wives? Sir Samuel Baker immortalised himself by discovering Lake Albert Nyanza, one of the main sources of the Nile, but in attaining this goal, which other explorers had in vain essayed to reach, he was not alone. The companion of his triumph, as of his trials and hardships, was Lady Baker, a woman who, although delicately reared, was as brave in presence of danger as she was resourceful in trials and difficulties. More than once her husband owed his life to her intrepidity and presence of mind, when confronted by the treacherous savages of equatorial Africa; and, if he achieved success where others failed, it was in no slight measure due to her tact, her energy and perseverance in what seemed at times a forlorn hope. "She had learned Arabic with him in a year of necessary but wearisome delay; her mind travelled with his mind as her feet had followed his footsteps." And, when after preliminary toils without number, after braving dangers from climate, disease and ruthless savages, they finally stood on the shore of that unknown sea which was then first beheld by English eyes, she could, in contemplating their achievements of which Albert Nyanza was the crowning glory, exclaim with exaltation and truth, "*Quorum pars magna fui.*"

When Livingstone lost, in the unexplored valley of the Zambezi, the faithful wife who had been his inspiring companion in his wanderings in darkest Africa, he lost completely that enthusiasm for deeds of high emprise that before had been one of his leading characteristics. Writing to his distinguished friend, Sir Roderick Murchison, he mournfully declares: "I must confess this heavy stroke quite takes the heart out of me. Everything that has happened only made me more determined to overcome all difficulties; but after this sad stroke I feel crushed and void of strength..... I shall do my duty still, but it is with a darkened horizon that I again set about it."

The noted English naturalist, Frank Buckland, in speaking of the aid afforded by his gifted mother to her distinguished husband, Dr Buckland, writes as follows: "During the long period that Dr Buckland was engaged in writing the book which I now have the honour of editing, my mother sat up night after night, for weeks and months consecutively, writing to my father's dictation; and this often until the sun's rays, shining through the shutters at early morn, warned the husband to cease from thinking and the wife to rest her weary hand.

"Not only with the pen did she render material assistance, but her natural talent in the use of her pencil enabled her to give accurate illustrations and finished drawings, many of which are perpetuated in Dr Buckland's works. She was also particularly clever and neat in mending broken fossils. There are many specimens in the Oxford Museum, now exhibiting their natural forms and beauty, which were restored by her perseverance to shape from a mass of broken and almost comminuted fragments. It was her occupation also to label the specimens, which she did in a particularly neat way; and there is hardly a fossil or a bone in the Oxford Museum which has not her handwriting upon it.

"Notwithstanding her devotion to her husband's pursuits, she did not neglect the education of her children, but occupied her mornings in superintending their instruction in sound and useful knowledge. The sterling value of her labours they now, in after life, fully appreciate, and feel most thankful that they were blessed with so good a mother."[15]

What has been said of the influence and cooperation of the women already named may, with equal truth, be affirmed of numberless others of recent as well as of earlier date. It is particularly true of the wife of the naturalist Heller and of the great astronomer, Kepler. It is true of the wife of the illustrious mathematician, the Marquis de l'Hopital. She not only shared her husband's talent for mathematics, but was of special assistance to him in preparing for the press his important *Analyse des Infiniment Petits*. It is true of the wife of Asaph Hall, the illustrious discoverer of the satellites of Mars. Often he was on the point of abandoning the quest of these diminutive moons — which no one had ever seen but which his calculations led him to believe really existed — but he was encouraged by Mrs Hall to continue his observations, with the result that his labours and vigils were at last rewarded by the startling discovery of Deimos and Phobos.

And there is Mme Pasteur, who, in her way, was quite as important a factor in the scientific career of her immortal husband as were the women just mentioned in the lives of their husbands, to whose triumphs they so materially contributed.

One of the great Frenchman's biographers has truly declared that "it is impossible rightly to appreciate Pasteur's life without some understanding of the immense assistance which he received in his home. Whether in discussing forms of crystals, watching over experiments, shielding her husband from all the daily fret of life, or busy at the customary evening task of writing to his dictation, Madame Pasteur was at once his most devoted assistant and incomparable companion. His surroundings at home were entirely subordinated to his scientific life, and his family shared with him both his trials and his triumphs. At the time when Pasteur was engrossed with the study of anthrax, and, after many difficulties and disappointments, had at length succeeded in preparing a vaccine against it, he at once hurried from the laboratory to communicate his great discovery first to his wife and daughter."[16]

It was particularly during his long and arduous researches on the disease of silkworms that Pasteur found his wife's aid of incalculable value. For Mme Pasteur and her daughter then constituted themselves veritable silkworm rearers. They collected mulberry leaves, sorted larvae, and were unremitting in their labours during the continuance of this memorable investigation. And not only in the silk-producing districts of Southern France were they thus occupied, but also in a special laboratory in École Normale, after their return to Paris.

And, when in the midst of these researches, on the successful outcome of which hinged one of the greatest sources of national wealth, the indefatigable savant was stricken with paralysis and his life was for a while despaired of, it was again his devoted helpmate that afforded him solace in suffering and exercised a supervision over those experiments which the great man was still conducting almost in the presence of death.

That Pasteur's life was prolonged for a quarter of a century after the terrible attack of hemiplegia in 1868, that he was able to unravel the deep mysteries of microbian life, that he was able to make discoveries whose economical value to France was, in the estimation of Professor Huxley, more than sufficient to liquidate the immense indemnity of five billion francs exacted from his country by Germany at the termination of the Franco-Prussian war, that he was able, especially during these fruitful twenty-five years, to render his "scientific life like a luminous trail in the great night of the infinitely little in those ultimate abysses of being where life is born," was, in great measure, due to the unceasing care, the untiring vigilance and the sympathetic collaboration of one of the most devoted of wives and most noble and whole-souled of women.

What has been said of the influence and helpfulness of Mme Pasteur can be asserted with even greater truth of Elizabeth Agassiz and of Caroline Herschel. For these two women, apart from the assistance they gave to a loved husband and an idolised brother, in the labours that made them so famous, both achieved distinction for their contributions to the sciences which they individually cultivated with such splendid results. And had they elected to devote all their time to scientific research, instead of giving the greater part of it to those to whom they were so devotedly attached, who can tell how much more brilliant would have been their achievements and how much greater would have been the fame they would have won for themselves. Both of them were dowered in an eminent degree with taste and talent for science, and had they chosen to make it the sole object of their life work, there can be no doubt that their personal contributions to natural history and astronomy would have been far greater than they were. As it was, they were so overshadowed by those for whom they laboured with such unselfishness and loyalty that the real value of their work is too often forgotten when there is question of the scientific triumphs of Louis Agassiz and Sir William Herschel.

But they willed it so. They gladly effaced themselves that those whom they loved with such a deep and abiding love might shine the more brightly in the firmament of science. They preferred to spend and be spent in strengthening the great workers and leaders with whose lives their own were so thoroughly identified — "Inspiring them with courage, keeping faith in their own ideas alive, in days of darkness

'When all the world seems adverse to desert.' "

Both of these noble women had the same quality in common — absolute devotion and unswerving faith in those to whose success and happiness they had dedicated their lives. They sought nothing for themselves, they thought nothing of themselves. They both had, to borrow the idea of another, an intense power of sympathy, a generous love of giving themselves to the service of others, which enabled them to transfuse the force of their own personality into the objects to which they dedicated their powers.

In the preface of the joint work of Mr and Mrs Agassiz entitled *A Journey in Brazil*, that delightful volume which throws such a flood of light on the fauna and flora of the Amazon valley, occur the following significant words regarding the share each had in producing the book: "Our separate contributions have become so closely interwoven that we should hardly know how to disconnect them." So was it with all their undertakings. There was the same common interest, the same unity of purpose, the same unselfish devotion to the cause of science during those long years of toil which were so prolific in results of supreme importance. Reading between the lines in *A Journey in Brazil*, and in *Louis Agassiz, His Life and Correspondence*, written by Mrs Agassiz, we can easily fancy that the great naturalist owed as much, if not more, to his wife's never-failing sympathy and inspiration as to her active cooperation in his work, and we are ready to apply to her the words of Longfellow when he sings:

> And whenever the way seemed long,
> Or his heart began to fail,
> She would sing a more wonderful song
> Or tell a more wonderful tale.

As to Caroline Herschel as a helper and sustainer of her illustrious brother, too much cannot be said. "In the days when he gave up a lucrative career that he might devote himself to astronomy, it was owing to her thrift and care that he was not harassed by the rankling vexations of money matters. She had been his helper and assistant when he was a leading musician; she became his helper and assistant when he gave himself up to astronomy. By sheer force of will and devoted affection she learned enough of mathematics and of methods of calculation, which to those unlearned seem mysteries, to be able to commit to writing his researches. She became his assistant in the workshop; she helped him to grind and polish his mirrors; she stood beside his telescope in the nights of midwinter, to write down his observations when the very ink was frozen in the bottle. She kept him alive by her care; thinking nothing of herself, she lived for him. She loved him and believed in him, and helped him with all her heart and with all her strength. She might have become a distinguished woman on her own account, for with the seven-foot Newtonian sweeper given her by her brother she discovered eight comets first and last. But the pleasure of seeking and finding for herself was scarcely tested. She 'minded the heavens' for her brother; she worked for him, not for herself, and the unconscious self-denial with which she gave up 'her own pleasure in the use of her sweeper' is not the least beautiful picture in her life."[17]

While recounting the achievements of women who directly or indirectly contributed to our knowledge of the earth and what it contains we cannot forget what the world owes to the gracious and glorious Isabella of Castile. For it is to her probably as much as to Columbus that a new continent was discovered at the close of the fifteenth century. For, while the doctors of Salamanca — most of whom were what Galileo called "paper philosophers," men who fancied that a correct knowledge of the physical universe was to be obtained by a collation of ancient texts — were denouncing the great navigator as an idle dreamer, and quoting the ill-founded notions of Pliny and Aristotle to prove the impossibility of his carrying out his project, Isabella was quietly revolving in her own mind the reasons which Columbus had adduced in favour of his great enterprise. Having satisfied herself that his views were sufficiently probable to justify action, she was prepared to make any sacrifices to have his plans executed. The result of her decision is but another illustration of the value of woman's quick intuition, as against the slow reasoning processes of philosophers and men of science.

Again, while considering what women have accomplished for the advancement of science by inspiration and collaboration, we must not lose sight of what they have done by suggestion. For, as John Stuart Mill well observes: "It no doubt often happens that a person who has not widely and accurately studied the thoughts of others on a subject has by natural sagacity a happy intuition which he can suggest but cannot prove, which yet, when matured, may be an important addition to knowledge: but, even then, no justice can be done to it until some other person, who does possess the previous acquirements, takes it in hand, tests it, gives it a scientific or practical form, and fits it into its place among the existing truths of philosophy or science. Is it supposed that such felicitous thoughts do not occur to women? They occur by hundreds to every woman of intellect; but they are mostly lost for want of a husband or friend who has the other knowledge which can enable him to estimate them properly and bring them before the world; and, even when they are brought before it, they usually appear as his ideas, not their real author's. Who can tell how many of the original thoughts put forth by male writers belong to a woman by suggestion, to themselves only by verifying and working out? If I may judge by my own case, a very large proportion indeed."[18]

Nor should we forget those active and energetic women — and their number is much greater than is ordinarily supposed — whose husbands, although often endowed with genius of the highest order, were indolent by temperament and disorderly and unmethodical by nature. Such men would, in the majority of cases, have run to seed had not their genius been given special force and impulse by their vigorous and methodical helpmates. Sir William Hamilton, the most learned philosopher of the Scottish school, is a striking instance in point; for it was due almost entirely to the stimulation he received from his ever active wife that he was always kept keyed up to his fullest working capacity as a philosopher and became recognised the world over as one of the commanding intellects of his age.

"Lady Hamilton," writes Professor Veitch in his *Memoir of Sir William Hamilton,* "had a power of keeping her husband up to what he had to do. She contended wisely against a sort of energetic indolence which characterised him, and which, while he was always labouring, made him apt to put aside the task actually before him, sometimes diverted by subjects of inquiry suggested in the course of study on the matter in hand, sometimes discouraged by the difficulty of reducing to order the immense mass of materials he had accumulated in connection with it. Then her resolution and cheerful disposition sustained and refreshed him, and never more so than when, during the last twelve years of his life, his bodily strength was broken and his spirit, though languid, yet ceased not from mental toil. The truth is that Sir William's marriage, his comparatively limited circumstances, and the character of his wife supplied to a nature that would have been contented to spend its mighty energies in work that brought no reward but in the doing of it, and that might never have been made publicly known or available, the practical force and impulse which enabled him to accomplish what he actually did in literature and philosophy. It was this influence, without doubt, which saved him from utter absorption in his world of rare, noble and elevated but ever-increasingly unattainable ideas. But for it the serene sea of abstract thought might have held him becalmed for life; and, in the absence of all utterance of definite knowledge of his conclusions, the world might have been left to an ignorant and mysterious wonder about the unprofitable scholar."[19]

What has been so far said, important as it is, does not tell the whole story of woman's influence on men of science, and consequently on the progress of science. We should not have an adequate conception of women as inspirers and collaborators if we did not advert to certain faculties which they usually possess in a more eminent degree than the most of men. It is a well-known fact that in many of the affairs of life women are more practical, have more tact, and possess keener and quicker perceptions than men. They are, too, more ideal, more romantic and more enthusiastic.

Men of science in their investigations usually proceed by the slow and laborious process of collecting facts and collating phenomena, either by observation or experiment, or both, and, from the observed facts and phenomena, they formulate a law which explains and correlates them. This is known as induction, a method which proceeds from facts to ideas.

Women, on the contrary, are rather disposed to proceed from ideas to facts; to explain phenomena from ideas which already exist in the mind, without having recourse to the slow process of induction. This is the deductive method, and is the very reverse of that employed by the average man of science. It would, however, be a mistake to maintain that the inductive method is always employed, for such is not the case. More than a half a century ago the historian, Buckle, in a notable lecture delivered in the Royal Institution of Great Britain, directed attention to the fact that some of the greatest scientific discoveries had been made by the deductive method.

One of these was Newton's epoch-making discovery of universal gravitation. While sitting in a garden he saw an apple fall, and this simple fact caused him to advance from idea to idea, and to be carried, by what Tyndall loved to call "the scientific use of the imagination," into the distant realms of space. And, heedless of the operations of nature, neither observing nor experimenting, the great philosopher, by pure *a priori* reasoning, "completed the most sublime and majestic speculation that it ever entered into the heart of man to conceive." "It was," as Buckle well observes, "the triumph of an idea. It was the audacity of genius." It was also the triumph of the deductive method in the solution of a problem that one not a genius could have worked out only by the long and toilsome process of induction.

Similarly, the great law of metamorphosis in plants, "according to which the stamens, pistils, corollas, bracts, petals and so forth, of every plant, are simply modified leaves," was discovered not by an inductive investigator, but by a poet. "Guided by his brilliant imagination, his passion for beauty and his exquisite conception of form which supplied him with ideas," Germany's greatest poet, Goethe, by reasoning deductively, was able to generalise a law which lesser minds could never have arrived at except through the application of the inductive method.

So also was it in the science of crystallography. Its foundations were laid, not by a mineralogist nor a mathematician, as one would suppose, but by one of strong imagination and marked poetic temperament. Like Goethe, Haüy was led by his ideas of beauty and symmetry to work deductively on the problem before him. Descending from ideas to facts, he finally succeeded, after a long series of subsequent labours, in reading "the riddle which had baffled his able but unimaginative predecessors."

It is the possession of this deductive faculty, so characteristic of men of genius — their ability to reach conclusions directly, as great mathematicians perceive inferences which those less gifted reach only after pages of elaborate calculations — which enable women, "not indeed to make scientific discoveries, but to exercise the most momentous and salutary influence over the method by which scientific discoveries are made." For, as Buckle points out, men of science are too inclined to employ the inductive method to the exclusion of the deductive.[20] They have become slaves to the tyranny of facts, and, as such, are incompetent to further the progress of science as they would by using both methods instead of one. And their slavery would be still more complete and ignominious were it not for the great though unconscious service to science rendered by women who have kept alive the deductive habit of thought. "Their turn of thought, their habits of mind, their conversation, their influence, insensibly extending over the whole surface of society and frequently penetrating its intimate structure, have, more than all other things put together, tended to raise us up into an ideal world, lift us from the dust in which we are too prone to grovel, and develop in us those germs of imagination which even the most sluggish and apathetic understandings in some degree possess."

From the foregoing observations it is manifest that the best results to science are secured when men and women work together — men supplying the slow, logical reasoning power, women the vivid, far-reaching imagination; men generalising from facts, women from ideas; men working chiefly by induction, women principally by deduction. For thus collaborating, each with his or her predominant faculties, the two combined possess in a measure the elements which go to make up a man or woman of genius and which enable them to achieve far more for the advancement of science than would otherwise be possible.

No one has ever given more eloquent expression to this truth than John Stuart Mill, who was as keen as an observer as he was profound as a thinker. Writing on the subject under discussion, he does not hesitate to say: "Hardly anything can be of greater value to a man of theory and speculation who employs himself, not in collecting materials of knowledge by observation, but in working them up by processes of thought into comprehensive truths of science and laws of conduct, than to carry on his speculations in the companionship and under the criticism of a really superior woman. There is nothing comparable to it for keeping his thoughts within the limits of real things and the actual facts of nature. A woman seldom runs wild after an abstraction..... Women's thoughts are thus as useful in giving reality to those of thinking men as men's thoughts in giving width and largeness to those of women. In depth, as distinguished from breadth, I greatly doubt if even now women, compared with men, are at any disadvantage."[21]

We have already learned, from his own avowal, how much Mill was beholden to his wife for her active cooperation in the production of those works of his which have exerted so profound an influence on many phases of modern thought. A more striking illustration of the value of woman's assistance, but in the domain of biology, is found in the biography of the late Professor Huxley. By those who know this distinguished man of science — so remarkable for his intellectual vigour — only from his writings, the impression would be gleaned that he was one of the most independent of thinkers, and that his utterances on all subjects were absolutely personal and entirely unmodified by suggestion or criticism from any quarter.

How far this view is from being correct is found in the statement by his son that his father "invariably submitted his writings to the criticism of his wife before they were seen by any other eye. To her judgement was due the toning down of many a passage which erred by excess of vigour, and the clearing up of phrases which would be obscure to the public. In fact, if any essay met with her approval, he felt sure it would not fail of its effect when published."[22] She was not only his "help and stay for forty years; in his struggles ready to counsel, in adversity to comfort," but, over and above this, she was "the critic whose judgement he valued above almost any, and whose praise he cared most to win" — the other self who made his life work possible.[23]

An intelligent, sympathetic pair of this kind — and this, as we have seen, is but one of a multitude which illuminates and beautifies the history of science — are competent to achieve wonders. They are like "the two-celled heart beating with one full stroke" —

Two plummets dropt for one to sound the abyss
Of science, and the secrets of the mind.

The woman is then truly, as De Lamennais in Scriptural phrases has it, "Man's companion, man's assistant, bone of his bone and flesh of his flesh," and, in her sublime and endearing character so complete in every relation of life, she fully answers to the beautiful characterisation which Adam, in *Paradise Lost*, gives of his beloved Eve:

So absolute she seems,
And in herself complete, so well to know
Her own, that what she wills to do or say
Seems wisest, virtuosest, discreetest, best.
* * * *
Authority and reason on her wait,
* * * *
..... and, to consummate all,
Greatness of mind and nobleness their seat
Build in her loveliest, and create an awe
About her, as a guard angelic plac'd.

## Notes & References

1) To one woman, Mary E Poupard, of London, England, were granted in a single year no less than three patents for horseshoes — two of the patents being for sectional and segmental horseshoes.

2) Sis oppido meminens quod olim Martia Hortensio, Terentia Tullio, Calpurnia Plinio, Pudentilla Apuleio, Rusticana Symmacho legentibus meditantibusque candelas and candelabra tenuerunt. Lib II, Epist 10.

3) "Verum hoc — seu gratitudini seu ineptiae ascribendum — non sileo, me quantulucunque conspicis, per illam esse, nec unquam ad hoc, si quid est nominis aut gloriae fuisse venturum, nisi virtutum tenuissman sementem, quasi pectore in hoc natura locaverat, nobilissimis his affectibus coluisset." Francisci Petrarchae, *Colloquiorum Liber quem Secretum Suum Inscripsit*, pp 105-06, Berne, 1603.
In his canzone beginning with the words *Perche la vita e breve*, Petrarch declares to his inspirer —

Thus if in me is nurst
Any good fruit, from you the seed came first;
To you, if such appear, the praise is due,
Barren myself till fertilised by you.

4) *The Life of St Francis of Assisi*, by Paul Sabatier, p 166, New York, 1894.

5) *Ibid*, p 167.

6) *Ibid*, p 307.

7) *The Women of the Renaissance*, p 394, New York, 1901.

8) *Women of Florence*, by Isodoro del Lungo, p xxvii, London, 1907.

9) This passage from the dedication is so important that I reproduce the Latin original: "Omnino vitam, aut, si quid mihi carius est, vobis autem debeo, tibi autem, o diva Melusinis, omne presertim Mathematicis studium, ad quod me excitavit tum tuus in earn amor, tum summa artis illius, quam tenes, peritia, immo vero nunquam satis admiranda in tuo tamque regii et nobilis generis sexu Encyclopaedia." *François Viete, Inventeur de l'Algebre Moderne*, p 20, par Frederic Ritter, Paris, 1895.

10) "E nell' amore della figlia il grande astronomo trovò non soltanto un conforto a suoi affanni, ma anche una guida benefica alla quale sembrò egli abandonarsi con cieca tenerezza figliale." *La Storia del Feminismo*, p 509, by G L Arrighi, Florence, 1911.

11) *Galileo Galilei e Suor Celeste*, by Antonio Favaro, p 256 *et seq*, Florence, 1891.

12) An English writer, discussing this subject, pertinently observes: "For, after all, is it not the personal incidents and commonplaces of life that gather interest as the centuries roll on, while its more pretentious events often drop into mere literary lumber? How much more interesting Dr Johnson's incidental admission, 'I have a strong inclination, Sir, to do nothing today,' is to us now than many of his more formal utterances. And, in reality, is it the personal element alone that is in the long run perennial? The wise may prate as they will about the importance of maintaining the continuity of history and of handing on the torch of science. The world cares for none of these things; they interest only some few political economists and laborious men. What does the crowd and poor little Tom Jones and his nestful, for instance, care about the fact that Cheops was — at any rate by courteous tradition — a mighty man of valour of such an era and land? But little Tom Jones and the rest of us would become mightily interested in this misty monster of many traditions, could we learn in some magical way all he thought, hated and loved in his inmost heart of hearts." *The National Review*, p 461, June

1889.

13) The Duke of Peiresc, in a letter to Gassendi, regarding Galileo, refers to certain letters — tres belles epistres — of the great philosopher, "à une sienne fille religieuse sur le sujet mesme des matieres traictees en son dernier livre." This shows that Sister Celeste was kept fully informed by her father respecting the nature and contents of his various works while he was preparing them for the press. It implies, likewise, that she was not only interested in them in a general way, but that she was able to read them intelligently and appreciate them as well.

How fondly Galileo treasured the letters written him by this daughter of predilection is made known to us by Sister Celeste herself, when she tells him in one of her letters "Resto confusa sentendo ch'ella conservi le mie lettere, e dubito che il grande affeto que mi porta gliele dimonstri piu compita di quello che sono." *Op cit*, p 317.

14) *Op cit*, p 404.

15) In the dedication of his *Principles of Philosophy* he addresses his young friend and pupil in the following words: "Je puis dire avec verite que je ne jamais rencontre que le seul esprit de votre altesse auquel l'un et l'autre" — metaphysics and mathematics — "fût egalement facile; ce qui fait quo j'ai une tres juste raison de l'estimer incomparable."

16) *Geology and Mineralogy Considered with Reference to Natural Theology*, by William Buckland, p xxxvi, London, 1858.

17) *Pasteur*, by Mr and Mrs Percy Frankland, p 26 *et seq*, London, 1898. A French writer referring to this happy discovery expresses himself as follows: "Quand Pasteur trouva le vaccin de charbon, il remonta triomphant de son laboratoire et les larmes lui vinrent aux yeux en embrassant sa femme et sa fille auxquelles annoncait sa victoire." *Revue Encyclopedique*, p 20, Jan 15, 1895.

18) *Memoir and Correspondence of Caroline Herschel*, London, 1879, pp vi and vii, by Mrs John Herschel. Cf Ch IV of this volume

19) *The Subjection of Women*, pp 98-99, London, 1909.

The idea herein expressed is beautifully accentuated in the touching dedication to the author's work *On Liberty*, which reads as follows: To the beloved and deplored memory of her who was the inspirer, and in part the author, of all that is best in my writings — the friend and wife whose exalted sense of truth and right was my strongest incitement, and whose approbation was my chief reward — I dedicate this volume. Like all that I have written for many years, it belongs as much to her as to me; but the work as it stands has had, in a very insufficient degree, the inestimable advantage of her revision, some of the most important portions having been reserved for a more careful re-examination, which they are now never destined to receive. Were I but capable of interpreting to the world one-half the great thoughts and noble feelings which are buried in her grave, I should be the medium of a greater benefit to it than is ever likely to arise from anything I can write, unprompted and unassisted by her all but unrivalled wisdom.

The chivalrous sentiments expressed in this generous tribute by one of the deepest thinkers of his time, to the memory of his noble and gifted life-companion, extravagant as they may seem, are but echoes of similar sentiments often voiced before by the world's greatest leaders of thought and science.

20) *Memoir of Sir William Hamilton*, by John Veitch, p 136 *et seq*, Edinburgh, 1869.

It is frequently said that women, unlike men, are indifferent to fame. This may be true so far as they are personally concerned; but it is certainly not true of them in regard to their husbands, or the men for whom they have a genuine affection. This is abundantly proved by the lives of Mme Huber, Mme Pasteur, Caroline Herschel and Lady Hamilton, not to name others who have been mentioned in the foregoing pages. After Sir William Hamilton, at the age of fifty-six, had been stricken by hemiplegia on the right side, as the result of overwork, his faithful wife became for twelve years eyes, hands and even mind for him. She read and consulted books for him, and helped him to prepare his lectures and the works which have given him such celebrity. "Everything that was sent to the press and all the courses of lectures were written by her, either to dictation or from copy." And when we remember that the lectures and books were of the most abstruse character and that Lady Hamilton was associated with her husband in his recondite work throughout his long and brilliant career, we must confess that her conduct was not only heroic to a degree, but also that the fame of the one she loved was to her a matter of the deepest concern.

21) "Induction is, indeed, a mighty weapon laid up in the armoury of the human mind, and by its aid great deeds have been accomplished and noble conquests have been won. But in that armoury there is another weapon, I will not say of stronger make, but certainly of keener edge; and, if that weapon had been oftener used during the present and preceding century, our knowledge would be far more advanced than it actually is. If the imagination had been more cultivated, if there had been a closer union between the spirit of poetry and the spirit of science, natural philosophy would have

made greater progress, because natural philosophers would have taken a higher and more successful aim, and would have enlisted on their side a wider range of human sympathies." Buckle, *The Influence of Women on the Progress of Knowledge.*

22) *The Subjection of Women*, ut sup, p 87.

23) *Life and Letters of Thomas Henry Huxley*, by his son Leonard Huxley, Vol I, p 324, New York, 1900.

# The Future of Women in Science : Summary and Epilogue

SAINT-EVREMOND, THE first great master of the genteel style in French literature, who was equally noted as a brilliant courtier, a graceful wit, a professed Epicurean, and who exerted so marked an influence on the writings of Voltaire and the essayists of Queen Anne's time, gives us in one of his desultory productions an entertaining disquisition on *La femme qui ne se trouve point et ne se trouvera jamais* — the woman who is not and never will be found. The caption of this singular essay admirably expresses the idea that the majority of mankind has, even until the present day, held respecting woman in science. For them she was non-existent. Nature, in their view, had disqualified her for serious and, above all, for abstract science. Never, therefore, in the opinion of these solemn wiseacres, had been found or could be found a woman who had achieved distinction in science.

The foregoing chapters show how ill-founded is such a view regarding woman in times past. For that half of humanity which has produced such scientific luminaries as Aspasia, Laura Bassi, Maria Gaetana Agnesi, Sophie Germain, Mary Somerville, Caroline Herschel, Sónya Kovalevsky, Agnes S. Lewis, Margaret Dunlop Gibson, Eleanor Ormerod and Mme Curie — to mention no others — is far from exhibiting any evidence of intellectual disqualification and still farther from warranting any one from declaring that the successful pursuit of science is entirely beyond the mental powers of womankind.

The preceding pages, likewise, afford an answer to those who insist on woman's incapacity for scientific pursuits, and point to the small number of those that have attained eminence in any of the branches of science; who continue to assert that the women named are but exceptions to the rule of the hopeless inferiority of their sex, and that no conclusions can be deduced from the paucity of women who have risen above the intellectual level of their less fortunate or less highly dowered sisters. They further show that, until the last few decades, woman's environment was rarely if ever favourable to her pursuit of science. From the days of Aspasia until the latter half of the nineteenth century she was discriminated against by law, custom and public opinion. Save only in Italy, she was excluded from the universities and from learned societies in which she might have had an opportunity of developing her intellect. In other countries her social ostracism in all that pertained to mental development was so complete and universal that she rarely had an opportunity of making a trial of her powers or exhibiting her innate capacity. The consequence was that her mind remained in a condition of comparative atrophy — a condition that gave rise to that long prevalent belief in woman's intellectual inferiority to man and her natural incapacity for everything that is not light or frivolous.

Practically all that women have achieved in science, until very recent years, has been accomplished in defiance of that conventional code which compelled them to confine their activities to the ordinary duties of the household. The lives and achievements of the eminent mathematicians, Sophie Germain and Mary Somerville, are good illustrations of the truth of this assertion. It was only their persistence in the study of their favourite branch of science, in spite of the opposition of their family and friends, and in spite of what was considered taboo for their sex by the usages and ordinances of society, that they were able to attain that eminence in the most abstruse of the sciences which won for them the plaudits of the world. Both were virtually self-made women. Deprived of the advantages of a college or university education, and denied the stimulus afforded by membership in learned scientific associations, they nevertheless succeeded by their own unaided efforts in winning a place of highest honour in the Walhalla of men of science.

M Alphonse de Candolle, in his great work, *Histoire des Sciences et des Savants depuis Deux Siecles,* devotes only two pages to the consideration of woman in science. She is, to him, a negligible quantity. And, although a professed man of science, he repeats, without any scientific warrant whatever, all the gratuitous statements of his predecessors regarding the superficial character of the female mind, "a mind," he will have it, which "takes pleasure in ideas that are readily seized by a kind of intuition;" a mind "to which the slow methods of observation and calculation by which truth is surely arrived at are not pleasing. Truths themselves," the Swiss savant continues, "independent of their nature and possible consequences — especially general truths which have no relation to a particular person — are of small moment to most women. Add to this a feeble independence of opinion, a reasoning faculty less intense than in man, and, finally, the horror of doubt, that is, a state of mind in which all research in the sciences of observation must begin and often end. These reasons are," according to de Candolle, "more than sufficient to explain the position of women in scientific pursuits."[1]

They certainly are more than sufficient to explain their position if we choose to accept the author's method of determining one's attainments in the realm of science. His chief test of one's eminence in science is the number of learned societies to which one belongs. For De Candolle, membership in one or more such bodies is *prima facie* evidence of special distinction in some branch of science. But "We," he declares, "do not see the name of any woman on the lists of learned men connected with the principal academies. This is not due entirely to the fact that the customs and regulations have made no provision for their admission, for it is easy to assure one's self that no person of the feminine sex has ever produced an original scientific work which has made its mark in any science and commanded the attention of specialists in science. I do not think it has ever been considered desirable to elect a woman a member of any of the great scientific academies with restricted membership."[2]

When De Candolle insisted on membership in learned societies as a necessary indication of scientific eminence, he must have known, what everybody knew, that such exclusive societies as the French Academy of Sciences and the Royal Society of Great Britain have always been dead set against the admission of women members. It is difficult to imagine that the learned author of the *History of Science and Scientists* was entirely ignorant of the exclusion from the French Academy of Maria Gaetana Agnesi solely because she was a woman. And he must have been aware that, had it not been for her sex, Sophie Germain would have been accorded a fauteuil in the same society for her remarkable investigations in one of the difficult departments of mathematical physics. He must likewise have been cognisant of the attitude of such organisations as the Royal Society toward women, no matter how meritorious their achievements in science.

According to De Candolle's criterion, such women as Mme Curie, Sónya Kovalevsky, Eleanor Ormerod, Agnes S Lewis, Margaret Dunlop Gibson have accomplished nothing worthy of note because, forsooth, their names are not found on the rolls of membership of the Royal Society or the French Academy of Sciences — associations whose constitutions have been purposely so framed as to exclude women from membership. It would, indeed, be difficult to instance a more unfair or a more unscientific test of woman's eminence in science, and that, too, proposed by one who is supposed to be actuated in his judgements by rigorously scientific methods. Had any of the women named belonged to the male sex, there never would have been any question of their fitness to become members of the societies in question. This is particularly true of Mme Curie, who, in the estimation of the world, has done more to enhance the prestige of French science than any man of the present generation — a statement that is sufficiently justified by the fact that she is the only one so far who has twice, in competition with the greatest of the world's men of science, succeeded in carrying away the great Nobel prize.[3]

Not only have men, from time immemorial, been wont to point to woman's incapacity for science as evidenced by the small number of those who have achieved distinction in any of its branches, but they have also taken a special pleasure in directing attention to the fact that no woman has ever given to the world any of the great creations of genius, or been the prime-mover in any of the far-reaching discoveries which have so greatly contributed to the weal, the advancement and the happiness of our race.

No one, probably, has expressed himself on this subject in a more positive or characteristic fashion than the noted litterateur and philosopher, Count Joseph de Maistre. Writing from St Petersburg to his daughter, Constance, he says: "Voltaire, according to what you affirm — for as to me, I know nothing, as I have not read all his works, and have not read a line of them during the last thirty years — says that women are capable of doing all that men do, etc. This is merely a compliment paid to some pretty woman, or, rather, it is one of the hundred thousand and thousand silly things which he said during his lifetime. The very contrary is the truth. Women have produced no *chef d'oeuvre* of any kind whatsoever. They have been the authors neither of the *Iliad*, nor the *Aeneid*, nor the *Jerusalem Delivered*, nor *Phedre*, nor *Athalie* nor *Rodogune*, nor *The Misanthrope*, nor *Tartufe*, nor *The Joueur*, nor *The Pantheon*, nor *The Church of St Peter's*, nor the *Venus de' Medici*, nor the *Apollo Belvidere*, nor the *Principia*, nor the *Discourse on Universal History*, nor *Telemachus*. They have invented neither algebra nor the telescope, nor achromatic glasses nor the fire engine, nor hose-machines, etc."[4]

All this is true, but what does it prove? It does not prove, as is so frequently assumed, woman's lesser brain power or inferior intelligence. It does not prove — as the learned Frenchman and those who are similarly minded would have us believe — her incapacity for the highest flights of genius in every sphere of intellectual effort. Such assumptions are entirely negatived by woman's past achievements in all departments of art, literature and science.

Far from making the inference that De Maistre wished his daughter to draw from his letter, we should, from what we know of woman's ability as disclosed in the foregoing chapters, hesitate to set a limit to her powers, or to declare apodictically that she could not have been the author of works of as great merit as most of those — if not all of them — mentioned as among men's supreme achievements. The simple fact that Mme Curie and Sónya Kovalevsky were able, in sciences usually considered beyond female intelligence, to wrest from their male competitors the most coveted prizes within the gift of the Nobel Prize Commission and the French Academy of Sciences, demonstrates completely that woman's assumed incapacity for even the most recondite scientific pursuits is a mere figment of the masculine imagination.

What women have done "that at least, if nothing else," as John Stuart Mill aptly observes, "it is proved they can do. When we consider how sedulously they are all trained away from, instead of being trained toward, any of the occupations or objects reserved for men, it is evident that I am taking very humble ground for them, when I rest their case on what they have actually achieved. For, in this case, negative evidence is worth little, while any positive evidence is conclusive. It cannot be inferred to be impossible that a woman should be a Homer, or an Aristotle, or a Michaelangelo, or a Beethoven, because no woman has yet actually produced works comparable to theirs in any of those lines of excellence. This negative fact at most leaves the question uncertain and open to psychological discussion. But it is quite certain that a woman can be a Queen Elizabeth or a Deborah or a Joan of Arc, since this is not inference but a fact."[5]

In like manner it is quite certain that, in spite of all kinds of disabilities and prejudices and adverse legislation, there have been a large number of women who, in every department of intellectual activity, have achieved marked distinction and won imperishable renown for their proscribed sex. It is a fact, which admits of no question, that, notwithstanding their being debarred from all the educational advantages so generously lavished upon the dominant sex, women have since the days of Sappho and Hypatia shown themselves the equals and often the superiors of men in the highest and noblest spheres of mental achievement.

Such being the case, what, we may ask, would have been the result had women, from that splendid Heroic Period of which Homer sings until the present, enjoyed all the opportunities of mental development of which men have systematically claimed the exclusive privilege?[6] What would now be their condition if, from the days of the Muses — who were but learned women apotheosised — women had never been deprived of their intellectual birthright and had been permitted to continue in the path so auspiciously blazed by Corinna — the victor over Pindar — and Arete, the splendour of Greece and the possessor of the mind of Socrates and the tongue of Homer? What would not now be their intellectual efflorescence, if Plato's dream of twenty-three centuries ago of giving women equal rights with men in all things of the mind could have been realised; if those ardent female disciples of his, who so lovingly followed him through the streets of Athens — "the home of the intellectual and the beautiful" — and hung on his lips during his matchless discourses in the groves of the Academy and on the banks of the Ilyssus, could have continued that race of intellect and genius which was the admiration and the inspiration of all Hellas during the most brilliant period of its marvellous history?

Speculating only on what the gifted daughters of Greece might have achieved, we may easily believe that they would have kept pace with their most highly gifted countrymen, and that, following in the footsteps of Sappho and the other Muses of the "Terrestrial Nine," they would have been worthy rivals of Homer, Pindar and Aeschylus, and would have occupied a prominent place in that brilliant galaxy of genius composed of such luminaries as Anaxagoras, Sophocles, Euclid, Archimedes, Theophrastus,

Polygnotus, Diophantus, Pausanias and Thucydides.

To those who base their opinions on what so long has been the absurdly anomalous condition of women and who, in formulating their theories of human progress, completely ignore the fundamental laws of heredity, such conjectures will seem extravagant, if not chimerical. But, when one bears in mind the universal fact that offspring, whatever the sex, inherits its characteristics and its powers from both parents alike; that the soul, unlike the body, has no sex, and that, so far as legitimate indications from the teachings of biology and psychology can serve as a guide, there is no valid reason for asserting the mental superiority of man over woman, one will be obliged to confess that these surmises are far from being either fanciful or preposterous.

It is then the veriest sophism to predicate woman's incapacity for science and for intellectual achievements of the highest order on what she has not accomplished in the past, or on the comparatively limited number of her contributions to the advancement of knowledge; for up till the present she has, for the most part, been but a dwarf of the gynaeceum,

> Cramp'd under worse than South-sea isle taboo.

Had men been compelled to labour under similar conditions, it is doubtful if they would have accomplished any more than women have now to their credit.

Considering woman's past achievements in science, as well as in other departments of knowledge; considering her present opportunities for developing her long-hampered faculties, and considering, especially, the many new social and economic adjustments which have been made within the last half century, in consequence of the greatly changed conditions of modern life, it requires no prophetic vision to forecast what share the gentler sex will have in the future advancement of science. That it will be far greater than it has been hitherto there can be no reasonable doubt. That the number of savantes of the type of Maria Gaetana Agnesi, Sónya Kovalevsky and Mme Curie will be greatly enlarged there is every reason to believe. That among these coming votaries of science there will be more than one woman who, even in the most abstruse sciences, will stand Upon an even pedestal with man, seems to be assured by the achievements of many who are now so materially adding to the sum of human knowledge.

Is it probable that the future will bring forth women whose achievements in science will rank with those of Euler, Faraday, Liebig, Leverrier, Champollion and Geoffry Saint-Hillaire? It would be a rash man who would answer in the negative. We cannot, as De Maistre seems to do, reason from what they have not done — when everything was against them — to what they may do when conditions shall, in every way, be as favourable to them as they always have been to the dominant sex.

Still rasher would be the man who would attempt to prove the negative of this question. Mere *a priori* arguments, based on preconceived bias or on the vague and groundless impression that woman is essentially and hopelessly the intellectual inferior of man, have no more value than gratuitous opinions. The unprejudiced seeker after truth will insist on a demonstration based on incontrovertible facts. He will appeal to history to learn what the sex has already accomplished, and to science to inquire if there be anything in the female brain to differentiate it from that of the male, or to preclude woman from attaining the highest rank in the activities of the intellect.

The result of such an investigation will, I think, cause even the most biased person to suspend judgement, if it does not induce him to align himself with those who, finding no differences in the mental endowments of the sexes, have reached the conclusion that the day will come, and, mayhap, in the near future, when the achievements of women will be on a par with those of man. The facts stated in the preceding chapters seem, not unreasonably, to point to such a conclusion, if, indeed, they do not warrant it as a necessary inference.

A few considerations germane to this discussion will illustrate the danger of forming hasty judgements regarding questions like the one under discussion.

During the last hundred years no country in the world has done more for the education of the masses than the United States. Everything that money could purchase and ingenuity suggest has been adopted to develop the minds and stimulate the latent talents and genius of our youth. From the primary schools to the highest and best equipped universities, a special premium has been put on success in study, and the highest rewards have awaited those who should make any notable contribution towards the advancement of knowledge. But, notwithstanding all the educational advantages our people have enjoyed and all the encouragement they have received to achieve something of supreme excellence, our great country with its teeming millions attracted from the most gifted nations of the Old World has not yet produced a single man who has attained the highest rank in either literature or art or science. Far from having a pre-eminent master of song like Homer or Dante, we have not even a poet approaching Goethe or Tasso or Camoens. We have no Cervantes, no Milton, no Racine, no Moliere. America has produced no Raphael or Michaelangelo; no Mozart or Wagner or Tschaikovsky. Nor has it given us a Descartes, a Leibnitz, a Newton or a Darwin. Would any one, from this complete absence in America of representatives of the highest order in literature, art and science, ever dream of concluding that we shall never have such favourite sons of genius and such giants of intellect? Does our comparative intellectual sterility in the past, and in a country which seemed specially adapted to foster genius and attainments of the highest order, justify any one in inferring that the days of great geniuses, like the days of demigods, are gone never to return?

And yet the number of men in our broad commonwealth who, during the past hundred years, have enjoyed such signal opportunities for attaining distinction in every domain of intellectual effort is incomparably greater than that of all the women so favoured since the earliest days of human history. If, from the first flowering of Greek culture to the present day, as many millions of women had enjoyed all the transcendent advantages of education as have been in the United States so lavishly accorded to the same number of millions of men, who will say that very many of them would not have attained a much higher rank in science, as well as in art and literature, than has yet been reached by any man that America has yet produced? Who even, on the evidence now available, would be warranted in denying that at least some of these millions of women might have attained the very highest rank in every department of intellectual achievement?

Gray, in his *Elegy Written in a Country Churchyard*, muses on the potential statesmen and the "mute, inglorious Miltons" of those countless multitudes who, for lack of opportunity to develop their inborn gifts, were condemned to pass their lives in obscurity and die, "to Fortune and to Fame unknown." But how much more truthfully could his words have been applied to that much larger number of women of rare mental powers to whose eyes knowledge

> Her ample page
> Rich with the spoils of time did ne'er unroll,

and whose God-given genius was ruthlessly suppressed from the cradle to the grave?

We are still in ignorance as to many of the conditions which are essential to the development of genius and which contribute to its loftiest flights. We have yet to learn how far the efflorescence of the human mind is aided and modified by heredity, environment, atmosphere, as well as by education, encouragement and other stimuli equally potent.

But we do know that Germany, in spite of its famed universities and its feverish intellectual activity in many departments of knowledge, had to wait many long dreary centuries before it could point to a Goethe, a Schiller, a Humboldt, a Bach, or a Beethoven. We know that France — so long the reputed centre of culture — has so far produced no great epic poet, no Cervantes, no Murillo. But shall we affirm that she will never give to the world imperishable works like *Paradise Lost, Don Quixote* or the *Immaculate Conception*? We know that Athens, which during the most brilliant period of its history counted only fifty-four hundred free-born citizens — less than the population of a small modern town — was able to produce within a very brief epoch more men of supreme distinction than all the rest of Europe from the Age of Pericles until the dawn of the Renaissance. Hers is still the art of the world, the literature of the world, the philosophy of the world, the culture of the world. For twenty-five centuries her canons of taste and beauty have guided poets, orators, artists; and her matchless productions have been the inspiration, as they have been the despair, of the greatest geniuses of our modern world.

Had the women of Greece not been put under constraint just as they were beginning to exhibit the splendid results of their intellectual activities; had they been encouraged to develop to the utmost their richly-dowered minds, as were the men, a far larger number of them, no doubt, would have been as successful in carrying off coveted prizes in the intellectual arena as was Corinna in her contests with Pindar. And they would, likewise, as we may easily conceive, have greatly added to the number of masterpieces of Greek intellect in science as well as in art and letters.

But the opportunity for women to test their powers, which was so wantonly snatched from their sisters in the Hellenic world, seems again to be offered to their sex. This opportunity, as has been stated, is due chiefly to their persistence in claiming the same right as men to intellectual development as well as to the countless proofs they have given that their demands are founded on reason and justice. What shall be the outcome of the new opportunity for woman to prove her capacity as compared with man's in things of the intellect remains to be seen, but, from indications she has during recent years given of her powers in every branch of scientific inquiry, there can be little doubt that it will be of such character as to place woman on a higher intellectual plane than she has yet occupied. In physical strength and in the rougher conflicts with the world she will doubtless always remain "the lesser man," but, once she feels in full possession of liberty

To burgeon out of all
Within her,

she will duly justify her advocates who throughout the centuries have been

Maintaining that with equal husbandry
The woman were an equal to the man.

Not the least of the contributing factors to woman's intellectual growth, and especially to her future achievements in science, are the recent adjustments for women in social and economical conditions brought about chiefly by far-reaching changes in the industrial world. Even so late as the last half of the nineteenth century the energies of women, when they were not engaged in the kitchen or the nursery, were spent on the domestic loom, spinning wheel and the knitting needle. All the various processes from carding the wool to making it into clothing for all the members of the family were in the hands of the housewife. Ready-made clothing was far from being as common and inexpensive as it is now. Canned foods and cereals, which do away with so much of the drudgery of the kitchen, were unknown. Electricity, which has proved to be such a remarkable aid in every modern home, was little more than a mysterious force that was utilised in the electric telegraph. Most of the domestic labour-saving machines were still in their infancy and possessed by but few people. Large fortunes were confined to only a favoured few in our great metropolises. The mass of the people was preoccupied with the struggle for existence.

But science, the spirit of invention and the advent of the age of machinery have completely changed the conditions of life which obtained but a generation ago. They have not only opened up for women countless occupations that were undreamed of in their mother's time, but have also given to tens of thousands of them the necessary means and leisure to indulge their tastes for study and research and enabled an ever increasing number of them to realise their aspirations for achieving distinction in the divers departments of scientific research.

As an instance of this marked change in the intellectual activity of women, we need only consider what an important part they now take in our present prodigious literary output, as compared with their share in similar work but a few decades ago. As authors, as writers and readers in the editorial rooms of our leading periodicals, as contributors to learned journals and reviews dealing with every branch of science, even the most abstruse, they now occupy a conspicuous place and are doing work that is quite as creditable as that of men.

And it is no longer necessary, in deference to public sentiment, for them to write under a pseudonym, for it is no longer considered unfeminine, as it was in the time of the Bronte sisters, for women to acknowledge themselves the authors of books or of articles in magazines. If they elect to devote their lives to literary or scientific work, they will not be deterred from so doing by what Mrs Grundy may say, or by the fear that some feeble imitator of Moliere may dub them as *Precieuses Ridicules*. The value of their productions, like those of men, is gauged solely by merit and not by any narrow-minded considerations of the author's sex.

So also will it be in all other occupations where women choose to gain their livelihood by devoting themselves to scientific pursuits rather than to manual labour or to secretarial work in the counting-room. There are positions open for them in colleges, universities and the government service where, as professors or experts in every branch of science, their talents have full liberty of action and where they have the same opportunity of achieving distinction in their chosen life-work as have their male colleagues.

In Germany there are today a million more women than men. It is the same in England. In France the number of women who are widows or unmarried or divorcees or mothers with full-grown children aggregates no less than four and a half millions. A similar condition obtains in other parts of Europe. A large percentage of this number is without home ties and, as the old fields of labour are no longer open to women, they are forced to find new ones. They naturally demand the privilege of exercising their talents in occupations which are most congenial to them. Many have no inclination for any of the avocations in the industrial or commercial world, but have a very decided inclination as well as talent for scientific pursuits. Hence the ever-increasing number of women who seek employment in chemical and biological laboratories, in museums and astronomical observatories, as well as aspire to professorships of science in schools and colleges. From this large number of votaries of science some are sure to achieve distinction in their calling and to contribute materially to the advancement of knowledge. In the course of time the number of those, like Mme Curie, Mme Coudreau, Mary Kingsley, Sónya Kovalevsky, Eleanor Ormerod, Caroline Herschel, Zelia Nuttall, Harriet Boyd Hawes, Donna Eersilia Bovatillo, Sophie Pereyaslawewa — to name only a few — who will become prominent as chemists, explorers, naturalists, mathematicians, entomologists, astronomers, archaeologists, biologists will be vastly increased, for women will find a greater stimulus for such work and more numerous demands for their service in the constantly expanding sphere of scientific research.

Many women will, doubtless, become specialists in some specific branch of science, particularly if they have a genuine love for it, or be fired by an ambition to achieve fame as discoverers. But it is not probable that they will ever specialise to the same extent as men do. For men scientific work has to a large extent become a *metier*, and success, as in industry, depends on a division of labour. Hence it is that their field of investigation is daily becoming more and more circumscribed. This is observable in all the sciences, but especially in such all-embracing sciences as chemistry, biology, and archaeology. A man now does well if he master a single branch of any of these sciences, and is hailed as exceptionally fortunate if he succeed in making some notable discovery in his limited field of research. So great, indeed, has been the activity of scientific men in every department of science during the last half century, and so thoroughly have they explored the most hidden recesses of nature, that it, at times, seems as if there were but little left to discover. A prominent scientist recently well expressed the difficulty of making any striking additions to our knowledge of nature by asserting that all great discoveries would hereafter be made in the sixth place of decimals. This statement is well illustrated by the delicate experiments that were required to isolate such rare elements as radium, polonium, helium and neon, which occur only in infinitesimal quantities.

While men of science will be forced to continue as specialists as long as the love of fame, to consider no other motives of research, continues to be a potent influence in their investigations, it is probable that women will have less love for the long and tedious processes involved in the more difficult kinds of specialisation. They will, it seems likely, be more inclined to acquire a general knowledge of the whole circle of the sciences — a knowledge that will enable them to take a comprehensive survey of nature. And it will be fortunate for themselves, as well as for the men who must perforce remain specialists, if they elect to do so. For nothing gives falser views of nature as a whole, nothing more unfits the mind for a proper apprehension of higher and more important truths, nothing more incapacitates one for the enjoyment of the masterpieces of literature or the sweeter amenities of life, than the narrow occupation of a specialist who sees nothing in the universe but electrons, microbes and protozoa.

But just at the critical moment, when men of science would rather discover a process than a law, when they are so preoccupied with the infinitely little that they lose sight of the cosmos as a whole; when their attention is so riveted on particular phenomena that they will no longer have aptitude for rising from effects to causes; when they cease to have any interest in general ideas and stray away from the guidance of the true philosophic spirit; when, like Plato's cave men, they have so long groped in darkness that their powers of vision are impaired, then it is that woman, "The herald of a brighter race," comes to the rescue and holds up to their astonished gaze the picture of an ideal world whose existence they had almost forgotten. For women, as a rule, love science for its own sake, and, unlike the specialists in question, they are, in its pursuit, rarely actuated by any selfish or mercenary interests, or by the hope of financial reward. Precise and never-ending observations with the microscope and spectroscope, which at best give them but a superficial knowledge of certain details of science, while it leaves them in ignorance of the greater and better part of it, do not appeal to them. They prefer general ideas to particular facts, and love to roam over the whole realm of science rather than confine themselves to one of its isolated corners.

"Women," writes M Étienne Lamy, the distinguished French Academician, "group themselves at the centre of human knowledge, whereas men disperse themselves towards its outer boundaries. While men are always pushing analysis to its utmost limits, women are seeking a synthesis. While men are becoming more technical, women are becoming more intellectual. They are better placed to observe the correlations of the different sciences, and to subordinate them to the common and unique source of truth from which they all descend. We seem, indeed, to be approaching a time when women will become the conservers of general ideas."[7]

In the preceding chapter reference was made to the fact that women are naturally inclined to adopt the deductive method in their search for truth when men would employ only the inductive method. This disposition of theirs to arrive at conclusions by a kind of intuition, coupled with their more pronounced idealism, is sure to react favourably on men, and prevent them from becoming so involved in mere facts and phenomena as to cause them to forget that it is as important to reason well as to observe well — that the fundamental principles of a true philosophy are quite as necessary for the eminent man of science as they are to the trustworthy historian or commanding statesman.

From what has been said, it is clear that man's ideal of the woman of the future will be quite different from what it was but a little more than a century ago, when Dr Johnson could say that "any acquaintance with books," among women, "was distinguished only to be censured." It will be quite different from the ideal woman, as portrayed by poets and novelists, for centuries past. For among the thousands of women painted by our leading writers of fiction, poets and dramatists there are few, if any, outside of those sketched by Tennyson in *The Princess*, who are distinguished for their learning or for their love of intellectual pursuits. Even Portia, Shakespeare's most learned woman, was, according to her own confession, but

> An unlessoned girl, unschooled, unpracticed.

And the heroines of the novelist, far from being women who had a thirst for knowledge, or were eager

> To sound the abyss
> Of science and the secrets of the mind,

were those only whose chief attractions were physical graces and charms, affectionate natures, brilliant wit together with "sweet laughs for bird-notes and blue eyes for a heaven."

Now, however, that women after ages of struggle are beginning to experience a sense of intellectual freedom before unknown, and to exult in the fact that

> Knowledge is now no more a fountain sealed;

now that they are, for the first time, beginning, in every civilised nation, to realise their age-long aspirations for unimpeded opportunity in all the activities of the intellect; now that they are no longer

> Dismiss'd in shame to live
> No wiser than their mothers, household stuff,
> Live chattels,.....
> ..... laughing-stocks of Time,

we may expect soon to see a marked change in the character of the ideal woman as depicted in literature and as desired by the intelligent portion of mankind.

What woman's liberation from intellectual bondage and her

freedom to devote herself to scientific pursuits mean for the future of humanity it is difficult at present adequately to forecast. That it will contribute immensely to the betterment of social conditions and to the elevation of the masses of humanity, there can be no doubt. Setting free the imprisoned energies of one half of our race, means more than doubling mankind's capacity for advancement. For the failure to utilise woman's vast energies, pining for an outlet, acted as a drag on man's own potentialities, and thus retarded to an untold extent the world's advancement. In times past, as has aptly been said, "an enormous part of the brain power of mankind has been spent or wasted in smiting the Philistines hip and thigh, and an enormous part of the brain power of womankind has been spent in cajoling Sampson."

It will mean that the women of the future will be more suitable companions for the rapidly increasing number of highly educated men of science; that having their intellects developed *pari passu* with those of men, they will be able to sympathise with the noblest aims of their husbands and assist them in their most important undertakings, as did the wives of Huber, Lavoisier, Pasteur, Huxley, Louis Agassiz and others scarcely less renowned in the annals of science. It will mean that they will not only share in the joys and the sorrows of their life-companions, but that they will also have a part in their thoughts, their studies, their labours, their achievements. For one should bear in mind that the first essential to a perfect union of hearts is a perfect harmony of minds. Where neither husband nor wife is educated, the virtues may suffice for companionship, but where the man is educated and the woman ignorant, there are sooner or later estrangements and the wife becomes little better than an old Japanese conception of her, "a cook without pay," or a pasha's toy for an idle hour. Chrysalde in Moliere's *L'École des Femmes,* declares:

> Qu'il est assez ennuyeux, que je crois,
> D'avoir toute sa vie une bête avec soi.

A briefer and truer statement of the evils of unequal intellectual mating was never penned.[8] Men of intelligence are no longer, like

Rousseau, satisfied with an ignorant domestic for a wife, and still less are they disposed with Schopenhauer to regard woman as an incurable Philistine, and as a mere intermediary between a child and a man. They have learned by sad experience that it is contrary both to justice and public policy to impose artificial restrictions on the acquisition of knowledge by women, or to close to the vigorous and capable representatives of their sex careers which are open to the weakest and most incompetent men. History has taught them that the fall of Greece and Rome was owing to the failure of these nations to make due provision for the mental development of women.

And women know that it was because of the inability of the wives of the Athenians to enter into the thoughts of their highly educated husbands and to sympathise with their aims and appreciate their achievements that caused the men to leave them in their solitude and seek in the companionship of the hetaerae the intellectual atmosphere which was wanting in their own homes. They know, too, that the lack of knowledge in the wife and the absence of virtue in the hetaerae, which brought such disasters on the most learned and most cultured of nations are still evils to be guarded against, and that one of the means over and above moral rule and revealed truth of safe-guarding their own interests and preserving the sanctity of the home is to make themselves by knowledge and culture the intellectual equals of their consorts.

They realise also that if they are to attain the highest measure of success as wives and mothers, a broad and thorough education — a knowledge of science, as well as familiarity with art and literature and the teachings of religion — is essential to them for their children's sake. It is said that

> The hand that rocks the cradle rules the world,

but how much truer is it that "The domestic hearth is the first of schools, and the best of lecture-rooms; for here the heart will cooperate with the mind, the affections with the reasoning power."[9] It is only when the mothers of this, the woman's century, shall dispute with men the primacy of erudition — when they shall prove their mastery of those newer sciences by which our age sets such great store — when they shall possess

Seraphic intellect and force
To seize and throw the doubts of man;

that their grown-up sons will have the same confidence in their intelligence as they now have in their hearts. Then only will mothers be properly equipped for developing the character of their children; for inspiring them with a love of the true, the beautiful and the good; for stimulating their talents and aiding them to attain to all the sublimities of knowledge; for assisting them in doubt and despondency and firing them with an ambition to strive for supreme excellence in all that makes for the nobility of manhood and the glory of womanhood; for making them, as Beatrice made Dante after he was renewed and purified in the waters of Eunoe, "fit to mount up to the stars."

*Puro e disposto a salire alle stelle.*

The romantic idea of treating woman as a clinging vine, and thus eliminating half the energies of humanity, is rapidly disappearing and giving place to the idea that the strong are for the strong — the intellectually strong; that the evolution of the race will be complete only when men and women shall be associated in perfect unity of purpose, and shall, in fullest sympathy, collaborate for the attainment of the highest and the best. Then, indeed, will man's helpmate become to him and to his children

More rich than pearls of Ind or gold of Ophir,
And in her sex more wonderful and rare.

Then will men and women for the first time fully supplement each other in their aspirations and endeavours and realise somewhat of that oneness of heart and mind which was so beautifully adumbrated in Plato's androgyn. Then will the world witness the return of another Golden Age — the Golden Age of Science — the Golden Age of cultured, noble, perfect womanhood. Then to all who really think and love will be manifest the clearness and power of vision of England's great poet laureate when in matchless numbers he sings :

The woman's cause is man's; they rise or sink
Together, dwarfed or godlike, bond or free.

* * * *

For woman is not undevelopt man
But diverse: could we make her as the man,
Sweet Love were slain; his dearest bond is this,
Not like to like, but like in difference.
Yet in the long years liker must they grow;
The man be more of woman, she of man;
He gain in sweetness and in moral height,
Nor lose the wrestling thews that throw the world;
She mental breadth, nor fail in childward care,
Nor lose the childlike in the larger mind;
Till at the last she set herself to man,
Like perfect music unto noble words;
And as these twain, upon the skirts of Time,
Sit side by side, full-summ'd in all their powers,
Dispensing harvest, sowing the To-be,
Self-rev'rent each, and reverencing each,
Distinct in individualities,
But like each other ev'n as those who love,
Then comes the statelier Eden back to men;
Then reign the world's great bridals chaste and calm;
Then springs the crowning race of human-kind.
May these things be!

## Notes & References

1) *Ibid*, p 39; Vol II, p 458.
2) *Histoire des Sciences et des Savants*, p 271, Geneve-Bale, 1885.
3) *Ibid*, p 270.
4) A writer in the English magazine, *Nature*, under date of January 12, 1911, when the European press was discussing Mme Curie's claims to membership in the French Academy of Sciences, makes the following sane observations on the admission of women to the various academies of the French Institute:

   There may be room for difference of opinion as to the wisdom or expediency of permitting women to embark on the troubled sea of politics, or of allowing them a determinate voice in the settlement of questions which may affect the existence or the destiny of a nation; but surely there ought to be no question that in the peaceful walks of art, literature and science, there should be the freest possible scope extended to them, and that, as human beings, every avenue to distinction and success should unreservedly be open to them. All academies tend to be conservative and to move slowly; they are the homes of privilege and of vested interest. Some of them incline to be reactionary. They were created by men for men and for the most part at a time when women played little or no part in those occupations which such societies were intended to foster and develop But the times have changed. Women have gradually won for themselves their rightful position as human beings. We have now to recognise that academies as seats of learning were made for humanity and that, as members of the human race, women have the right to look upon their heritage and property no less than men. This consummation may not at once be reached, but, as it is based upon reason and justice, it is certain to be attained eventually.

A fortnight later the same magazine contained a second article, in which the matter is treated in an equally manly fashion.

"As scientific work," the writer pertinently observes, "must ultimately be judged by its merits, and not by the nationality or sex of its author, we believe that the opposition to the election of women into scientific societies will soon be seen to be unjust and detrimental to the progress of natural knowledge. By no pedantic reasoning can the rejection of a candidate for membership of a scientific society be justified, if the work done places the candidate in the leading position among other competitors. Science knows no nationality and should recognise no distinction of sex, colour or creed among those who are contributing to its advancement. Believing that this is the conclusion to which consideration of the question must inevitably lead, we have confidence that the doors of all scientific societies will eventually be open to women on equal terms with men."

5) *Lettres et Opuscules Inedits du Comte Joseph de Maistre*, Vol I, p 194, Paris, 1851.

It was this same brusque and original writer who asserted that "science was a most dangerous thing for women; that no woman should study science under penalty of becoming ridiculous and unhappy; that a coquette can more readily get married than a savante." And he it was who declared that women who attempted to emulate men in the pursuit of science are monkeys and *donne barbute* — bearded women — and who designated Mme de Staël as "*la science en jupons, une impertinente femelette*" — science in petticoats, a silly, impertinent female.

He, however, met an opponent worthy of his steel in the person of the eloquent bishop of Orleans, Mgr Dupanloup In a lengthy and brilliant critique of De Maistre's views he shows them to be untenable, if not ridiculous. "I by no means," he writes, "agree with M de Maistre that '*la science en jupons*,' as he calls it, or talents of any kind whatsoever, militates in the slightest against a woman being a good wife or a good mother. Quite the contrary." And considering woman as the companion and aid of man — *socia et adjutorium* — he expresses a view which is quite the opposite of that championed by his distinguished adversary for, in words precise and pregnant, he asserts that the education of women cannot be too consistent, too serious, and too solid — "*L'education des femmes ne saurait être trop suivie, trop serieuse et trop forte.*" *La Femme Studieuse*, p 160, Paris, 1895.

6) *The Subjection of Women*, p 81, London, 1909.

7) The late Mr Gladstone asserts that "It would be hard to discover any period of history or country of the world, not being Christian, in which they" — women — "stood so high as with the Greeks of the Heroic Age" — when the position of the Greek woman was so remarkable and "so elevated, both absolutely and in comparison with what it became in the Historic Ages of Greece and Rome amidst their elaborate civilisation." *Studies on Homer and the Homeric Age*, Vol II, p 479 *et seq*, Oxford, 1858. Cf also the same author's *Juventus Mundi*, p 405 *et seq*, London, 1869.

8) *La Femme de Demain*, pp 45-46, Paris, 1912.

9) Dr Johnson expressed the same sentiment when he declared that a man of sense should meet a suitable companion in a wife. "It was a miserable thing," he asserted in characteristic fashion, "when the conversation could only be such as whether the mutton should be boiled or roasted, and a probable dispute about that."

Sidney Smith, in a forceful and trenchant essay *On the Education of Women*, written for the *Edinburgh Review* a century ago, gives it as his deliberate opinion that "The instruction of women improves the stock of natural talents, and employs more minds for the instruction and amusement of the world; it increases the pleasures of society by multiplying the topics upon which the two sexes take a common interest; and makes marriage an intercourse of understanding as well as of affection by giving dignity and importance to the female character. The education of women favours public morals; it provides for every season of life as well as for the brightest and the best; and leaves a woman when she is stricken by the hand of time, not as she now is, destitute of everything and neglected by all, but with the full power and the splendid attractions of knowledge — diffusing the elegant pleasures of polite literature, and receiving the just homage of learned and accomplished men." As to the oft repeated commonplace of noodledom that higher education puts an end to domestic economy and deteriorates the noblest qualities of womanhood, the same clear-headed writer asks: "Can anything..... be more perfectly absurd than to suppose that the care and perpetual solicitude which a mother feels for her children, depends upon her ignorance of Greek or mathematics; and that she would desert an infant for a quadratic equation — that Cimmerian ignorance can aid parental affection, or the circle of the arts and sciences produce its destruction — that the moment you suffer women to eat of the tree of knowledge the rest of the family will very soon be reduced to the same kind of aerial and unsatisfactory diet?"

Still more insistent on the necessity of the broadest and deepest education for woman — education in science as well as in art and literature — is the Most Rev Archbishop, J L Spalding, who by his writing and lectures has done so much for the cause of the higher education of both men and women. In an eloquent and pregnant discourse, pronounced in the Church of the Gesù in Rome, in March 1900, he told his vast audience — composed of the elite of the Eternal City — that:

If we are to have a race of enlightened, noble, and brave men, we must give to woman the best education it is possible for her to receive. She has the same right as man to become all that she may be, to know whatever may be known, to do whatever is fair and just and good. In souls there is no sex. If we leave half the race in ignorance, how shall we hope to lift the other half into the light of truth and love? Let woman's mental power increase, let her influence grow, and more and more she will stand by the side of man as a helper in all his struggles to make the will of God prevail. From the time the Virgin Mother held the Infant Saviour in her arms, to this hour, woman has been the great lover of Christ and the unweary helper of His little ones; and the more we strengthen and illumine her, the more we add to her sublime faith and devotion the power of knowledge and culture, the more efficaciously shall she work to purify life, to make justice, temperance, chastity, and love prevail. She is more unselfish, more capable of enthusiasm for spiritual ends, she has more sympathy with what is beautiful, noble, and godlike than man; and the more her knowledge increases, the more shall she become a heavenly force to help spread God's kingdom on earth.

# Bibliography

## *Partial List of the Works Quoted or Referred to in the Text*


Agassiz, Mrs L: *Louis Agassiz, His Life and Correspondence*, Boston, 1893.
Agnesi, Maria Gaetana: *Instituzioni Analitiche*, Milan, 1748.
—— *Propositiones Philosophicae*, Milan, 1738.
Anzoletti, Luisa: *Maria Gaetana Agnesi*, Milan, 1900.
Arrighi, G L: *Storia del Feminismo*, Florence, 1911.
Asse, Eugene: *Lettres de la Marquise du Chatelet*, Paris, 1882.
Athenaeus: *The Deipnosophists; or the Banquet of the Learned*, Bohn Edition, London, 1907.

Becq de Fouquieres, L: *Aspasie de Milet*, Paris, 1872.
Bede, Venerable: *Historia Ecclesiastica Gentis Anglorum*, London, 1848.
Bievre, Le Comte de: *Histoire des Deux Aspasies*, Paris, 1736.
Bigoni, D Guido: *Ipazia, Alessandrina*, Venice, 1887.
Birch, Una, *Anna van Schurman*, London, 1909.
Blackwell, Elizabeth, *Pioneer Work in Opening the Medical Profession to Women*, London, 1895.
Bonnet, J: *Vie d'Olympie Morata* Paris, 1850.
Bremmer, C S: *Education of Girls and Women*, London, 1897.
Briand, E M: *Histoire de Sainte Radegonde, Reine de France*, Paris, 1898.
Broca, Paul: *Memoires d'Anthropologie*, Paris, 1871-88.
Burckhardt, J: *Die Cultur der Renaissance in Italien*, Leipzig, 1899.

Candolle, Alphonse de: *Histoire des Sciences et des Savants depuis Deux Siecles*, Geneve, 1885.
Castiglione, Baldassare: *Libro del Cortigiano*, Milan, 1890.
Chabaud, L: *Mesdames de Maintenon, de Genlis et Campan. Leur Role dans l'Éducation Chretienne de la Femme*, Paris, 1901.
Chatelet, Marquise du: *Principes Mathematiques de la Philosophie Naturelle*, Paris, 1759.
Chiappelli, A: *Saggi e Note Critche*, Bologna, 1895.
Coudreau, Mme H: *Voyage au Maycurú*, Paris, 1903.

—— *Voyage au Rio Curuá*, Paris, 1903.
Curie, Pierre, Mme: *Traite de Radioactivite*, Paris, 1910.

Delambre, J B: *Histoire de l'Astronomie*; Paris, 1817.
Denifle, H: *Die Entstehung der Universitäten des Mittelalters*, Berlin, 1885.
Denifle et Chatelain: *Chartularium Universitatis Parisiensis*, Paris, 1889-91.
Diogenes Laertius: *Lives and Opinions of the Ancient Philosophers*, Bohn Edition, London.
Dieulafoy, Mme: *At Susa the Ancient Capital of the Kings of Persia: Narrative of Travel through Western Persia and Excavations Made on the Site of the Lost City of the Lilies, 1884-1886*, Philadelphia, 1890.
—— *La Perse, la Chaldee et la Susane*, Paris, 1887.
D'istria, Dora: *Des Femmes par une Femme*, Paris, 1865.
Donaldson, James: *Woman: Her Position in Ancient Greece and Rome among the Early Christians*, London, 1907.
Drane, Augusta T: *Christian Schools and Scholars*, London, 1881.
Druon, H: *Oeuvres de Synesius, Évêque de Ptolemais, Traduites entierement pour la premiere Fois en Francais et Precedees d'une Étude Biographique et Litteraire*, Paris, 1878.
Dupanloup, Mgr: *La Femme Studieuse*, Paris, 1895.

Eckstein, Lena: *Women under Monasticism*, Cambridge, 1896.
Ellis, Havelock: *Man and Woman*, London, 1898.

Fantuzzi, Giovanni: *Notizie degli Scrittori Bolognesi*, Bologna, 1781.
Favaro, Antonio: *Galileo Galilei e Suor Celeste*, Florence, 1891.
Fenelon, François de Salignac de la Mothe: *De L'education Des Filles*, Paris, 1881.
Finot, J: *Problems of the Sexes*, New York, 1913.
Fiorelli, Jos: *Pompeinarum Antiquitatum Historia*, Naples, 1860-64.
Frankland, Mr and Mrs Percy: *Pasteur*, London, 1898.

Gasquet, Aidan: *Henry VIII and the English Monasteries*, London, 1895.
Gibson, Margaret Dunlop: *How the Codex was Found*, Cambridge, 1893.
Gicosa, Piero: *Magistri Salernitani Nondum Editi*, Turin, 1891.
Girard, Paul: *L'Éducation Athenienne an Ve et IVe Siecle avant Jesus Christ*, Paris, 1889.
Gladstone, W E: *Studies on Homer and the Homeric Age*, Oxford, 1858.
Grimaux, E: *Lavoisier, 1743-1794, d'apres sa Correspondence, Ses Manuscrits, Ses Papiers de Famille et d'Autres Documents Inedits*, Paris, 1896.

Hall, G Stanley: *Adolescence*, New York, 1904.

Hamel, F: *An Eighteenth Century Marquise. A Study of Émilie du Chatelet and Her Times*, New York, 1911.
Harless, C F: *Die Verdienste der Frauen um Naturwissenschaft and Heilkunde. Gottingen*, 1830.
Harrison, Jane E: *Prolegomena to the Study of Greek Religion*, Cambridge, 1903.
Hawes, Harriet Boyd: *Gournia, Vasilike and Other Prehistoric Sites on the Isthmus of Hierapetra, Crete*, Philadelphia, 1908.
Henschel, G E T: *Daremberg, C, e Renzi de, S Collectio Salernitana, Ossia Documenti Inediti e Trattati di Medicina Appertenti alla la Scuola Medica Salernitana*, Naples, 1852-59.
Herrade de Lansberg: *Hortus Deliciarum*, Strasburg, 1901.
Herschel, Mrs John: *Memoirs and Correspondence of Caroline Herschel*, London, 1879.
Hertzen Et Rossi: *Inscriptiones Urbis Romae Latinae*, Berlin, 1882.
Hildegardis, S: *Causae et Curae*, Leipzig, 1903.
—— *Opera Omnia*, Edition Migne, Paris, 1882.
—— *Nova S Hildegardis Opera*, Editor: J B Card, Pitra, Paris, 1882.
Hill, Georgiana: *Women in English Life*, London, 1896.
Hunt, Caroline L: *The Life of Helen H Richards*, Boston, 1912.
Huxley, L H: *Life and Letters of Thomas H Huxley*, New York, 1900.

Jerome: *St Epistolae*, Edition Vallarsi, Verona, 1734-42.
Jex-Blake, Sophia: *Medical Women*, Edinburgh, 1886.
Jourdain, Charles: *Excursions Historiques et Philosophiques à travers le Moyen Age*, Paris, 1888.

Kendall, Phebe M: *Maria Mitchell, Life, Letters and Journals*, Boston, 1896.
Kingsley, Mary H: *Travels in West Africa*, London, 1897.
—— *West African Studies*, London, 1899.
Kirchhoff, A: *Die Akademische Frau*, Berlin, 1897.

Labe, Louise: *Oeuvres de*, Paris, 1871.
Lagrange, F: *Histoire de Sainte Paule*, Paris, 1870.
Laigle, Mathilde: *Le Livre de Trois Vertus de Christine de Pisan et son Milieu Historique et Literaire*, Paris, 1912.
Lalande, Jerome: *Bibliographie Astronomique*, Paris, 1803.
Lamy, E: *La Femme de Demain*, Paris, 1912.
Lange, Helene: *Higher Education of Women in Europe*, New York, 1890.
Laurie, S C: *Historical Survey of Pre-Christian Education*, London, 1900.
Lecky, W E: *History of European Morals*, New York, 1905.
Lefevre, M: *La Femme à travers l'Histoire*, Paris, 1902.
Leffler, Anna Carlotta: *Sónya Kovalevsky, Her Recollections of Childhood, with a Biography*, New York, 1895.

Lepinska, Melanie, Mlle: *Histoire des Femmes Medecins*, Paris, 1900.
Lewis, Agnes Smith: *In the Shadow of Sinai*, Cambridge, 1898.
Ligier, Hermann: *De Hypatia Philosopha et Eclectismi Fine*, Dijon, 1879.
Lourbet, J: *La Femme devant la Science*, Paris, 1896.
Lungo, Isodoro del: *Women of Florence*, London, 1907.

Macpherson, Geraldine: *Memoirs of the Life of Anna Jameson*, London, 1878.
Maistre, Comte Joseph de: *Lettres et Opuscules Inedits*, Paris, 1851.
Marini, Gaetano: *Archiatri Pontifici*, Rome, 1784.
Mason, O T: *Woman's Share in Primitive Culture*, London, 1895.
—— *Origin of Inventions*, London, 1895.
Maulde la Claviere, R de: *The Women of the Renaissance*, New York, 1901.
Mazzuchelli, Giammaria: *Gli Scrittori d'Italia*, Brescia, 1758.
Medici, Michele: *Compendio Storico della Scuola Anatomica di Bologna*, Bologna, 1857.
Menagius, Aegidius: *Historia Mulierum Philosopharum*, Amsterdam, 1692.
Meyer, H F: *Geschichte der Botanik*, Konigsburg, 1856.
Michaelis, A: *A Century of Archaeological Discoveries*, New York, 1908.
Mill, John Stuart: *The Subjection of Women*, London, 1909.
Mobius, P J: *Über die Anlage zur Mathematik*, Leipzig, 1907.
Montagu, Lady Mary Wortley: *Letters and Works of*, Bohn Edition, London, 1887.
Montalambert, Comte de: *Monks of the West*, London, 1896.
Myers, Annie N: *Woman's Work in America*, New York, 1891.

Nuttall, Zelia: *The Fundamental Principles of Old and New World Civilisations*, Cambridge, Mass, 1901.
Nolhac, Pierre de: *Petrarque et l'Humanisme*, Paris, 1892.

Oelsner, Elise: *Die Leistungen der deutschen Frau in den letzen vierhundert Jahren auf wissenschaftlichem Gebiete*, Guhrau, 1894.
Osanam, A F: *Documents Inedits pour servir à l'Histoire Litteraire de l'Italie*, Paris, 1850.

*Plato's Dialogues*, Jowett's Translation, London, 1892.
Poestion, J C: *Griechische Dicterinnen*, Wien, 1876.
—— *Griechische Philosophinnen*, Norden, 1885.

Rashdall, H: *The Universities of Europe in the Middle Ages*, Oxford, 1895.
Rebiere, A: *Les Femmes dans la Science*, Paris, 1897.
Reich, Emil: *Woman through the Ages*, London, 1908.
Renaud, A: *Histoire Nouvelle des Arts et des Sciences*, Paris, 1878.
Renzi de, Salvatore: *Storia Documentata della Scuola Medica di Salerno*,

Naples, 1857.
Rodocanachi, E: *La Femme Italienne à l'Epoque de la Renaissance*, Paris, 1907.
Rousselot, Paul: *Histoire de l'Éducation des Femmes en France*, Paris, 1883.

Sabbadini, Remigio: *Vita di Guarino Veronese*, Genoa, 1891.
Sainte-Beuve, Charles-Augustin de: *Nouvelle Galerie de Femmes Celebres*, Paris, 1872.
Sarti, Mauri, et Fattorini, Mauri: *De Claris Archigymnasii Bononiensis Professoribus a Saeculo XI usque ad Saeculum XIV*, Bologna, 1888-96.
Schliemann, H: *Ilios, the City and Country of the Trojans*, New York, 1881.
Schmidt, A: *Sur l'Age de Pericles*, 1877-79.
Sertillanges, A D: *Feminisme et Christianisme*, Paris, 1908.
Simon, Jules: *La Femme du Vingtieme Siecle*, Paris, 1892.
Socrates, Scholasticus: *Ecclesiastical History*, London, 1848.
Somerville, Mary: *Personal Recollections from Early Life to Old Age*, Boston, 1874.
Stanton, Theodore: *The Woman Question in Europe*, New York, 1884.
Stupuy, H: *Oeuvres Philosophiques de Sophie Germain*, Paris, 1896.
Symonds, J A: *A Short History of the Renaissance in Italy*, London, 1893.

Thierry, A: *Saint Jerome, La Societe Chretienne à Rome et l'Émigration Romaine en Terre Sainte*, Paris, 1867.
Tiraboschi, G: *Storia della Letteratura Italiana*, Milan, 1822.

Veitch, J: *Memoir of Sir William Hamilton*, Edinburgh, 1869.
Vives, Joannes Ludovicus: *De Tradendis Disciplinis*, Colon, Agr, 1536.

Wallace, R: *Eleanor Ormerod, Economic Entomologist, Autobiography and Correspondence*, London, 1904.
Wharton, H T: *Sappho*, London, 1898.
Wolf, J C: *Mulierum Graecarum Quae Oratione Prosa Usae Sunt Fragmenta et Elogia Graece et Latine*, London, 1739.
—— *Poetriarum Octo, Erinnae, Myrus, Myrtidis, Corinnae, Telesillae, Praxillae, Nossidis, Anytae, Fragmenta et Elogia*, Hamburg, 1734.
—— *Sapphus, Poetriae Lesbiae, Fragmenta et Elogia*, Hamburg, 1733.
Woodward, W H: *Vittorino da Feltre and Other Humanist Educators*, Cambridge, England, 1905.
Wright, T. *Womankind in Western Europe*, London, 1869.

Zuccante, Giusseppe. *Fra il Pensiero Antico e il Moderno*, Milan, 1905.